Symmetry and Mesoscopic Physics

Symmetry and Mesoscopic Physics

Editors

Rashid G. Nazmitdinov
Vyacheslav Yukalov

MDPI • Basel • Beijing • Wuhan • Barcelona • Belgrade • Manchester • Tokyo • Cluj • Tianjin

Editors

Rashid G. Nazmitdinov
Dubna State University
Russia

Vyacheslav Yukalov
Joint Institute for Nuclear Research
Russia

Editorial Office
MDPI
St. Alban-Anlage 66
4052 Basel, Switzerland

This is a reprint of articles from the Special Issue published online in the open access journal *Symmetry* (ISSN 2073-8994) (available at: https://www.mdpi.com/journal/symmetry/special_issues/mesoscopic_physics).

For citation purposes, cite each article independently as indicated on the article page online and as indicated below:

LastName, A.A.; LastName, B.B.; LastName, C.C. Article Title. *Journal Name* **Year**, *Volume Number*, Page Range.

ISBN 978-3-0365-2759-8 (Hbk)
ISBN 978-3-0365-2758-1 (PDF)

Contents

About the Editors

Rashid G. Nazmitdinov received the M.Sc. degree in theoretical physics from the Samarkand State University, Samarkand, Uzbekistan in 1974. In 1979 and 1990 he received the Ph.D. degree and the Dr.Sci. degree in physics and mathematics from the Higher Attestation Committee of USSR, respectively. He was a Junior Reseacher from 1981 to 1982, before becoming a Senior Researcher, with the Institute of Applied Physics of the Tashkent State University, Tashkent, Uzbekistan, from 1982 to 1989. Since 1989, he has been a Senior Scientist at the Joint Institute for Nuclear Research in Dubna, Russia, where he currently holds the position of a Leading Scientist. He has authored about 160 papers in refereed journals. His current research interests include mesoscopic physics, nuclear structure, transport in nanosystems, classical and quantum chaos.

Vyacheslav Yukalov received the M.Sc. degree in theoretical physics and Ph.D. degree in theoretical and mathematical physics from the Moscow State University, Moscow, Russia, in 1970 and 1974, respectively. He has also received the Dr.Hab. degree in theoretical physics from the University of Poznan, Poznan, Poland, and the Dr.Sci. degree in physics and mathematics, from the Higher Attestation Committee of Russia. He was a Graduate Assistant with the Moscow State University from 1970 to 1973, before becoming an Assistant Professor, Senior Lecturer, and an Associate Professor with the Moscow Engineering Physics Institute, Moscow, Russia, from 1973 to 1984. Since 1984, he has been a Senior Scientist and Department Head at the Joint Institute for Nuclear Research in Dubna, Russia, where he currently holds the position of a Leading Scientist. He has authored about 500 papers in refereed journals and 4 books and edited a number of books and special issues. His current research interests include statistical physics, quantum theory, dynamical theory, complex systems, nonlinear and coherent phenomena, self-organization, decision theory, and applied mathematics.

Preface to "Symmetry and Mesoscopic Physics"

Nature loves to play dice with us, providing this or that riddle along the way of our understanding of its diversity. It is like the faces of an n-dimensional facially regular polyhedron, the numerous faces of which from time to time indicate different problems for us. Depending on our luck, we can get the Ariadne's thread, which could lead us to the laws of nature based on beauty and harmony. Since ancient times people have realized that very often harmony is associated with symmetry. In this issue, we tried to present some of the results of the manifestation of symmetries and symmetry breaking in finite quantum systems that have peculiarities compared to macroscopic samples. Some authors find the effects of symmetry and asymptotic symmetry breaking in the analysis of Bose-Einstein condensate. Other authors have shown how symmetries, due to spin-orbit interaction in quantum dots and graphene, can explain the effects of electron transport in these systems. The problem of scaling symmetry, related to fractals, is studied as well, which allows us to understand the mechanisms of fabrication of complex materials with predefined physical properties and functionalities.

Rashid G. Nazmitdinov, Vyacheslav Yukalov
Editors

Article

Particle Fluctuations in Mesoscopic Bose Systems

Vyacheslav I. Yukalov [1,2]

[1] Bogolubov Laboratory of Theoretical Physics, Joint Institute for Nuclear Research, Dubna 141980, Russia; yukalov@theor.jinr.ru

[2] Instituto de Fisica de São Carlos, Universidade de São Paulo, CP 369, São Carlos 13560-970, São Paulo, Brazil

Received: 10 April 2019; Accepted: 25 April 2019; Published: 1 May 2019

Abstract: Particle fluctuations in mesoscopic Bose systems of arbitrary spatial dimensionality are considered. Both ideal Bose gases and interacting Bose systems are studied in the regions above the Bose–Einstein condensation temperature T_c, as well as below this temperature. The strength of particle fluctuations defines whether the system is stable or not. Stability conditions depend on the spatial dimensionality d and on the confining dimension D of the system. The consideration shows that mesoscopic systems, experiencing Bose–Einstein condensation, are stable when: (i) ideal Bose gas is confined in a rectangular box of spatial dimension $d > 2$ above T_c and in a box of $d > 4$ below T_c; (ii) ideal Bose gas is confined in a power-law trap of a confining dimension $D > 2$ above T_c and of a confining dimension $D > 4$ below T_c; (iii) the interacting Bose system is confined in a rectangular box of dimension $d > 2$ above T_c, while below T_c, particle interactions stabilize the Bose-condensed system, making it stable for $d = 3$; (iv) nonlocal interactions diminish the condensation temperature, as compared with the fluctuations in a system with contact interactions.

Keywords: Bose systems; asymptotic symmetry breaking; Bose–Einstein condensation; particle fluctuations; stability of Bose systems

---

1. Introduction

The theory of Bose systems has recently attracted high attention triggered by experimental studies of cold trapped atoms (see, e.g., the books and review articles [1–19]). Special attention has been payed to the study of particle fluctuations, mainly considering three-dimensional macroscopic Bose systems or harmonically-trapped atoms. The importance of this problem has been emphasized after the appearance of a number of papers claiming the occurrence of thermodynamically-anomalous particle fluctuations in the whole region below the condensation temperature T_c even for equilibrium three-dimensional interacting systems (a list of the papers containing such claims has been summarized in [20]). The origin of the arising fictitious anomalies and the ways of avoiding them have been discussed in detail in reviews [16–18].

It would not be strange if anomalously strong fluctuations would be found at the point of a second-order phase transition. This would be natural, since at the point of a phase transition, the system is unstable and fluctuations in a system can drastically increase. It is exactly the system instability that drives the phase transition and forces the system to transfer to another state. However, as soon as the transition to the other state has happened, the real system becomes stable and has to exhibit thermodynamically normal fluctuations. It is therefore more than strange how thermodynamically-anomalous fluctuations could arise in realistic three-dimensional interacting systems.

Moreover, Bose–Einstein condensation is necessarily accompanied by the spontaneous breaking of global gauge symmetry. From the mathematical point of view, the similar breaking of continuous symmetry occurs under magnetic phase transitions [21]; hence, anomalous fluctuations of the order parameter should appear in magnets below T_c. However, thermodynamically-anomalous fluctuations imply the system instability [22]. Therefore, if such fluctuations would really arise in the whole range below T_c, then neither superfluids nor magnets would exist. Fortunately, it has been shown [23,24] that thermodynamically-anomalous fluctuations in interacting three-dimensional equilibrium systems, discussed in theoretical papers, are just calculational artifacts caused, briefly speaking, by the use of a second-order approximation for calculating fourth-order terms.

The aim of the present paper is to extend the investigation of particle fluctuations in Bose systems in several aspects: First, we consider mesoscopic systems that are finite, although containing many particles $N \gg 1$. Taking into account a finite number of particles requires modifying the definition of the Bose function by introducing a finite cutoff responsible for the existence of a minimal wave vector prescribed by the system geometry. Second, we analyze particle fluctuations above, as well as below T_c for the Bose systems of arbitrary dimensionality, which allows us to find the critical spatial dimension above which the system is stable. Third, we consider two types of Bose systems, confined either in a rectangular box or in a power-law trap. Fourth, the influence of nonlocal interactions on particle fluctuations is analyzed, as compared to that of local interactions.

Throughout the paper, the system of units is employed where the Planck and Boltzmann constants are set to one, $\hbar = 1$ and $k_B = 1$.

2. Particle Fluctuations and Stability

Here and in what follows, we consider mesoscopic systems that are finite, containing a finite number of particles N, although with this number is rather large, $N \gg 1$.

Observable quantities are given by statistical averages $\langle \hat{A} \rangle$ of Hermitian operators $\hat{A}$. Fluctuations of the observable quantities are characterized by the variance:

$$\mathrm{var}(\hat{A}) \equiv \langle \hat{A}^2 \rangle - \langle \hat{A} \rangle^2 .$$

The observable is called extensive when:

$$\langle \hat{A} \rangle \propto N \qquad (N \gg 1) , \tag{1}$$

which is equivalent to the condition:

$$\frac{\langle \hat{A} \rangle}{N} \simeq const \qquad (N \gg 1) . \tag{2}$$

Fluctuations are termed thermodynamically normal if the inequalities:

$$0 \leq \frac{\mathrm{var}(\hat{A})}{|\langle \hat{A} \rangle|} < \infty \tag{3}$$

are valid for any N, which can also be represented as the condition:

$$\frac{\mathrm{var}(\hat{A})}{|\langle \hat{A} \rangle|} \simeq const \qquad (N \gg 1) . \tag{4}$$

When these conditions do not hold, the fluctuations are called thermodynamically anomalous. Sometimes, instead of the terms thermodynamically normal or thermodynamically anomalous, one says, for short, that fluctuations are just normal or anomalous.

Particle fluctuations, describing the fluctuations of the number of particles, characterized by the number-of-particles operator $\hat{N}$, are quantified by the relative variance:

$$\frac{\mathrm{var}(\hat{N})}{N} = \frac{1}{N} \left(\langle \hat{N}^2 \rangle - \langle \hat{N} \rangle^2 \right) , \tag{5}$$

where $N = \langle \hat{N} \rangle$. The fluctuations are normal when:

$$0 \leq \frac{\text{var}(\hat{N})}{N} < \infty \tag{6}$$

for any N, or in other words, when:

$$\frac{\text{var}(\hat{N})}{N} \simeq const \qquad (N \gg 1) . \tag{7}$$

The strength of particle fluctuations characterizes the system stability, since these fluctuations are directly connected to the isothermal compressibility:

$$\kappa_T \equiv -\frac{1}{V}\left(\frac{\partial V}{\partial P}\right)_{TN} = \frac{1}{\rho N}\left(\frac{\partial N}{\partial \mu}\right)_{TV} \tag{8}$$

by the equality:

$$\kappa_T = \frac{\text{var}(\hat{N})}{\rho T N} \qquad \left(\rho \equiv \frac{N}{V}\right) , \tag{9}$$

with ρ being the average particle density. The system stability requires that:

$$0 \leq \kappa_T < \infty \tag{10}$$

for any N, which yields Conditions (6) and (7). The above relations give us one of the ways for calculating the relative variance:

$$\frac{\text{var}(\hat{N})}{N} = \rho T \kappa_T = \frac{T}{N}\left(\frac{\partial N}{\partial \mu}\right)_{TV} . \tag{11}$$

3. Ideal Gas in a Rectangular Box

Bose systems in a rectangular box are not merely an interesting object allowing for detailed calculations, but it can also be realized experimentally inside box-shaped traps [25–27].

3.1. Modified Bose Function

The grand Hamiltonian of a gas in a rectangular box of volume V reads as:

$$H = \hat{H} - \mu\hat{N} = \int \psi^\dagger(\mathbf{r})\left(-\frac{\nabla^2}{2m} - \mu\right)\psi(\mathbf{r})\, d\mathbf{r} , \tag{12}$$

where the integration is over the given volume V. Assuming periodic continuation of the box, the field operators can be expanded in plane waves,

$$\psi(\mathbf{r}) = \sum_k a_k \varphi_k(\mathbf{r}) , \qquad \varphi_k(\mathbf{r}) = \frac{1}{\sqrt{V}}\, e^{i\mathbf{k}\cdot\mathbf{r}} , \tag{13}$$

which gives:

$$H = \sum_k \omega_k a_k^\dagger a_k \qquad \left(\omega_k = \frac{k^2}{2m} - \mu\right) . \tag{14}$$

The total number of particles is the sum:

$$N = N_0 + N_1 , \qquad N_1 = \sum_{k \neq k_0} n_k , \tag{15}$$

where N_0 is the number of condensed particles, while N_1 is the number of uncondensed particles, with the momentum distribution:

$$n_k \equiv \langle a_k^\dagger a_k \rangle = \left(e^{\beta\omega_k} - 1 \right)^{-1} . \tag{16}$$

Here, $\beta = 1/T$ is the inverse temperature. For a large number of particles, the sums over momenta can be represented as the integrals,

$$\sum_k n_k \to V \int n_k \, \frac{d\mathbf{k}}{(2\pi)^d} ,$$

where d is spatial dimensionality. In the case of isotropic functions under the integrals, it is possible to pass to spherical coordinates. However, it is necessary to take into account that for a finite system, the values of the wave vectors start not from zero, but from a finite minimal momentum k_0 that can be estimated as:

$$k_0 = \frac{2\pi}{L} = \frac{2\pi}{aN^{1/d}} , \tag{17}$$

with the box volume:

$$V = L^d , \qquad L = aN^{1/d} ,$$

where a is the mean interparticle distance. Thus, the integration over the momenta takes the form:

$$\int \frac{d\mathbf{k}}{(2\pi)^d} \to \frac{2}{(4\pi)^{d/2}\Gamma(d/2)} \int_{k_0}^\infty k^{d-1} \, dk , \tag{18}$$

where the lower limit is given by the cutoff prescribed by the minimal quantity k_0. Then, the number of uncondensed particles becomes proportional to the modified Bose function:

$$g_n(z) \equiv \frac{1}{\Gamma(n)} \int_{u_0}^\infty \frac{zu^{n-1}}{e^u - z} \, du , \tag{19}$$

with $z \equiv \exp(\beta\mu)$ being the fugacity and where the lower limit is given by the cutoff:

$$u_0 = \frac{k_0^2}{2mT} = \frac{\varepsilon_0}{T} \tag{20}$$

defined by the minimal energy:

$$\varepsilon_0 = \frac{2\pi^2}{ma^2} \, N^{-2/d} . \tag{21}$$

Since the minimal energy (21) tends to zero for large N, it is admissible to keep in mind that:

$$u_0 \ll 1 \qquad (N \gg 1) . \tag{22}$$

In this way, the relative variance (11) can be expressed through the derivative of the modified Bose function (19). The latter differs from the standard Bose function by the existence of a nonzero lower integration limit defined by the minimal wave vector.

3.2. Fluctuations above the Condensation Temperature

At temperatures above the condensation point, there are no condensed particles, so that the total number of particles reads as:

$$N = \frac{V}{\lambda_T^d} \, g_{d/2}(z) \qquad (T \geq T_c) , \tag{23}$$

where:

$$\lambda_T \equiv \sqrt{\frac{2\pi}{mT}}$$

is the thermal wavelength. Hence, the relative variance (11) is:

$$\frac{\text{var}(\hat{N})}{N} = \frac{z}{\rho\lambda_T^d}\,\frac{\partial g_{d/2}(z)}{\partial z} \qquad (T > T_c)\,, \tag{24}$$

where $\rho \equiv N/V$ is the particle density.

Estimating the Bose function above T_c, where $z < 1$, we find:

$$g_n(z) = -\,\frac{z}{(1-z)\Gamma(1+n)}\left[u_0^n - \frac{n u_0^{1+n}}{(1+n)(1-z)}\right] \qquad (n < 0\,,\; z < 1)\,. \tag{25}$$

In particular,

$$g_{-1/2}(z) = -\,\frac{z}{\sqrt{\pi}(1-z)}\left(u_0^{-1/2} + \frac{u_0^{1/2}}{1-z}\right) \qquad (z < 1) \tag{26}$$

and:

$$g_0(z) = -\,\frac{z}{1-z} \qquad (z < 1)\,. \tag{27}$$

Calculating the derivatives of the modified Bose functions requires being attentive, since some of the derivatives are different from those for the standard Bose functions. Generally, we have:

$$\frac{\partial g_n(z)}{\partial z} = \frac{1}{z}\,g_{n-1}(z) + \frac{u_0^{n-1}}{\Gamma(n)(1-z+u_0)}\,. \tag{28}$$

Using the smallness of u_0, we can write:

$$\frac{u_0^{n-1}}{1-z+u_0} \simeq \frac{1}{1-z}\left(u_0^{n-1} - \frac{u_0^n}{1-z}\right) \qquad (z < 1)\,.$$

Therefore, for $n < 1$, we find:

$$\frac{\partial g_n(z)}{\partial z} = -\,\frac{u_0^n}{(1-z)^2\Gamma(1+n)} \qquad (n < 1\,,\; z < 1)\,; \tag{29}$$

while for $n > 1$, keeping the main terms, we get:

$$\frac{\partial g_n(z)}{\partial z} = \frac{1}{z}\,g_{n-1}(z) \qquad (n > 1\,,\; z < 1)\,. \tag{30}$$

We shall also need the derivatives:

$$\frac{\partial g_{1/2}(z)}{\partial z} = -\,\frac{2u_0^{1/2}}{\sqrt{\pi}(1-z)^2} \qquad (d = 1\,,\; z < 1)\,,$$

$$\frac{\partial g_1(z)}{\partial z} = -\,\frac{u_0}{(1-z)^2} \qquad (d = 2\,,\; z < 1)\,,$$

$$\frac{\partial g_{3/2}(z)}{\partial z} = \frac{1}{z}\,g_{1/2}(z) \qquad (d = 3\,,\; z < 1)\,.$$

For the relative variance, depending on the space dimensionality, we obtain:

$$\frac{\text{var}(\hat{N})}{N} = -\,\frac{2z}{\sqrt{\pi}(1-z)^2\rho\lambda_T}\left(\frac{\varepsilon_0}{T}\right)^{1/2} \qquad (d = 1\,,\; T > T_c)\,,$$

$$\frac{\text{var}(\hat{N})}{N} = -\,\frac{z}{(1-z)^2\rho\lambda_T^2}\left(\frac{\varepsilon_0}{T}\right) \qquad (d = 2\,,\; T > T_c)\,,$$

$$\frac{\text{var}(\hat{N})}{N} = \frac{1}{\rho\lambda_T^3} \, g_{1/2}(z) \qquad (d = 3 \; , \; T > T_c) \; . \tag{31}$$

The negative values of the variance for $d = 1$ and $d = 2$ show that these low-dimensional systems are unstable. However, the gas is stable in three dimensions.

As follows from the derivative:

$$\frac{\partial g_{d/2}(z)}{\partial z} = \frac{1}{z} \, g_{(d-2)/2}(z) \qquad (d > 2 \; , \; z < 1) \; , \tag{32}$$

the system is stable for $d > 2$. That is, the critical spatial dimension, above which the uncondensed gas in a rectangular box is stable, is $d_c = 2$, so that the stability condition is:

$$d \; > \; d_c = 2 \qquad (T > T_c) \; . \tag{33}$$

3.3. Condensation Temperature of a Gas in a Rectangular Box

At the temperature of Bose condensation, the chemical potential becomes zero, $\mu = 0$, because of whuich $z = 1$. The total number of particles:

$$N = \frac{V}{\lambda_T^d} \, g_{d/2}(1) \qquad (T = T_c) \tag{34}$$

defines the critical temperature:

$$T_c = \frac{2\pi}{m} \left[\frac{\rho}{g_{d/2}(1)} \right]^{2/d} \; . \tag{35}$$

For different dimensionalities, we have:

$$g_{1/2}(1) = \frac{2}{\sqrt{\pi}} \, u_0^{-1/2} \qquad (d = 1) \; ,$$

$$g_1(1) = -\ln u_0 \qquad (d = 2) \; ,$$

$$g_{3/2}(1) = \zeta(3/2) \qquad (d = 3) \; .$$

This gives the critical temperatures for a one-dimensional system:

$$T_c = \frac{\pi\rho}{\sqrt{2m}} \, \varepsilon_0^{1/2} \qquad (d = 1) \; , \tag{36}$$

and for a two-dimensional system:

$$T_c = \frac{2\pi\rho}{m \ln(T_c/\varepsilon_0)} \qquad (d = 2) \; . \tag{37}$$

Iterating the latter equation and taking into account that:

$$\frac{T_c}{\varepsilon_0} \ll \exp\left(\frac{2\pi\rho}{m\varepsilon_0} \right) \; , \qquad \varepsilon_0 \propto N^{-2/d} \qquad (N \gg 1) \; ,$$

we obtain:

$$T_c = \frac{2\pi\rho}{m \ln(2\pi\rho/m\varepsilon_0)} \qquad (d = 2) \; . \tag{38}$$

For a three-dimensional box, we have the known result:

$$T_c = \frac{2\pi}{m} \left[\frac{\rho}{\zeta(3/2)} \right]^{2/3} \qquad (d = 3) \; . \tag{39}$$

The critical temperatures at a large $N \gg 1$ scale as:

$$T_c \propto \frac{1}{N} \qquad (d = 1) \; ,$$
$$T_c \propto \frac{1}{\ln N} \qquad (d = 2) \; ,$$
$$T_c \propto const \qquad (d = 3) \; . \tag{40}$$

For $d \leq 2$, the critical temperature diminishes to zero as N increases. It remains finite for $d > 2$. Recall that, as is found above, the system is unstable for $d \leq 2$. Thus, the Bose gas in a box is stable in the case where the critical temperature remains finite for large N.

3.4. Fluctuations below Critical Temperature

Below the critical temperature, there appears the Bose–Einstein condensate, so that the number-of-particle operator becomes the sum of the number of condensed particles N_0 and the number-of-particle operator $\hat{N}_1$ of uncondensed particles,

$$\hat{N} = N_0 + \hat{N}_1 \qquad (T < T_c) \; . \tag{41}$$

When the condensate function η is introduced by means of the Bogolubov shift [28–30] of the field operator:

$$\psi(\mathbf{r}) \rightarrow \eta(\mathbf{r}) + \psi_1(\mathbf{r}) \; ,$$

then particle fluctuations are defined by the fluctuations of uncondensed particles (see the detailed explanations in [3,9,16–18]),

$$\text{var}(\hat{N}) = \text{var}(\hat{N}_1) \; .$$

The average number of uncondensed particles is:

$$N_1 = \frac{V}{\lambda_T^d} \, g_{d/2}(1) \qquad (T < T_c) \; . \tag{42}$$

Therefore, the relative particle variance reads as:

$$\frac{\text{var}(\hat{N})}{N} = \frac{1}{\rho\lambda_T^d} \lim_{z\rightarrow 1} \frac{\partial g_{d/2}(z)}{\partial z} \qquad (T < T_c) \; . \tag{43}$$

For a mesoscopic system, we have:

$$\lim_{z\rightarrow 1} \frac{\partial g_n(z)}{\partial z} = g_{n-1}(1) + \frac{u_0^{n-2}}{\Gamma(n)} \; . \tag{44}$$

Notice that the last term here would be absent for a macroscopic system. In particular,

$$\lim_{z\rightarrow 1} \frac{\partial g_{1/2}(z)}{\partial z} = g_{-1/2}(1) + \frac{u_0^{-3/2}}{\sqrt{\pi}} \; .$$

Since:

$$g_{-1/2}(1) = -\,\frac{1}{3\sqrt{\pi}} \, u_0^{-3/2} \; ,$$

we find:

$$\lim_{z\rightarrow 1} \frac{\partial g_{1/2}(z)}{\partial z} = \frac{2}{3\sqrt{\pi}} \, u_0^{-3/2} \; .$$

We also need the limits:

$$\lim_{z\to 1} \frac{\partial g_1(z)}{\partial z} = \frac{1}{u_0} ,$$

and:

$$\lim_{z\to 1} \frac{\partial g_{3/2}(z)}{\partial z} = g_{1/2}(1) + \frac{2}{\sqrt{\pi}}\, u_0^{-1/2} .$$

Using $g_{1/2}(1)$ from the previous subsection, we get:

$$\lim_{z\to 1} \frac{\partial g_{3/2}(z)}{\partial z} = \frac{4}{\sqrt{\pi}}\, u_0^{-1/2} .$$

Generally, from the expression:

$$\lim_{z\to 1} \frac{\partial g_{d/2}(z)}{\partial z} = g_{(d-2)/2}(1) + \frac{u_0^{(d-4)/2}}{\Gamma(d/2)} \tag{45}$$

we see that the last term here increases with N by a power-law, when $d < 4$, while it increases logarithmically for $d = 4$,

$$\lim_{z\to 1} \frac{\partial g_2(z)}{\partial z} = 1 - \ln u_0 \qquad (d = 4) .$$

For the relative variance (43), we obtain:

$$\frac{\mathrm{var}(\hat{N})}{N} = \frac{2}{3\sqrt{\pi}\rho\lambda_T} \left(\frac{T}{\varepsilon_0}\right)^{3/2} \qquad (d = 1) ,$$

$$\frac{\mathrm{var}(\hat{N})}{N} = \frac{1}{\rho\lambda_T^2} \left(\frac{T}{\varepsilon_0}\right) \qquad (d = 2) ,$$

$$\frac{\mathrm{var}(\hat{N})}{N} = \frac{4}{\sqrt{\pi}\rho\lambda_T^3} \left(\frac{T}{\varepsilon_0}\right)^{1/2} \qquad (d = 3) ,$$

$$\frac{\mathrm{var}(\hat{N})}{N} = \frac{1}{\rho\lambda_T^4} \left(\frac{T}{\varepsilon_0}\right) \qquad (d = 4) . \tag{46}$$

Keeping in mind that $\varepsilon_0 \propto N^{-2/d}$, the scaling of these expressions with respect to N is as follows:

$$\frac{\mathrm{var}(\hat{N})}{N} \propto N^3 \qquad (d = 1) ,$$

$$\frac{\mathrm{var}(\hat{N})}{N} \propto N \qquad (d = 2) ,$$

$$\frac{\mathrm{var}(\hat{N})}{N} \propto N^{1/3} \qquad (d = 3) ,$$

$$\frac{\mathrm{var}(\hat{N})}{N} \propto \ln N \qquad (d = 4) . \tag{47}$$

This shows that for all dimensions below and including four, particle fluctuations are anomalous, corresponding to an unstable systems. In that sense, the dimension four is critical, implying that the stability condition for a condensed gas in a box is:

$$d > d_c = 4 \qquad (T < T_c) . \tag{48}$$

4. Ideal Gas in a Power-Law Trap

Power-law traps are the most often used devices for trapping particles. Here, we study particle fluctuations and the related stability of mesoscopic clouds in such traps.

4.1. Modified Semiclassical Approximation

The general form of confining potentials, employed in power-law traps, can be represented as:

$$U(\mathbf{r}) = \sum_{\alpha=1}^{d} \frac{\omega_\alpha}{2} \left| \frac{r_\alpha}{l_\alpha} \right|^{n_\alpha} , \tag{49}$$

where:

$$l_\alpha \equiv \frac{1}{\sqrt{m\omega_\alpha}}$$

is the effective trap radius in the α direction. As a whole, a trap can be characterized by the effective trap frequency ω_0 and effective length l_0 connected by the relations:

$$\omega_0 \equiv \left(\prod_{\alpha=1}^{d} \omega_\alpha \right)^{1/d} = \frac{1}{ml_0^2} , \qquad l_0 \equiv \left(\prod_{\alpha=1}^{d} l_\alpha \right)^{1/d} = \frac{1}{\sqrt{m\omega_0}} . \tag{50}$$

In the limit $n_\alpha \to \infty$, we return to a rectangular box.

When the effective trap frequency is much lower than temperature,

$$\frac{\omega_0}{T} \ll 1 , \tag{51}$$

it is possible to resort to the semiclassical approximation that, however, needs to be modified for considering mesoscopic systems [18,31].

In the semiclassical approximation, one defines the density of states:

$$\rho(\varepsilon) = \frac{(2m)^{d/2}}{(4\pi)^{d/2}\Gamma(d/2)} \int_{\mathbb{V}_\varepsilon} [\varepsilon - U(\mathbf{r})]^{d/2-1} \, d\mathbf{r} ,$$

in which:

$$\mathbb{V}_\varepsilon \equiv \{\mathbf{r} : \; U(\mathbf{r}) \leq \varepsilon\}$$

is the volume available for particle motion.

For trapped particles, an important notion is the confining dimension [18,31]:

$$D \equiv d + \sum_{\alpha=1}^{d} \frac{2}{n_\alpha} . \tag{52}$$

The density of states for the power-law potential (49) reduces to:

$$\rho(\varepsilon) = \frac{\varepsilon^{D/2-1}}{\gamma_D \Gamma(D/2)} , \tag{53}$$

where we use the notation:

$$\gamma_D \; \equiv \; \frac{\pi^{d/2}}{2^{D/2}} \prod_{\alpha=1}^{d} \frac{\omega_\alpha^{1/2+1/n_\alpha}}{\Gamma(1+1/n_\alpha)} .$$

In the normal state above T_c, the number of particles is given by the formula:

$$N = \frac{T^{D/2}}{\gamma_D} \, g_{D/2}(z) \qquad (T \geq T_c) . \tag{54}$$

We again meet the Bose function that has to be modified according to the definition (19) by using the integral cutoff:

$$u_0 = \frac{\varepsilon_0}{T} \qquad (\varepsilon_0 \sim \omega_0) \; , \tag{55}$$

with ε_0 being the lowest energy level in the trap, which is of the order of ω_0.

4.2. Condensation Temperature of a Gas in a Power-Law Trap

At the critical temperature T_c, we have $\mu = 0$ and $z = 1$. Then, Equation (54) yields:

$$T_c = \left[\frac{\gamma_D N}{g_{D/2}(1)} \right]^{2/D} \; . \tag{56}$$

The modified Bose function, depending on the confining dimension, takes the forms:

$$g_{D/2}(1) = \frac{2}{(2-D)\Gamma(D/2)} \left(\frac{T}{\varepsilon_0} \right)^{1-D/2} \qquad (D < 2) \; ,$$

$$g_1(1) = \ln \frac{T}{\varepsilon_0} \qquad (D = 2) \; ,$$

$$g_{D/2}(1) = \zeta \left(\frac{D}{2} \right) \qquad (D > 2) \; .$$

If $D < 2$, the spatial dimension can only be $d = 1$, when:

$$\gamma_D = \frac{\sqrt{\pi}}{\Gamma(D)} \left(\frac{\omega_0}{2} \right)^{D/2} \qquad (d = 1) \; .$$

Then, the critical temperature is:

$$T_c = \frac{\sqrt{\pi}}{\Gamma(D)} \left(1 - \frac{D}{2} \right) \Gamma \left(\frac{D}{2} \right) \left(\frac{\omega_0}{2\varepsilon_0} \right)^{D/2} N\varepsilon_0 \qquad (D < 2 \; , \; d = 1) \; . \tag{57}$$

The confining dimension equals two, $D = 2$, when $d = 1$ and $n = 2$, so that:

$$\gamma_2 = \omega_0 \qquad (D = 2 \; , \; d = 1 \; , \; n = 2) \; .$$

This yields the critical temperature:

$$T_c = \frac{N\omega_0}{\ln(T_c/\varepsilon_0)} \qquad (D = 2 \; , \; d = 1) \; . \tag{58}$$

For large N, one has:

$$\frac{T_c}{\varepsilon_0} \ll \exp \left(\frac{\omega_0}{\varepsilon_0} \, N \right) \; ,$$

since for a one-dimensional harmonic oscillator, $\varepsilon_0 = \omega_0/2$. Because of this, the critical temperature (58) can be simplified to:

$$T_c = \frac{N\omega_0}{\ln(2N)} \; . \tag{59}$$

For the confining dimension larger than two, the critical temperature is:

$$T_c = \left[\frac{\gamma_D N}{\zeta(D/2)} \right]^{2/D} \qquad (D > 2) \; . \tag{60}$$

In the case of harmonic traps, when $n_\alpha = 2$, hence $D = 2d$ and $\gamma_D = \omega_0^d$, the critical temperature becomes:

$$T_c = \left[\frac{N}{\zeta(d)}\right]^{1/d} \omega_0 \qquad (n_\alpha = 2\,,\; d > 1)\;.$$

4.3. Scaling with Respect to the Particle Number

As is explained in Section 2, extensive observables are proportional to the number of particles N, when this number is large. This definition prescribes the scaling of the system characteristics with respect to N. As a representative of an observable quantity, we may take, e.g., internal energy:

$$\langle \hat{H} \rangle = \langle H \rangle + \mu N\;. \tag{61}$$

This is an extensive quantity satisfying the condition:

$$\frac{\langle \hat{H} \rangle}{N} \simeq const \qquad (N \gg 1)\;. \tag{62}$$

For the considered case of a gas in a power-law trap, we have:

$$\frac{\langle \hat{H} \rangle}{N} = \frac{D g_{1+D/2}(z)}{2N\gamma_D}\, T^{1+D/2}\;. \tag{63}$$

The function $g_{1+D/2}(z)$ is finite for all $D > 0$ and all z. Hence, the condition (62) implies:

$$N\gamma_D \simeq const \qquad (N \gg 1);. \tag{64}$$

To make the consideration slightly less cumbersome, let us set the powers $n_\alpha = n$ for the trapping potential. Then, the confining dimension is:

$$D = \left(1 + \frac{2}{n}\right) d\;. \tag{65}$$

γ_D becomes:

$$\gamma_D = \frac{\pi^{d/2}}{\Gamma^d(1+1/n)} \left(\frac{\omega_0}{2}\right)^{D/2}\,,$$

which tells us that:

$$\gamma_D \propto \omega_0^{D/2} \qquad (N \gg 1)\;.$$

Therefore, ω_0 scales as:

$$\omega_0 \propto \frac{1}{N^{2/D}} \qquad (N \gg 1)\;. \tag{66}$$

Using this scaling and the fact that $\omega_0 \sim \varepsilon_0$, we see that the critical temperatures from the previous subsection behave as:

$$T_c \propto \frac{1}{N^{2/D-1}} \qquad (D < 2)\,,$$

$$T_c \propto \frac{1}{\ln N} \qquad (D = 2)\,,$$

$$T_c \propto const \qquad (D > 2)\;. \tag{67}$$

4.4. Fluctuations above the Condensation Temperature

Particle fluctuations above the condensation temperature are described by the formula:

$$\frac{\mathrm{var}(\hat{N})}{N} = \frac{T^{D/2}}{N\gamma_D}\, z\, \frac{\partial g_{D/2}(z)}{\partial z} \qquad (T > T_c)\,, \tag{68}$$

where $z < 1$. For the modified Bose function, we have:

$$\frac{\partial g_m(z)}{\partial z} = \frac{1}{z}\, g_{m-1}(z) + \frac{1}{(1-z)\Gamma(m)} \left(u_0^{m-1} - \frac{u_0^m}{1-z} \right) \qquad (z < 1) \,, \tag{69}$$

with the value:

$$g_{m-1}(z) = -\, \frac{z}{(1-z)\Gamma(m)} \left[u_0^{m-1} - \frac{m-1}{m(1-z)}\, u_0^m \right] \qquad (m < 1 \,,\, z < 1) \tag{70}$$

for $m < 1$. Summarizing, we have the derivatives:

$$\frac{\partial g_m(z)}{\partial z} = -\, \frac{u_0^m}{(1-z)^2\Gamma(1+m)} \qquad (m < 1 \,,\, z < 1) \,,$$

$$\frac{\partial g_1(z)}{\partial z} = -\, \frac{u_0}{(1-z)^2} \qquad (m = 1 \,,\, z < 1) \,,$$

$$\frac{\partial g_m(z)}{\partial z} = \frac{1}{z}\, g_{m-1}(z) \qquad (m > 1 \,,\, z < 1) \,.$$

From here, we find the relative variance:

$$\frac{\mathrm{var}(\hat{N})}{N} = -\, \frac{zT^{D/2}}{(1-z)^2 N\gamma_D \Gamma(1+D/2)} \left(\frac{\varepsilon_0}{T} \right)^{D/2} \qquad (D < 2 \,,\, T > T_c) \,,$$

$$\frac{\mathrm{var}(\hat{N})}{N} = -\, \frac{zT}{(1-z)^2 N\gamma_2} \left(\frac{\varepsilon_0}{T} \right) \qquad (D = 2 \,,\, T > T_c) \,,$$

$$\frac{\mathrm{var}(\hat{N})}{N} = \frac{T^{D/2}}{N\gamma_D}\, g_{D/2-1}(z) \qquad (D > 2 \,,\, T > T_c) \,, \tag{71}$$

characterizing particle fluctuations above the critical temperature. For $D \leq 2$, the variance is negative, which means instability. The system is stable only for $D > 2$, giving the stability condition:

$$d + \sum_{\alpha=1}^{d} \frac{2}{n_\alpha} > 2 \qquad (T > T_c) \,. \tag{72}$$

4.5. Fluctuations below the Condensation Temperature

Below the condensation temperature, where $\mu = 0$ and $z = 1$, the number of uncondensed particles reads as:

$$N_1 = \frac{T^{D/2}}{\gamma_D}\, g_{D/2}(1) \qquad (T \leq T_c) \,. \tag{73}$$

The variance of the total number of particles coincides with that of the uncondensed particles, which leads to:

$$\frac{\mathrm{var}(\hat{N})}{N} = \frac{T^{D/2}}{N\gamma_D} \lim_{z\to 1} \frac{\partial g_{D/2}(z)}{\partial z} \qquad (T < T_c) \,. \tag{74}$$

For the derivative in the right-hand side of the above formula, we have:

$$\lim_{z\to 1} \frac{\partial g_{D/2}(z)}{\partial z} = g_{D/2-1}(1) + \frac{u_0^{D/2-2}}{\Gamma(D/2)} \,. \tag{75}$$

Employing the values:

$$g_{D/2-1}(1) = \frac{2}{(4-D)\Gamma(D/2-1)}\, u_0^{D/2-2} \qquad (D < 4) \,,$$

$$g_1(1) = -\ln u_0 \qquad (D = 4) \; ,$$

we get the derivatives:

$$\lim_{z\to 1} \frac{\partial g_{D/2}(z)}{\partial z} = \left[\frac{2}{(4-D)\Gamma(D/2-1)} + \frac{1}{\Gamma(D/2)} \right] u_0^{D/2-2} \qquad (D < 4) \; ,$$

$$\lim_{z\to 1} \frac{\partial g_2(z)}{\partial z} = -\ln u_0 \qquad (D = 4) \; ,$$

$$\lim_{z\to 1} \frac{\partial g_{D/2}(z)}{\partial z} = g_{D/2-1}(1) = \zeta\left(\frac{D}{2} - 1 \right) \qquad (D > 4) \; .$$

In that way, we come to the relative variances:

$$\frac{\text{var}(\hat{N})}{N} = \frac{T^{D/2}}{N\gamma_D} \left[\frac{2}{(4-D)\Gamma(D/2-1)} + \frac{1}{\Gamma(D/2)} \right] \left(\frac{T}{\varepsilon_0} \right)^{2-D/2} \qquad (D < 4) \; ,$$

$$\frac{\text{var}(\hat{N})}{N} = \frac{T^2}{N\gamma_4} \ln\left(\frac{T}{\varepsilon_0} \right) \qquad (D = 4) \; ,$$

$$\frac{\text{var}(\hat{N})}{N} = \frac{T^{D/2}}{N\gamma_D} \zeta\left(\frac{D}{2} - 1 \right) \qquad (D > 4) \; . \tag{76}$$

Keeping in mind that $\varepsilon_0 \propto N^{-2/D}$ results in the scaling:

$$\frac{\text{var}(\hat{N})}{N} \propto N^{(4-D)/D} \qquad (D < 4) \; ,$$

$$\frac{\text{var}(\hat{N})}{N} \propto \ln N \qquad (D = 4) \; ,$$

$$\frac{\text{var}(\hat{N})}{N} \propto const \qquad (D > 4) \; . \tag{77}$$

This tells us that the system is stable only for $D > 4$. Therefore, the stability condition is:

$$d + \sum_{\alpha=1}^{d} \frac{2}{n_\alpha} > 4 \qquad (T < T_c) \; . \tag{78}$$

Notice that in the case of the often considered harmonic potential, when $n_\alpha = 2$, we have $D = 2d$ and $\gamma_D = \omega_0^d$. Then, the stability condition (78) reduces to the condition $d > 2$. The relative particle variance reads as:

$$\frac{\text{var}(\hat{N})}{N} = \frac{\zeta(d-1)}{\zeta(d)} \left(\frac{T}{T_c} \right)^d \qquad (n_\alpha = 2 \; , \; d > 2) \; .$$

5. Interacting Bose System above the Condensation Temperature

The grand Hamiltonian for a system of interacting Bose particles is:

$$H = \int \psi^\dagger(\mathbf{r}) \left(-\frac{\nabla^2}{2m} - \mu \right) \psi(\mathbf{r}) \, d\mathbf{r} \; +$$

$$+ \; \frac{1}{2} \int \psi^\dagger(\mathbf{r}) \psi^\dagger(\mathbf{r}') \Phi(\mathbf{r} - \mathbf{r}') \psi(\mathbf{r}') \psi(\mathbf{r}) \, d\mathbf{r} d\mathbf{r}' \; . \tag{79}$$

For generality, we consider a nonlocal isotropic interaction potential $\Phi(\mathbf{r}) = \Phi(r)$, where $r \equiv |\mathbf{r}|$. The integration is assumed to be over a rectangular box of volume V confining the system.

In the Hartree–Fock approximation, the Hamiltonian takes the form:

$$H_{HF} = E_{HF} + \int \psi^\dagger(\mathbf{r}) \left(-\frac{\nabla^2}{2m} - \mu \right) \psi(\mathbf{r})\, d\mathbf{r} +$$

$$+ \int \Phi(\mathbf{r} - \mathbf{r}') \left[\rho(\mathbf{r}')\psi^\dagger(\mathbf{r})\psi(\mathbf{r}) + \rho(\mathbf{r}', \mathbf{r})\psi^\dagger(\mathbf{r}')\psi(\mathbf{r}) \right] d\mathbf{r} d\mathbf{r}' , \tag{80}$$

where:

$$E_{HF} = -\frac{1}{2} \int \Phi(\mathbf{r} - \mathbf{r}') \left[\rho(\mathbf{r})\rho(\mathbf{r}') + \mid \rho(\mathbf{r}, \mathbf{r}') \mid^2 \right] d\mathbf{r} d\mathbf{r}'$$

and the notations are used for the single-particle density matrix:

$$\rho(\mathbf{r}, \mathbf{r}') = \langle\, \psi^\dagger(\mathbf{r}')\psi(\mathbf{r}) \rangle \tag{81}$$

and the particle density:

$$\rho(\mathbf{r}) = \rho(\mathbf{r}, \mathbf{r}) = \langle\, \psi^\dagger(\mathbf{r})\psi(\mathbf{r}) \rangle\, . \tag{82}$$

Employing the expansion of the field operators over plane waves, as in Equation (13), we get the Hamiltonian:

$$H_{HF} = E_{HF} + \sum_k \omega_k a_k^\dagger a_k\, , \tag{83}$$

in which:

$$E_{HF} = -\frac{1}{2}\, \rho \Phi_0 N - \frac{1}{2V} \sum_{kp} n_k n_p \Phi_{k+p}\, ,$$

Φ_k is a Fourier transform of $\Phi(\mathbf{r})$, and:

$$\Phi_0 = \int \Phi(\mathbf{r})\, d\mathbf{r}\, . \tag{84}$$

The momentum distribution is given by the expression (16), with the spectrum:

$$\omega_k = \frac{k^2}{2m} + \rho\Phi_0 + \frac{1}{V} \sum_p n_p \Phi_{k+p} - \mu\, . \tag{85}$$

The function n_p possesses a maximum at $p \to 0$, because of which it is possible to use the approximation [32,33]:

$$\sum_p n_p \Phi_{k+p} \cong \Phi_k \sum_p n_p \tag{86}$$

giving:

$$\omega_k = \frac{k^2}{2m} + \rho(\Phi_0 + \Phi_k) - \mu\, . \tag{87}$$

Introducing the effective interaction radius by the relation:

$$r_{eff}^2 \equiv \frac{\int \Phi(\mathbf{r}) r^2 d\mathbf{r}}{\int \Phi(\mathbf{r}) d\mathbf{r}} = \frac{4\pi}{\Phi_0} \int_0^\infty \Phi(\mathbf{r}) r^4\, d\mathbf{r} \tag{88}$$

shows that the long-wave limit of Φ_k is:

$$\Phi_k \simeq \left(1 - \frac{1}{6}\, k^2 r_{eff}^2 \right) \Phi_0\, . \tag{89}$$

Then, the spectrum (87) can be represented as:

$$\omega_k \simeq \frac{k^2}{2m^*} - \mu_{eff} \qquad (k \to 0)\, , \tag{90}$$

with the effective mass:

$$m^* \equiv \frac{m}{1 - \rho\Phi_0 r_{eff}^2/3} \tag{91}$$

and effective chemical potential:

$$\mu_{eff} \equiv \mu - 2\rho\Phi_0 \; . \tag{92}$$

In this approximation, the number of particles acquires the same form (23), however with the notations:

$$\lambda_T \equiv \sqrt{\frac{2\pi}{m^* T}} \; , \qquad z \equiv \exp(\beta\mu_{eff}) \; . \tag{93}$$

Using again the modified Bose function (19) and following the same analysis as in Section 3, we come to the conclusion that the system is stable for $d > 2$, when $T > T_c$. The difference is that now instead of mass m, there is the effective mass m^*, and at the critical temperature, we have:

$$\mu_{eff} = 0 \; , \qquad \mu = 2\rho\Phi_0 \qquad (T = T_c) \; .$$

Therefore, the critical temperature becomes:

$$T_c = \frac{2\pi}{m^*}\left[\frac{\rho}{g_{d/2}(1)}\right]^{2/d} \; . \tag{94}$$

As an example, let us consider the realistic three-dimensional case. Using the Robinson representation (see the details in review [18]), we can find the behavior of the effective chemical potential at high temperatures:

$$\mu_{eff} = T\ln\left(\rho\lambda_T^3\right) \qquad (T \gg T_c) \tag{95}$$

and at the temperature approaching the critical point from above,

$$\mu_{eff} \simeq -T\,\frac{\zeta^2(3/2)}{4\pi}\left[1 - \left(\frac{T_c}{T}\right)^{3/2}\right]^2 \; . \tag{96}$$

Then, the isothermal compressibility:

$$\kappa_T = \frac{g_{1/2}(z)}{\rho^2 T \lambda_T^3} \tag{97}$$

at high temperatures is:

$$\kappa_T \simeq \frac{1}{\rho T} \qquad (T \gg T_c) \; , \tag{98}$$

while close to the critical point, it is:

$$\kappa_T \simeq \frac{0.921}{\rho T}\left[1 - \left(\frac{T_c}{T}\right)^{3/2}\right]^{-1} \; ; \tag{99}$$

respectively, particle fluctuations, described by the relative variance:

$$\frac{\mathrm{var}(\hat{N})}{N} = \rho T\kappa_T = \frac{g_{1/2}(z)}{\rho\lambda_T^3} \; , \tag{100}$$

at high temperatures behave as:

$$\frac{\mathrm{var}(\hat{N})}{N} \simeq 1 \qquad (T \gg T_c), \tag{101}$$

and close to the critical point, we get:

$$\frac{\mathrm{var}(\hat{N})}{N} \simeq 0.921 \left[1 - \left(\frac{T_c}{T} \right)^{3/2} \right]^{-1} . \tag{102}$$

Outside of the critical temperature itself, particle fluctuations are thermodynamically normal. The divergence of the compressibility at the critical point signifies a second-order phase transition. At the point of the phase transition, the system is not stable, and the fluctuations do not need to be finite.

6. Interacting Bose System below the Condensation Temperature

In Section 3, it is proven that the ideal Bose gas, confined in a box, is stable below the condensation temperature only for $d > 4$. In the present section, we show that interactions stabilize the system, making it stable already for $d = 3$.

6.1. Self-Consistent Approach

For describing a Bose system with the Bose–Einstein condensate, we employ the self-consistent approach [16–18,24,34,35], providing a gapless spectrum, correct thermodynamics, the validity of all conservation laws, and good agreement with Monte Carlo simulations and experimental data.

The energy Hamiltonian has the form:

$$\hat{H} = \int \hat{\psi}^\dagger(\mathbf{r}) \left(-\frac{\nabla^2}{2m} \right) \hat{\psi}(\mathbf{r}) \, d\mathbf{r} +$$

$$+ \frac{1}{2} \int \hat{\psi}^\dagger(\mathbf{r}) \hat{\psi}^\dagger(\mathbf{r}') \Phi(\mathbf{r} - \mathbf{r}') \hat{\psi}(\mathbf{r}') \hat{\psi}(\mathbf{r}) \, d\mathbf{r} d\mathbf{r}' . \tag{103}$$

The genuine Bose–Einstein condensation necessarily requires global gauge symmetry breaking [6,9,17,18]. Finite systems, strictly speaking, do not exhibit this symmetry breaking. However, a system with a large number of particles $N \gg 1$ enjoys asymptotic symmetry breaking [36] in the sense that the system properties asymptotically, with respect to N, are close to the system with broken symmetry. The global gauge symmetry can be broken by the Bogolubov shift [28–30]:

$$\hat{\psi}(\mathbf{r}) = \eta(\mathbf{r}) + \psi_1(\mathbf{r}) , \tag{104}$$

in which the condensate function $\eta(\mathbf{r})$ and the operator of uncondensed particles $\psi_1(\mathbf{r})$ are mutually orthogonal,

$$\int \eta^*(\mathbf{r}) \psi_1(\mathbf{r}) \, d\mathbf{r} = 0 \tag{105}$$

and the operator of uncondensed particles satisfies the condition:

$$\langle \psi_1(\mathbf{r}) \rangle = 0 . \tag{106}$$

The number of condensed particles is:

$$N_0 = \int |\eta(\mathbf{r})|^2 \, d\mathbf{r} , \tag{107}$$

while the number of uncondensed particles is given by the average:

$$N_1 = \langle \hat{N}_1 \rangle , \qquad \hat{N}_1 = \int \psi_1^\dagger(\mathbf{r}) \psi_1(\mathbf{r}) \, d\mathbf{r} . \tag{108}$$

The grand Hamiltonian reads as:

$$H = \hat{H} - \mu_0 N_0 - \mu_1 \hat{N}_1 - \hat{\Lambda} \, , \tag{109}$$

where:

$$\hat{\Lambda} = \int \left[\lambda(\mathbf{r}) \psi_1^\dagger(\mathbf{r}) + \lambda^*(\mathbf{r}) \psi_1(\mathbf{r}) \right] \, d\mathbf{r}$$

and μ_0, μ_1, and $\lambda(\mathbf{r})$ are Lagrange multipliers guaranteeing the validity of the normalizations (107) and (108), as well as condition (106).

The evolution equation for the condensate function can be written as:

$$i \, \frac{\partial}{\partial t} \, \eta(\mathbf{r}, t) = \left\langle \frac{\delta H}{\delta \eta^*(\mathbf{r}, t)} \right\rangle \tag{110}$$

and the equation for the operator of uncondensed particles as:

$$i \, \frac{\partial}{\partial t} \, \psi_1(\mathbf{r}, t) = \frac{\delta H}{\delta \psi_1^\dagger(\mathbf{r}, t)} \, . \tag{111}$$

Keeping in mind, as usual, the periodic continuation of the box, we expand the field operators in plane waves, as in (13), and assume the existence of the Fourier representation for the interaction potential:

$$\Phi_k = \int \Phi(\mathbf{r}) e^{-i\mathbf{k}\cdot\mathbf{r}} \, d\mathbf{r} \, , \qquad \Phi(\mathbf{r}) = \frac{1}{V} \sum_k \Phi_k e^{i\mathbf{k}\cdot\mathbf{r}} \, . \tag{112}$$

Then, we get the normal density matrix:

$$\rho_1(\mathbf{r}, \mathbf{r}') = \langle \psi_1^\dagger(\mathbf{r}') \psi_1(\mathbf{r}) \rangle = \frac{1}{V} \sum_{k \neq 0} n_k e^{i\mathbf{k}\cdot(\mathbf{r}-\mathbf{r}')} \tag{113}$$

and the anomalous matrix:

$$\sigma_1(\mathbf{r}, \mathbf{r}') = \langle \psi_1(\mathbf{r}') \psi_1(\mathbf{r}) \rangle = \frac{1}{V} \sum_{k \neq 0} \sigma_k e^{i\mathbf{k}\cdot(\mathbf{r}-\mathbf{r}')} , \tag{114}$$

in which:

$$n_k \equiv \langle a_k^\dagger a_k \rangle \, , \qquad \sigma_k \equiv \langle a_k a_{-k} \rangle \, . \tag{115}$$

The condensate function $\eta(\mathbf{r}) = \eta$ defines the condensate density:

$$\rho_0 \equiv \frac{N_0}{V} = |\eta|^2 \, . \tag{116}$$

The density of uncondensed particles is:

$$\rho_1 \equiv \frac{N_1}{V} = \rho_1(\mathbf{r}, \mathbf{r}) = \frac{1}{V} \sum_k n_k \, . \tag{117}$$

The diagonal anomalous matrix gives the anomalous average:

$$\sigma_1 \equiv \sigma_1(\mathbf{r}, \mathbf{r}) = \frac{1}{V} \sum_k \sigma_k \, . \tag{118}$$

The average density of particles is the sum:

$$\rho \equiv \frac{N}{V} = \rho_0 + \rho_1 \, . \tag{119}$$

Then, we use the Hartree–Fock–Bogolubov approximation and accomplish the Bogolubov canonical transformation:

$$a_k = u_k b_k + v^*_{-k} b^\dagger_{-k} \, , \qquad b_k = u^*_k a_k - v^*_k a^\dagger_{-k} \, ,$$

where u_k and v_k are chosen so as to diagonalize the Hamiltonian. In that way, we obtain the diagonalized Hamiltonian:

$$H_B = E_B + \sum_k \varepsilon_k b^\dagger_k b_k \, , \tag{120}$$

in which:

$$E_B = -\frac{1}{2} \, N\rho\Phi_0 \; - \; \rho_0 \sum_p (n_p + \sigma_p)\Phi_p \; - \; \frac{1}{2V} \sum_{kp} (n_k n_p + \sigma_k \sigma_p)\Phi_{k+p} \; + \; \frac{1}{2} \sum_k (\varepsilon_k - \omega_k) \, ,$$

the particle spectrum is:

$$\varepsilon_k = \sqrt{\omega_k^2 - \Delta_k^2} \, , \tag{121}$$

and where:

$$\omega_k = \frac{k^2}{2m} + \Delta + \rho_0(\Phi_k - \Phi_0) + \frac{1}{V} \sum_p n_p (\Phi_{k+p} - \Phi_p) \, ,$$

$$\Delta_k = \rho_0 \Phi_k + \frac{1}{V} \sum_p \sigma_p \Phi_{k+p} \, , \qquad \Delta \equiv \lim_{k \to 0} \Delta_k = \rho_0 \Phi_0 + \frac{1}{V} \sum_p \sigma_p \Phi_p \, . \tag{122}$$

For the expressions in (115), we find:

$$n_k = \frac{\omega_k}{2\varepsilon_k} \, \coth\left(\frac{\varepsilon_k}{2T} \right) - \frac{1}{2} \, , \qquad \sigma_k = -\, \frac{\Delta_k}{2\varepsilon_k} \, \coth\left(\frac{\varepsilon_k}{2T} \right) \, . \tag{123}$$

The chemical potentials are:

$$\mu_0 = \rho\Phi_0 + \frac{1}{V} \sum_k (n_k + \sigma_k)\Phi_k \, , \qquad \mu_1 = \rho\Phi_0 + \frac{1}{V} \sum_k (n_k - \sigma_k)\Phi_k \, . \tag{124}$$

In the long-wave limit, we can use the expansion:

$$\Phi_{k+p} \; \simeq \; \Phi_p + \frac{k^2}{2} \, \Phi''_p \qquad (k \to 0) \, ,$$

where:

$$\Phi''_p \; \equiv \; \frac{\partial^2 \Phi_p}{\partial p^2} \, .$$

Then, the spectrum (121) becomes of the phonon type:

$$\varepsilon_k \; \simeq \; ck \qquad (k \to 0) \, , \tag{125}$$

with the sound velocity:

$$c = \sqrt{\frac{\Delta}{m_{eff}}} \tag{126}$$

and with the notation for the effective mass:

$$m_{eff} \equiv \frac{m}{1 + \frac{m}{V} \sum_p (n_p - \sigma_p)\Phi''_p} \, . \tag{127}$$

Actually, Expression (126), which can be written as:

$$m_{eff}c^2 = \Delta \, ,$$

is the equation:

$$\frac{mc^2}{1+\frac{m}{V}\sum_p(n_p-\sigma_p)\Phi''_p} = \rho_0\Phi_0 + \frac{1}{V}\sum_p \sigma_p\Phi_p \, , \tag{128}$$

defining the sound velocity c.

To simplify the consideration, we can resort to the approximation (86), similarly, to which we can write:

$$\sum_p \sigma_p\Phi_{k+p} \;\cong\; \Phi_k \sum_p \sigma_p \, . \tag{129}$$

This gives:

$$\frac{1}{V}\sum_p (n_p-\sigma_p)\Phi''_p = (\rho_1-\sigma_1)\Phi''_0 \, ,$$

where:

$$\Phi''_0 = \lim_{p\to 0}\Phi''_p = -\,\frac{4\pi}{3}\int_0^\infty \Phi(\mathbf{r})r^4\, d\mathbf{r} \, .$$

In view of the notation for the effective interaction radius (88), we get:

$$\Phi''_0 = -\,\frac{1}{3}\,\Phi_0 r^2_{eff} \, .$$

Then, the effective mass (127) acquires the form:

$$m_{eff} = \frac{m}{1+(\sigma_1-\rho_1)\Phi_0 m r^2_{eff}/3} \, . \tag{130}$$

In the approximations (86) and (129), the chemical potentials (124) become:

$$\mu_0 = \rho\Phi_0 + (\rho_1+\sigma_1)\Phi_0 \, , \qquad \mu_1 = \rho\Phi_0 + (\rho_1-\sigma_1)\Phi_0 \, . \tag{131}$$

Furthermore, we have:

$$\omega_k = \frac{k^2}{2m} + \Delta + \rho(\Phi_k-\Phi_0) \qquad \Delta_k = (\rho_0+\sigma_1)\Phi_k \, , \qquad \Delta = (\rho_0+\sigma_1)\Phi_0 \, . \tag{132}$$

The spectrum (121) can be written as:

$$\varepsilon_k^2 = \left[\frac{k^2}{2m} + (\rho_1-\sigma_1)(\Phi_k-\Phi_0)\right]\left[\frac{k^2}{2m} + \rho(\Phi_k-\Phi_0) + (\rho_0+\sigma_1)(\Phi_k+\Phi_0)\right] \, . \tag{133}$$

The density of uncondensed particles is:

$$\rho_1 = \int \left[\frac{\omega_k}{2\varepsilon_k}\,\coth\left(\frac{\varepsilon_k}{2T}\right) - \frac{1}{2}\right]\frac{d\mathbf{k}}{(2\pi)^3} \, . \tag{134}$$

The anomalous average (118) can be represented in the form:

$$\sigma_1 = -\int \frac{\Delta_k}{2\varepsilon_k}\,\frac{d\mathbf{k}}{(2\pi)^3} - \int \frac{\Delta_k}{2\varepsilon_k}\left[\coth\left(\frac{\varepsilon_k}{2T}\right) - 1\right]\frac{d\mathbf{k}}{(2\pi)^3} \, . \tag{135}$$

When the first term here diverges, which happens for the local interaction, we can use dimensional regularization [18].

6.2. Particle Fluctuations

The number-of-particles variance can be found by involving the formula:

$$\frac{\text{var}(\hat{N})}{N} = 1 + \rho \int [g(\mathbf{r}) - 1] \, d\mathbf{r} \, , \tag{136}$$

in which:

$$g(\mathbf{r}_{12}) = \frac{1}{g^2} \langle \hat{\psi}^\dagger(\mathbf{r}_1) \hat{\psi}^\dagger(\mathbf{r}_2) \hat{\psi}(\mathbf{r}_2) \hat{\psi}(\mathbf{r}_1) \rangle \tag{137}$$

is the pair correlation function, with $\mathbf{r}_{12} \equiv \mathbf{r}_1 - \mathbf{r}_2$.

Accomplishing the Bogolubov shift (104), we use the Hartree–Fock–Bogolubov (HFB) decoupling for the expressions containing the operators ψ_1. Since, mathematically, the HFB approximation is of second order with respect to the products of the operators ψ_1, it is necessary to leave in the pair correlation function only the terms of second order with respect to these operators [3,16–18,23,24]. As a result, we obtain:

$$\int [g(\mathbf{r}) - 1] \, d\mathbf{r} = \frac{2}{\rho} \lim_{k \to 0} (n_k + \sigma_k) \, . \tag{138}$$

In this way, for the relative variance, we find:

$$\frac{\text{var}(\hat{N})}{N} = 1 + 2 \lim_{k \to 0} (n_k + \sigma_k) \, . \tag{139}$$

For small k, when ε_k tends to zero, we have:

$$n_k \simeq \frac{T\Delta_k}{\varepsilon_k^2} + \frac{\Delta_k}{12T} + \frac{T}{2\Delta_k} - \frac{1}{2} + \left(\frac{\Delta_k}{3T} - \frac{T}{\Delta_k} - \frac{\Delta_k^3}{90T^3} \right) \frac{\varepsilon_k^2}{8\Delta_k^2} \, ,$$

$$\sigma_k \simeq - \frac{T\Delta_k}{\varepsilon_k^2} - \frac{\Delta_k}{12T} + \frac{\Delta_k \varepsilon_k^2}{720T^3} \qquad (\varepsilon_k \to 0) \, . \tag{140}$$

Therefore:

$$\lim_{k \to 0} (n_k + \sigma_k) = \frac{1}{2} \left(\frac{T}{\Delta} - 1 \right) \, ,$$

with:

$$\Delta = m_{eff} c^2 = (\rho_0 + \sigma_1) \Phi_0 \, .$$

Thus, we come to the expression:

$$\frac{\text{var}(\hat{N})}{N} = \frac{T}{m_{eff} c^2} \, ; \tag{141}$$

respectively, the compressibility is:

$$\kappa_T = \frac{\text{var}(\hat{N})}{N \rho T} = \frac{1}{\rho m_{eff} c^2} \, . \tag{142}$$

Taking into account Formula (126) leads to the variance:

$$\frac{\text{var}(\hat{N})}{N} = \frac{T}{(\rho_0 + \sigma_1) \Phi_0} \, . \tag{143}$$

Note that Expression (143) is valid at zero temperature, as well. This is easy to check considering the quantities (123) at zero temperature,

$$n_k = \frac{\sqrt{\varepsilon_k^2 + \Delta_k^2}}{2\varepsilon_k} - \frac{1}{2} , \qquad \sigma_k = -\frac{\Delta_k}{2\varepsilon_k} \qquad (T = 0) .$$

From here, in the long-wave limit, we have:

$$n_k \simeq \frac{\Delta_k}{2\varepsilon_k} + \frac{\varepsilon_k}{4\Delta_k} - \frac{1}{2} \qquad (\varepsilon_k \rightarrow 0 , \; T = 0) .$$

Hence:

$$\lim_{k\rightarrow 0} (n_k + \sigma_k) = -\frac{1}{2} \qquad (T = 0)$$

and:

$$\frac{\text{var}(\hat{N})}{N} = 0 \qquad (T = 0) .$$

The above result for the relative variance (143) can be generalized for nonuniform systems [37] by involving the local-density approximation, which yields:

$$\frac{\text{var}(\hat{N})}{N} = \frac{T}{N} \int \frac{\rho(\mathbf{r})}{\Delta(\mathbf{r})} \, d\mathbf{r} , \tag{144}$$

where:

$$\Delta(\mathbf{r}) = [\rho_0(\mathbf{r}) + \sigma_1(\mathbf{r})]\Phi_0 . \tag{145}$$

Particle fluctuations in a three-dimensional Bose-condensed system of interacting particles are thermodynamically normal in both cases, when particles are in a box or in a nonuniform external potential.

7. Conclusions

Particle fluctuations in Bose systems were studied. Investigating the behavior of these fluctuations is important because they are directly connected with isothermal compressibility and define the system stability with respect to pressure variations. Thermodynamically-anomalous fluctuations signify system instability; while thermodynamically-normal fluctuations mean that the equilibrium system is stable. The obtained results are as follows.

The ideal Bose gas confined in a rectangular box is stable, depending on the temperature, in spatial dimensions:

$$d > 2 \qquad (T > T_c) ,$$

$$d > 4 \qquad (T < T_c) .$$

The stability of the ideal Bose gas in a power-law trap depends on the confining dimension:

$$D \equiv d + \sum_{\alpha=1}^{d} \frac{2}{n_\alpha} .$$

This gas is stable for the confining dimensions:

$$D > 2 \qquad (T > T_c) ,$$

$$D > 4 \qquad (T < T_c) .$$

Interactions stabilize Bose-condensed systems, so that an interacting system with Bose–Einstein condensate becomes stable at $d = 3$ for either a system in a box or in an external potential.

Nonlocal interactions with a stronger strength or with a larger interaction radius increase the effective mass, hence diminishing the condensation temperature.

Funding: This research received no external funding.

Acknowledgments: The author is grateful to E.P. Yukalova for useful discussions.

Conflicts of Interest: The author declares no conflict of interest.

References

1. Courteille, P.W.; Bagnato, V.S.; Yukalov, V.I. Bose-Einstein condensation of trapped atomic gases. *Laser Phys.* **2001**, *11*, 659–800.
2. Andersen, J.O. Theory of the weakly interacting Bose gas. *Rev. Mod. Phys.* **2004**, *76*, 599–639.
3. Yukalov, V.I. Principal problems in Bose-Einstein condensation of dilute gases. *Laser Phys. Lett.* **2004**, *1*, 435–461.
4. Bongs, K.; Sengstock, K. Physics with coherent matter waves. *Rep. Prog. Phys.* **2004**, *67*, 907–964.
5. Yukalov, V.I.; Girardeau, M.D. Fermi-Bose mapping for one-dimensional Bose gases. *Laser Phys. Lett.* **2005**, *2*, 375–382.
6. Lieb, E.H.; Seiringer, R.; Solovej, J.P.; Yngvason, J. *The Mathematics of the Bose Gas and its Condensation*; Birkhauser: Basel, Switzerland, 2005.
7. Posazhennikova, A. Weakly interacting dilute Bose gases in 2*D*. *Rev. Mod. Phys.* **2006**, *78*, 1111–1134.
8. Morsch, O.; Oberthaler, M. Dynamics of Bose-Einstein condensates in optical lattices. *Rev. Mod. Phys.* **2006**, *78*, 179–216.
9. Yukalov, V.I. Bose-Einstein condensation and gauge symmetry breaking. *Laser Phys. Lett.* **2007**, *4*, 632–647.
10. Letokhov, V. *Laser Control of Atoms and Molecules*; Oxford University Press: Oxford, UK, 2007.
11. Moseley, C.; Fialko, O.; Ziegler, K. Interacting bosons in an optical lattice. *Ann. Phys.* **2008**, *17*, 561–608.
12. Bloch, I.; Dalibard, J.; Zwerger, W. Many-body physics with ultracold gases. *Rev. Mod. Phys.* **2008**, *80*, 885–964.
13. Proukakis, N.P.; Jackson, B. Finite-temperature models of Bose-Einstein condensation. *J. Phys. B* **2008**, *41*, 203002.
14. Yurovsky, V.A.; Olshanii, M.; Weiss, D.S. Collisions, correlations, and integrability in atom waveguides. *Adv. At. Mol. Opt. Phys.* **2008**, *55*, 61–138.
15. Pethick, C.J.; Smith, H. *Bose-Einstein Condensation in Dilute Gases*; Cambridge University: Cambridge, UK, 2008.
16. Yukalov, V.I. Cold bosons in optical lattices. *Laser Phys.* **2009**, *19*, 1–110.
17. Yukalov, V.I. Basics of Bose-Einstein condensation. *Phys. Part. Nucl.* **2011**, *42*, 460–513.
18. Yukalov, V.I. Theory of cold atoms: Bose-Einstein statistics. *Laser Phys.* **2016**, *26*, 062001.
19. Yukalov, V.I.; Yukalova, E.P. Bose-Einstein condensation temperature of weakly interacting atoms. *Laser Phys. Lett.* **2017**, *14*, 073001.
20. Kocharovsky, V.V.; Kocharovsky, V.V.; Holthaus, M.; Ooi, C.R.; Svidzinsky, A.; Ketterle, W.; Scully, M.O. Fluctuations in ideal and interacting Bose–Einstein condensates: From the laser phase transition analogy to squeezed states and Bogolubov quasiparticles. *Adv. At. Mol. Opt. Phys.* **2006**, *53*, 291–411.
21. Patashinskii, A.Z.; Pokrovskii, V.L. *Fluctuation Theory of Phase Transitions*; Pergamon: Oxford, UK, 1979.
22. Yukalov, V.I. Theory of cold atoms: Basics of quantum statistics. *Laser Phys.* **2013**, *23*, 062001.
23. Yukalov, V.I. No anomalous fluctuations exist in stable equilibrium systems. *Phys. Lett. A* **2005**, *340*, 369–374.
24. Yukalov, V.I. Fluctuations of composite observables and stability of statistical systems. *Phys. Rev. E* **2005**, *72*, 066119.
25. Gaunt, A.L.; Tobias, F.; Schmidutz, T.F.; Gotlibovych, I.; Smith, R.P.; Hadzibabic, Z. Bose-Einstein condensation of atoms in a uniform potential. *Phys. Rev. Lett.* **2013**, *110*, 200406.
26. Navon, N.; Gaunt, A.L.; Smith, R.P.; Hadzibabic, Z. Critical dynamics of spontaneous symmetry breaking in a homogeneous Bose gas. *Science* **2015**, *347*, 167–170.
27. Lopes, R.; Eigen, C.; Navon, N.; Clement, D.; Smith, R.P.; Hadzibabic, Z. Quantum depletion of a homogeneous Bose-Einstein condensate. *Phys. Rev. Lett.* **2017**, *119*, 190404.
28. Bogolubov, N.N. *Lectures on Quantum Statistics*; Gordon and Breach: New York, NY, USA, 1967; Volume 1.

29. Bogolubov, N.N. *Lectures on Quantum Statistics*; Gordon and Breach: New York, NY, USA, 1970; Volume 2.
30. Bogolubov, N.N. *Quantum Statistical Mechanics*; World Scientific: Singapore, 2015.
31. Yukalov, V.I. Modified semiclassical approximation for trapped Bose gases. *Phys. Rev. A* **2005**, *72*, 033608.
32. Yukalov, V.I.; Yukalova, E.P. Bose-condensed atomic systems with nonlocal interaction potentials. *Laser Phys.* **2016**, *26*, 045501.
33. Yukalov, V.I. Dipolar and spinor bosonic systems. *Laser Phys.* **2018**, *28*, 053001.
34. Yukalov, V.I. Self-consistent theory of Bose-condensed systems. *Phys. Lett. A* **2006**, *359*, 712–717.
35. Yukalov, V.I. Representative statistical ensembles for Bose systems with broken gauge symmetry. *Ann. Phys.* **2008**, *323*, 461–499.
36. Birman, J.L.; Nazmitdinov, R.G.; Yukalov, V.I. Effects of symmetry breaking in finite quantum systems. *Phys. Rep.* **2013**, *526*, 1–91.
37. Yukalov, V.I. Particle fluctuations in nonuniform and trapped Bose gases. *Laser Phys. Lett.* **2009**, *6*, 688–695.

Article

Analysis of a Trapped Bose–Einstein Condensate in Terms of Position, Momentum, and Angular-Momentum Variance

Ofir E. Alon [1,2]

[1] Department of Mathematics, University of Haifa, Haifa 3498838, Israel; ofir@research.haifa.ac.il
[2] Haifa Research Center for Theoretical Physics and Astrophysics, University of Haifa, Haifa 3498838, Israel

Received: 30 September 2019; Accepted: 27 October 2019; Published: 1 November 2019

Abstract: We analyze, analytically and numerically, the position, momentum, and in particular the angular-momentum variance of a Bose–Einstein condensate (BEC) trapped in a two-dimensional anisotropic trap for static and dynamic scenarios. Explicitly, we study the ground state of the anisotropic harmonic-interaction model in two spatial dimensions analytically and the out-of-equilibrium dynamics of repulsive bosons in tilted two-dimensional annuli numerically accurately by using the multiconfigurational time-dependent Hartree for bosons method. The differences between the variances at the mean-field level, which are attributed to the shape of the BEC, and the variances at the many-body level, which incorporate depletion, are used to characterize position, momentum, and angular-momentum correlations in the BEC for finite systems and at the limit of an infinite number of particles where the bosons are 100% condensed. Finally, we also explore inter-connections between the variances.

Keywords: Bose–Einstein condensates; density; position variance; momentum variance; angular-momentum variance; harmonic-interaction model; MCTDHB

PACS: 03.75.Hh; 03.75.Kk; 67.85.Bc; 67.85.De; 03.65.-w

---

1. Introduction

Bose–Einstein condensates (BECs) made of ultra-cold atoms offer a wide platform to study many-body physics [1–5]. Here, there is a growing interest in the so-called particle limit [6–16], in which the interaction parameter (i.e., the product of the interaction strength times the number of particles) is kept fixed while the number of particles is increased to infinity. At the particle limit, the energy per particle, density per particle, and reduced density matrices [17] per particle computed at the many-body level of theory boil down to those obtained in mean-field theory [7–10,14,16], despite the fact that the respective many-boson wavefunctions are (much) different [13,15]. It turns out that variances of many-particle operators are a useful tool to characterize correlations (namely, differences between respective many-body and mean-field quantities) that exist even when the interacting bosons are 100% condensed [11,12].

The variance of a many-particle operator of a trapped BEC generally depends on the trap shape, strength and sign of the interaction and, in out-of-equilibrium problems, on time. Consequently, the difference between variances computed at the many-body and mean-field levels of theory also depends on these variables and, of course, on the observable under examination. The first examples [11,12] concentrated on one-dimensional problems and the position and momentum variances, and investigated conditions and mechanisms for the differences between the respective many-body and mean-field variances at the particle limit. In two spatial dimensions, further types of trap topologies come into play, and respective many-body and mean-field variances can exhibit additional phenomena, such as opposite anisotropy [18] and distinct (effective) dimensionality [19]. The many-body variance of a trap BEC has been applied to extract excitations [20], analyze the range of inter-particle interaction [21], examine the effects of asymmetry of a double-well potential [22], and to assess numerical convergence [23,24].

So far, only the position and momentum variances were studied for BECs in rather general traps. In [25,26], the angular-momentum variance is studied for BECs in two-dimensional isotropic traps, and scenarios were the mean-field angular-momentum variance has less [25] or more [26] symmetry (in terms of its conservation) than the many-body angular-momentum variance are identified. Going beyond these works, in the present work we study, analytically and numerically, the angular-momentum variance of a trapped BEC in a two-dimensional anisotropic trap for static and dynamic scenarios, and analyze the difference between the many-body and mean-field variances for finite systems and at the limit of an infinite number of particles. Furthermore, we also study the respective position and momentum variances, and thereby offer a comprehensive characterization of the BEC in terms of its variances. This would allow us to put forward inter-connections between the variances.

Let us elaborate on the strategy of exposition chosen in the paper. We first study the ground state of a many-particle model which is exactly solvable, i.e., integrable, both at the many-body and mean-field levels of theory. A couple of symmetries are also used in the analysis. These would allow us to obtain exact and transparent results for any number of particles and particularly to analyze the variances at the particle limit. The merit of analytical closed-form results and, in the context of interacting bosons, their explicit evaluation at the limit of an infinite number of particles is obvious. Then, as is the usual case in many realistic systems, we continue to explore a set-up which is not integrable, and more so, examine its out-of-equilibrium dynamics which is rather complicated already at the mean-field level of theory, let alone at the many-body level of theory. The later necessitates the state-of-the art numerical tools for the accurate integration of the Schrödinger equation and a careful interpolation of properties to the particle limit. All in all, we show below that the combination of analytics and numerics, i.e., of completely opposite methodologies, provides substantial and complementary novel knowledge on the position, momentum, and angular-momentum variances of anisotropic trapped BECs in two spatial dimensions.

The structure of the paper is as follows. In Section 2 we study the position, momentum, and angular-momentum variances of the ground state within an exactly solvable model, the anisotropic harmonic-interaction model. In Section 3 we study numerically the time-dependent variances of an out-of-equilibrium BEC sloshing in a tilted annulus. Summary and outlook are given in Section 4. Finally, Appendix A discusses translations of variances and inter-connections of the latter.

2. The Anisotropic Harmonic-Interaction Model

Solvable models of particles interacting by harmonic forces, or, briefly, the harmonic-interaction model (and its variants), have drawn including in the BEC literature much attention [27–41]. Here we consider the anisotropic two-dimensional harmonic-interaction model

$$\hat{H}(\mathbf{r}_1,\ldots,\mathbf{r}_N) = \sum_{j=1}^{N}\left[\left(-\frac{1}{2}\frac{\partial^2}{\partial x_j^2}+\frac{1}{2}\omega_x^2 x_j^2\right)+\left(-\frac{1}{2}\frac{\partial^2}{\partial y_j^2}+\frac{1}{2}\omega_y^2 y_j^2\right)\right]$$
$$+\lambda_0 \sum_{1\le j<k}^{N}\left[(x_j-x_k)^2+(y_j-y_k)^2\right], \tag{1}$$

where λ_0 is the interaction strength; positive values imply attraction and negative repulsion. Without loss of generality we take $\omega_y > \omega_x$, namely, that the trap is tighter along the y-axis than along the x-axis (the trap anisotropy satisfies $\frac{\omega_y}{\omega_x} > 1$). Here and hereafter $\hbar = m = 1$.

Transforming from Cartesian to Jacobi coordinates,

$$Q_{k,x} = \frac{1}{\sqrt{k(k+1)}} \sum_{j=1}^{k} \left(x_{k+1} - x_j\right), \quad Q_{k,y} = \frac{1}{\sqrt{k(k+1)}} \sum_{j=1}^{k} \left(y_{k+1} - y_j\right), \quad 1 \le k \le N-1,$$
$$Q_{N,x} = \frac{1}{\sqrt{N}} \sum_{j=1}^{N} x_j, \quad Q_{N,y} = \frac{1}{\sqrt{N}} \sum_{j=1}^{N} y_j, \tag{2}$$

the many-body solution for the ground state is given by

$$\Psi(\mathbf{Q}_1, \ldots, \mathbf{Q}_N) = \left(\frac{\omega_x}{\pi}\right)^{\frac{1}{4}} \left(\frac{\omega_y}{\pi}\right)^{\frac{1}{4}} \left(\frac{\Omega_x}{\pi}\right)^{\frac{N-1}{4}} \left(\frac{\Omega_y}{\pi}\right)^{\frac{N-1}{4}}$$
$$\times e^{-\frac{1}{2}\left(\Omega_x \sum_{j=1}^{N-1} Q_{j,x}^2 + \omega_x Q_{N,x}^2\right)} \times e^{-\frac{1}{2}\left(\Omega_y \sum_{j=1}^{N-1} Q_{j,y}^2 + \omega_y Q_{N,y}^2\right)}$$
$$= \Psi(\mathbf{r}_1, \ldots, \mathbf{r}_N) = \left(\frac{\omega_x}{\pi}\right)^{\frac{1}{4}} \left(\frac{\omega_y}{\pi}\right)^{\frac{1}{4}} \left(\frac{\Omega_x}{\pi}\right)^{\frac{N-1}{4}} \left(\frac{\Omega_y}{\pi}\right)^{\frac{N-1}{4}}$$
$$\times e^{-\frac{\alpha_x}{2} \sum_{j=1}^{N} x_j^2 - \beta_x \sum_{1 \le j < k}^{N} x_j x_k} \times e^{-\frac{\alpha_y}{2} \sum_{j=1}^{N} y_j^2 - \beta_y \sum_{1 \le j < k}^{N} y_j y_k}, \tag{3}$$

where

$$\Omega_x = \sqrt{\omega_x^2 + 2N\lambda_0}, \qquad \Omega_y = \sqrt{\omega_y^2 + 2N\lambda_0} \tag{4}$$

are the interaction-dressed frequencies of the relative-motion degrees-of-freedom, and

$$\alpha_x = \Omega_x + \beta_x, \qquad \beta_x = \frac{1}{N}\left(\omega_x - \Omega_x\right),$$
$$\alpha_y = \Omega_y + \beta_y, \qquad \beta_y = \frac{1}{N}\left(\omega_y - \Omega_y\right) \tag{5}$$

are parameters arising in the transformation from Jacoby coordinates back to Cartesian coordinates. Equation (4) prescribes the range of interactions for which the system is trapped, $\lambda_0 > -\frac{\omega_x^2}{2N}$, i.e., from moderate repulsion to any attraction. Clearly, the many-body solution (3) in two spatial dimensions factorizes to a product of respective one-dimensional many-body solutions.

All properties of the ground state can in principle be obtained from Ψ, such as the energy, densities, and reduced density matrices, see [29]. Here, as mentioned above, we concentrate on variances and their inter-connections. The many-particle position $\hat{X} = \sum_{j=1}^{N} x_j$, $\hat{Y} = \sum_{j=1}^{N} y_j$ variance per particle is given by

$$\frac{1}{N}\Delta_{\hat{X}}^2 = \frac{1}{2\omega_x}, \qquad \frac{1}{N}\Delta_{\hat{Y}}^2 = \frac{1}{2\omega_y}. \tag{6}$$

Due to the symmetry of center-of-mass separation in the Hamiltonian (1), the many-particle position variance per particle is independent both of the interaction strength and the number of bosons in the system. Similarly, the many-particle momentum $\hat{P}_X = \sum_{j=1}^{N} \frac{1}{i}\frac{\partial}{\partial x_j}$, $\hat{P}_Y = \sum_{j=1}^{N} \frac{1}{i}\frac{\partial}{\partial y_j}$ variance per particle is given by

$$\frac{1}{N}\Delta_{\hat{P}_X}^2 = \frac{\omega_x}{2}, \qquad \frac{1}{N}\Delta_{\hat{P}_Y}^2 = \frac{\omega_y}{2}, \tag{7}$$

reflecting the minimal uncertainty product $\frac{1}{N}\Delta_{\hat{X}}^2 \frac{1}{N}\Delta_{\hat{P}_X}^2 = \frac{1}{N}\Delta_{\hat{Y}}^2 \frac{1}{N}\Delta_{\hat{P}_Y}^2 = \frac{1}{4}$ of the interacting system in the anisotropic harmonic trap.

The many-particle angular-momentum $\hat{L}_Z = \sum_{j=1}^{N} \frac{1}{i}\left(x_j \frac{\partial}{\partial y_j} - y_j \frac{\partial}{\partial x_j}\right)$ variance per particle is, at least for bosons, a less familiar and more intricate quantity. After some lengthy but otherwise straightforward algebra it is given by

$$\frac{1}{N}\Delta^2_{\hat{L}_Z} = \frac{1}{4}\frac{(\Omega_y-\Omega_x)^2}{\Omega_y\Omega_x}\left(\frac{N-1}{N}\right)^2\left[\left(1+\frac{1}{N-1}\frac{\Omega_y}{\omega_y}\right)\left(1+\frac{1}{N-1}\frac{\Omega_x}{\omega_x}\right)\right. \\ \left.+\left(\frac{\Omega_y}{\omega_y}-1\right)\left(\frac{\Omega_x}{\omega_x}-1\right)\right]+\frac{1}{4N}\frac{[(\omega_y-\Omega_y)-(\omega_x-\Omega_x)]\,[(\omega_y+\Omega_y)-(\omega_x+\Omega_x)]}{\omega_y\omega_x}, \tag{8}$$

where we have made use of the bosonic permutational symmetry, the structure of Ψ,

$$\hat{L}_Z\Psi = -\frac{1}{i}\left\{[(\alpha_y-\beta_y)-(\alpha_x-\beta_x)]\left(\sum_{j=1}^{N} x_j y_j\right)+(\beta_y-\beta_x)\left(\sum_{j=1}^{N} x_j\right)\left(\sum_{k=1}^{N} y_k\right)\right\}\Psi, \tag{9}$$

and the inverse coordinate transformations

$$\begin{aligned} x_N &= \frac{1}{\sqrt{N}}Q_{N,x}+\sqrt{\frac{N-1}{N}}Q_{N-1,x}, \qquad y_N = \frac{1}{\sqrt{N}}Q_{N,y}+\sqrt{\frac{N-1}{N}}Q_{N-1,y}, \\ x_{N-1} &= \frac{1}{\sqrt{N}}Q_{N,x}-\frac{1}{\sqrt{N(N-1)}}Q_{N-1,x}+\sqrt{\frac{N-2}{N-1}}Q_{N-2,x}, \\ y_{N-1} &= \frac{1}{\sqrt{N}}Q_{N,y}-\frac{1}{\sqrt{N(N-1)}}Q_{N-1,y}+\sqrt{\frac{N-2}{N-1}}Q_{N-2,y} \end{aligned} \tag{10}$$

to evaluate the various integral terms contributing to (8).

The angular-momentum variance per particle of the ground state (3) depends on the dressed frequencies, Ω_x and Ω_y, and the number of particles N. Namely, unlike the respective position and momentum variances it depends explicitly on the interaction strength and the number of particles. $\frac{1}{N}\Delta^2_{\hat{L}_Z}$ is, of course, non-zero only for anisotropic traps [for isotropic traps, from $\omega_y=\omega_x$ we get $\Omega_y=\Omega_x$ and expression (8) then vanishes]. For non-interacting bosons, Equation (8) boils down to $\frac{1}{N}\Delta^2_{\hat{L}_Z}=\frac{1}{4}\frac{(\omega_y-\omega_x)^2}{\omega_y\omega_x}=\frac{1}{4}\frac{\left(\frac{\omega_y}{\omega_x}-1\right)^2}{\frac{\omega_y}{\omega_x}}$, the value for a single particle in the anisotropic trap $\frac{1}{2}\omega_x^2x^2+\frac{1}{2}\omega_y^2y^2$, which only depends on the trap anisotropy. Opposite to the non-vanishing of the angular-momentum variance, we note that the expectation value of the angular-momentum operator, $\frac{1}{N}\langle\Psi|\hat{L}_Z|\Psi\rangle$, vanishes for any anisotropy $\frac{\omega_x}{\omega_y}$, interaction strength λ_0, and number of particles N. This is straightforward to see since Ψ is even under reflection of all coordinates $X\rightarrow -X$ and separately of $Y\rightarrow -Y$, whereas $\hat{L}_Z$ is odd under reflection.

The anisotropic harmonic-interaction model (1) can be solved analytically at the mean-field level of theory as well, like in [29], also see [41]. Starting from the ansatz where each and every boson resides in one and the same orbital, the mean-field solution is given by

$$\begin{aligned} &\Phi^{GP}(\mathbf{r}_1,\ldots,\mathbf{r}_N) = \\ &= \left(\frac{\sqrt{\omega_x^2+2\Lambda}}{\pi}\right)^{\frac{N}{4}}\left(\frac{\sqrt{\omega_y^2+2\Lambda}}{\pi}\right)^{\frac{N}{4}} e^{-\frac{1}{2}\sqrt{\omega_x^2+2\Lambda}\sum_{j=1}^N x_j^2}\times e^{-\frac{1}{2}\sqrt{\omega_y^2+2\Lambda}\sum_{j=1}^N y_j^2} \\ &= \Phi^{GP}(\mathbf{Q}_1,\ldots,\mathbf{Q}_N) \\ &= \left(\frac{\sqrt{\omega_x^2+2\Lambda}}{\pi}\right)^{\frac{N}{4}}\left(\frac{\sqrt{\omega_y^2+2\Lambda}}{\pi}\right)^{\frac{N}{4}} e^{-\frac{1}{2}\sqrt{\omega_x^2+2\Lambda}\sum_{k=1}^N Q_{k,x}^2}\times e^{-\frac{1}{2}\sqrt{\omega_y^2+2\Lambda}\sum_{k=1}^N Q_{k,y}^2}, \end{aligned} \tag{11}$$

where $\Lambda = (N-1)\lambda_0$ is the interaction parameter and $\Lambda > -\frac{\omega_x^2}{2}$ the condition for a trapped solution. Like the many-body solution (3), the mean-field solution (11) in two spatial dimensions factorizes to a product of respective one-dimensional mean-field solutions.

The many-particle position variance computed at the mean-field level is given by

$$\frac{1}{N}\Delta^2_{\hat{X},GP} = \frac{1}{2\sqrt{\omega_x^2+2\Lambda}}, \qquad \frac{1}{N}\Delta^2_{\hat{Y},GP} = \frac{1}{2\sqrt{\omega_y^2+2\Lambda}}, \tag{12}$$

and seen to be dressed by the interaction. Similarly, the many-particle momentum variance computed at the mean-field level is dressed by the interaction and given by

$$\frac{1}{N}\Delta^2_{\hat{P}_X,GP} = \frac{\sqrt{\omega_x^2+2\Lambda}}{2}, \qquad \frac{1}{N}\Delta^2_{\hat{P}_Y,GP} = \frac{\sqrt{\omega_y^2+2\Lambda}}{2}. \tag{13}$$

Interestingly, because the mean-field solution (11) is made of Gaussian functions, it satisfies the minimal uncertainty product $\frac{1}{N}\Delta^2_{\hat{X},GP}\frac{1}{N}\Delta^2_{\hat{P}_X,GP} = \frac{1}{N}\Delta^2_{\hat{Y},GP}\frac{1}{N}\Delta^2_{\hat{P}_Y,GP} = \frac{1}{4}$ as well.

The many-particle angular-momentum variance computed at the mean-field level is given by

$$\frac{1}{N}\Delta^2_{\hat{L}_Z,GP} = \frac{1}{4}\frac{\left(\sqrt{\omega_y^2+2\Lambda}-\sqrt{\omega_x^2+2\Lambda}\right)^2}{\sqrt{\omega_y^2+2\Lambda}\sqrt{\omega_x^2+2\Lambda}}, \tag{14}$$

where we have made use of the structure and symmetries of Φ^{GP},

$$\hat{L}_Z\Phi^{GP} = -\frac{1}{i}\left(\sqrt{\omega_y^2+2\Lambda}-\sqrt{\omega_x^2+2\Lambda}\right)\sum_{j=1}^{N} x_j y_j \Phi^{GP}, \tag{15}$$

to arrive at the final expression.

The relation between the mean-field and many-body variances deserves a discussion. Their difference is used to define position, momentum, and angular-momentum correlations in the system. For the position and momentum variances, the following ratios hold,

$$\frac{\frac{1}{N}\Delta^2_{\hat{X},GP}}{\frac{1}{N}\Delta^2_{\hat{X}}} = \frac{1}{\sqrt{1+\frac{2\Lambda}{\omega_x^2}}}, \qquad \frac{\frac{1}{N}\Delta^2_{\hat{Y},GP}}{\frac{1}{N}\Delta^2_{\hat{Y}}} = \frac{1}{\sqrt{1+\frac{2\Lambda}{\omega_y^2}}},$$
$$\frac{\frac{1}{N}\Delta^2_{\hat{P}_X,GP}}{\frac{1}{N}\Delta^2_{\hat{P}_X}} = \sqrt{1+\frac{2\Lambda}{\omega_x^2}}, \qquad \frac{\frac{1}{N}\Delta^2_{\hat{P}_Y,GP}}{\frac{1}{N}\Delta^2_{\hat{P}_Y}} = \sqrt{1+\frac{2\Lambda}{\omega_y^2}}, \tag{16}$$

obviously for any number of particles N. These ratios simply imply that, since repulsion ($\Lambda < 0$) broadens the position density, the many-body position variance is smaller than the corresponding mean-field one for repulsive interaction, and vise verse for attraction ($\Lambda > 0$). Inversely, since repulsion narrows the momentum density, the many-body momentum variance is larger than the corresponding mean-field one for repulsive interaction, and vise versa for attraction. Furthermore, both the position and momentum variances per particle exhibit the same anisotropies as the respective densities for any interaction parameter Λ, namely, if $\frac{1}{N}\Delta^2_{\hat{X},GP} > \frac{1}{N}\Delta^2_{\hat{Y},GP}$ then $\frac{1}{N}\Delta^2_{\hat{X}} > \frac{1}{N}\Delta^2_{\hat{Y}}$ is satisfied and, analogously, if $\frac{1}{N}\Delta^2_{\hat{P}_X,GP} < \frac{1}{N}\Delta^2_{\hat{P}_Y,GP}$ then $\frac{1}{N}\Delta^2_{\hat{P}_X} < \frac{1}{N}\Delta^2_{\hat{P}_Y}$ is satisfied. We shall return to these relations and the anisotropy of the variance in the numerical example below.

We now extend the above discussion to the particle limit, in which the energy per particle, densities per particle, and reduced densities per particle at the mean-field and many-body levels of theory coincide, see in the context of the harmonic-interaction model [16]. Particularly, the system of bosons becomes 100% condensed. The results (16) for the position and momentum variances hold at the particle limit as well, owing to the center-of-mass separability for any number of particles, namely, $\frac{\lim_{N\to\infty}\frac{1}{N}\Delta^2_{\hat{X},GP}}{\lim_{N\to\infty}\frac{1}{N}\Delta^2_{\hat{X}}} = \frac{1}{\sqrt{1+\frac{2\Lambda}{\omega_x^2}}}$, $\frac{\lim_{N\to\infty}\frac{1}{N}\Delta^2_{\hat{Y},GP}}{\lim_{N\to\infty}\frac{1}{N}\Delta^2_{\hat{Y}}} = \frac{1}{\sqrt{1+\frac{2\Lambda}{\omega_y^2}}}$, $\frac{\lim_{N\to\infty}\frac{1}{N}\Delta^2_{\hat{P}_X,GP}}{\lim_{N\to\infty}\frac{1}{N}\Delta^2_{\hat{P}_X}} = \sqrt{1+\frac{2\Lambda}{\omega_x^2}}$, and $\frac{\lim_{N\to\infty}\frac{1}{N}\Delta^2_{\hat{P}_Y,GP}}{\lim_{N\to\infty}\frac{1}{N}\Delta^2_{\hat{P}_Y}} = \sqrt{1+\frac{2\Lambda}{\omega_y^2}}$. For the angular-momentum variance the limit has to be taken explicitly for each of the terms in (8). First are the frequencies (4), for which we have at the limit of an infinite number of bosons when Λ is held fixed

$$\lim_{N\to\infty}\Omega_x = \sqrt{\omega_x^2+2\Lambda}, \qquad \lim_{N\to\infty}\Omega_y = \sqrt{\omega_y^2+2\Lambda}. \tag{17}$$

Then, the angular-momentum variance takes on the appealing form

$$\lim_{N\to\infty}\frac{1}{N}\Delta^2_{\hat{L}_Z} = \frac{1}{4}\frac{\left(\sqrt{\omega_y^2+2\Lambda}-\sqrt{\omega_x^2+2\Lambda}\right)^2}{\sqrt{\omega_y^2+2\Lambda}\sqrt{\omega_x^2+2\Lambda}}\left[1+\left(\sqrt{1+\frac{2\Lambda}{\omega_y^2}}-1\right)\left(\sqrt{1+\frac{2\Lambda}{\omega_x^2}}-1\right)\right]. \tag{18}$$

Comparing (18) to the mean-field expression (14), it is instrumental to prescribe their ratio at the limit of an infinite number of particles (where, as mentioned above, the density per particle and other properties coincide),

$$\frac{\lim_{N\to\infty}\frac{1}{N}\Delta^2_{\hat{L}_Z,GP}}{\lim_{N\to\infty}\frac{1}{N}\Delta^2_{\hat{L}_Z}} = \frac{1}{1+\left(\sqrt{1+\frac{2\Lambda}{\omega_y^2}}-1\right)\left(\sqrt{1+\frac{2\Lambda}{\omega_x^2}}-1\right)}, \tag{19}$$

which is always smaller than 1 for interacting bosons in the anisotropic trap. Furthermore, we see that for attractive interaction the many-body variance can become much larger than the mean-field quantity in the anisotropic trap, signifying the growing necessity of the many-body treatment, even when the system is 100% condensed. This concludes our investigation of a solvable anisotropic many-boson model in which the variances of the momentum, position, and angular-momentum many-particle operators can be computed and investigated analytically and their values at the many-body and mean-field levels of theory compared and contrasted.

3. Bosons in an Annulus Subject to a Tilt

In most scenarios of interest, the position, momentum, and angular-momentum variance cannot be computed analytically. This in many cases is the situation when symmetries are lifted. Moreover, even when the variances can be computed for the ground state, like in the previous Section 2, their values for an out-of-equilibrium scenario are rarely within analytical reach. This would be the situation of the present investigation.

Bosons in rings, annuli, and shells have attracted considerable attention [42–66]. Here we consider weakly interacting bosons initially prepared in the ground state of a two-dimensional annulus. The annulus is then suddenly slightly tilted, leading to an out-of-equilibrium dynamics in an anisotropic setup. We build on and extend the study of bosons' dynamics in an annulus within an isotropic setup [19] (for which, e.g., the angular-momentum variance is 0). We analyze the BEC dynamics in terms of its time-dependent variances and other quantities of relevance, see Figures 1–7 below.

We consider the out-of-equilibrium dynamics governed by the time-dependent many-particle Schrödinger equation in two spatial dimensions, $\hat{H}(\mathbf{r}_1,\ldots,\mathbf{r}_N)\Psi(\mathbf{r}_1,\ldots,\mathbf{r}_N;t) = i\frac{\partial\Psi(\mathbf{r}_1,\ldots,\mathbf{r}_N;t)}{\partial t}$. The bosons are initially prepared in the ground state of the annulus, see Figure 1 in [19]. The trap potential is given by $\hat{V}(\mathbf{r}) = 0.05\mathbf{r}^4 + V_0 e^{-\frac{\mathbf{r}^2}{2}}$, with a barrier of heights $V_0 = 5$ and 10 throughout this work. The interaction between the bosons is repulsive and taken to be $\lambda_0 W(\mathbf{r}-\mathbf{r}') = \lambda_0 e^{-\frac{(\mathbf{r}-\mathbf{r}')^2}{2}}$, where the interaction strengths are $\lambda_0 = 0.02$ and 0.04 throughout this work. The form and extant of the interaction potential do not have a qualitative influence on the physics to be described below. At time $t = 0$ a linear term is added such that $V(\mathbf{r}) = 0.05\mathbf{r}^4 + V_0 e^{-\frac{\mathbf{r}^2}{2}} + 0.01x$. The physical meaning of the added potential is that a constant force pointing to the left is suddenly acting on the interacting bosons. Geometrically, the annulus can be considered to be slightly tilted to the left. Symmetry-wise, the isotropy of the potential is lifted and anisotropy sets in. All in all, the interacting bosons are not in their ground state any more and out-of-equilibrium dynamics emerges.

To compute the time-dependent many-boson wavefunction we use the multiconfigurational time-dependent Hartree for bosons (MCTDHB) method [67–69]. MCTDHB represents the wavefunction as a variationally optimal ansatz which is a linear-combination of all time-dependent permanents generated by distributing the N bosons over M time-adaptive orbitals. The quality of the wavefunction increases with M and convergence of quantities of interest is attained. The theory, applications, benchmarks, and extensions of MCTDHB are extensively discussed in the literature, see, e.g., Refs. [70–95]. Here we employ the numerical implementation in [96,97] both for preparing the ground state [98] (using imaginary-time propagation) and real-time dynamics. Finally, we mention that MCTDHB is the bosonic version of the nearly three-decades-established distinguishable-particle multiconfigurational time-dependent Hartree method frequently used (alongside its extensions) in molecular physics [99–105].

From the time-dependent wavefunction $\Psi(\mathbf{r}_1, \ldots, \mathbf{r}_N; t)$, here normalized to 1, we compute properties of interest. The reduced one-particle density matrix is defined as $\rho(\mathbf{r}, \mathbf{r}'; t) = N \int d\mathbf{r}_2 \cdots d\mathbf{r}_N \Psi^*(\mathbf{r}', \mathbf{r}_2, \ldots, \mathbf{r}_N; t)\Psi(\mathbf{r}, \mathbf{r}_2, \ldots, \mathbf{r}_N; t) = \sum_j n_j(t)\phi_j^*(\mathbf{r}'; t)\phi_j(\mathbf{r}; t)$, where $\{\phi_j(\mathbf{r}; t)\}$ are the natural orbitals and $\{n_j(t)\}$ the natural occupations. The number of particles residing outside the condensed mode $\phi_1(\mathbf{r}; t)$, i.e., the total number of depleted particles, is given by $\sum_{j>1} n_j(t) = N - n_1(t)$. Analogously, the reduced two-particle density matrix is given by $\rho(\mathbf{r}_1, \mathbf{r}_2, \mathbf{r}_1', \mathbf{r}_2'; t) = N(N-1) \int d\mathbf{r}_3 \cdots d\mathbf{r}_N \Psi^*(\mathbf{r}_1', \mathbf{r}_2', \mathbf{r}_3, \ldots, \mathbf{r}_N; t) \times \Psi(\mathbf{r}_1, \mathbf{r}_2, \mathbf{r}_3, \ldots, \mathbf{r}_N; t) = \sum_{jpkq} \rho_{jpkq}(t)\phi_j^*(\mathbf{r}_1'; t)\phi_p^*(\mathbf{r}_2'; t)\phi_k(\mathbf{r}_1; t)\phi_q(\mathbf{r}_2; t)$, from which the variance of a many-particle operator $\hat{A} = \sum_j \hat{a}(\mathbf{r})$ is computed,

$$\begin{aligned} \frac{1}{N}\Delta^2_{\hat{A}}(t) &= \frac{1}{N}\left(\langle\Psi(t)|\hat{A}^2|\Psi(t)\rangle - \langle\Psi(t)|\hat{A}|\Psi(t)\rangle^2\right) \\ &= \frac{1}{N}\Bigg\{ \sum_j n_j(t) \int d\mathbf{r}\phi_j^*(\mathbf{r}; t)\hat{a}^2(\mathbf{r})\phi_j(\mathbf{r}; t) - \left[\sum_j n_j(t) \int d\mathbf{r}\phi_j^*(\mathbf{r}; t)\hat{a}(\mathbf{r})\phi_j(\mathbf{r}; t)\right]^2 \\ &+ \sum_{jpkq} \rho_{jpkq}(t) \left[\int d\mathbf{r}\phi_j^*(\mathbf{r}; t)\hat{a}(\mathbf{r})\phi_k(\mathbf{r}; t)\right] \left[\int d\mathbf{r}\phi_p^*(\mathbf{r}; t)\hat{a}(\mathbf{r})\phi_q(\mathbf{r}; t)\right] \Bigg\}. \end{aligned} \quad (20)$$

To compute the various terms for the position, momentum, and angular-momentum variance numerically we work in coordinate representation and operate on orbitals first with coordinate derivatives and then with coordinate multiplications. Thus, for the position operator $\hat{a}(\mathbf{r}) = \hat{x}$ and $\hat{a}^2(\mathbf{r}) = \hat{x}^2$ and likewise for $\hat{a}(\mathbf{r}) = \hat{y}$, for the momentum operator $\hat{a}(\mathbf{r}) = \frac{1}{i}\frac{\partial}{\partial x}$ and $\hat{a}^2(\mathbf{r}) = -\frac{\partial^2}{\partial x^2}$ and likewise for $\hat{a}(\mathbf{r}) = \frac{1}{i}\frac{\partial}{\partial y}$, and for the angular-momentum operator $\hat{a}(\mathbf{r}) = \frac{1}{i}\left(x\frac{\partial}{\partial y} - y\frac{\partial}{\partial x}\right)$ and $\hat{a}^2(\mathbf{r}) = -x^2\frac{\partial^2}{\partial y^2} - y^2\frac{\partial^2}{\partial x^2} + 2yx\frac{\partial}{\partial y}\frac{\partial}{\partial x} + x\frac{\partial}{\partial x} + y\frac{\partial}{\partial y}$. For the numerical solution we use a grid of 64^2 points in a box of size $[-8, 8) \times [-8, 8)$ with periodic boundary conditions. Convergence of the results with respect to the number of grid points has been checked using a grid of 128^2 points.

We begin with the dynamics of $N = 10$ bosons in the annulus. Following the sudden tilt of the potential, the bosons start to flow to the left. To quantify their sloshing dynamics, Figure 1 shows the time-dependent center-of-mass, $\frac{1}{N}\langle\Psi|\hat{X}|\Psi\rangle(t)$, for the two barrier heights, $V_0 = 5$ and $V_0 = 10$, and the two interaction strengths, $\lambda_0 = 0.02$ and $\lambda_0 = 0.04$ [we mention that $\frac{1}{N}\langle\Psi|\hat{Y}|\Psi\rangle(t) = 0$ due to the $Y \to -Y$ reflection symmetry]. The dynamics of $\frac{1}{N}\langle\Psi|\hat{X}|\Psi\rangle(t)$ appears to be almost periodic and rather simple. We examine the amplitude and frequency of oscillations. It is useful to compare the amplitude of the center-of-mass motion with the radius of the (un-tilted) annulus. The radius of the density at its maximal value, R, is determined numerically using a computation with a resolution of 256^2 grid points as $R = 1.75(0)$ for $V_0 = 5, \lambda_0 = 0.02$, and $R = 2.06(2)$ for $V_0 = 10, \lambda_0 = 0.02$ [19]. From Figure 1 we see that the amplitude is about 13–25% of the radius, implying a mild sloshing of the density along the tilted annulus. The amplitude increases with the radius of the annulus and decreases with the interaction strength, where the latter implies that it is more difficult to compress the BEC for a stronger interaction. The decrease of the frequency of oscillations with R (V_0) and increase with λ_0 are compatible with angular excitations, also see [19]. Last but not least, convergence with M is clearly seen. In fact, here already $M = 1$ orbitals accurately describe the center-of-mass dynamics for short and intermediate times, and $M = 3$ orbitals for all times.

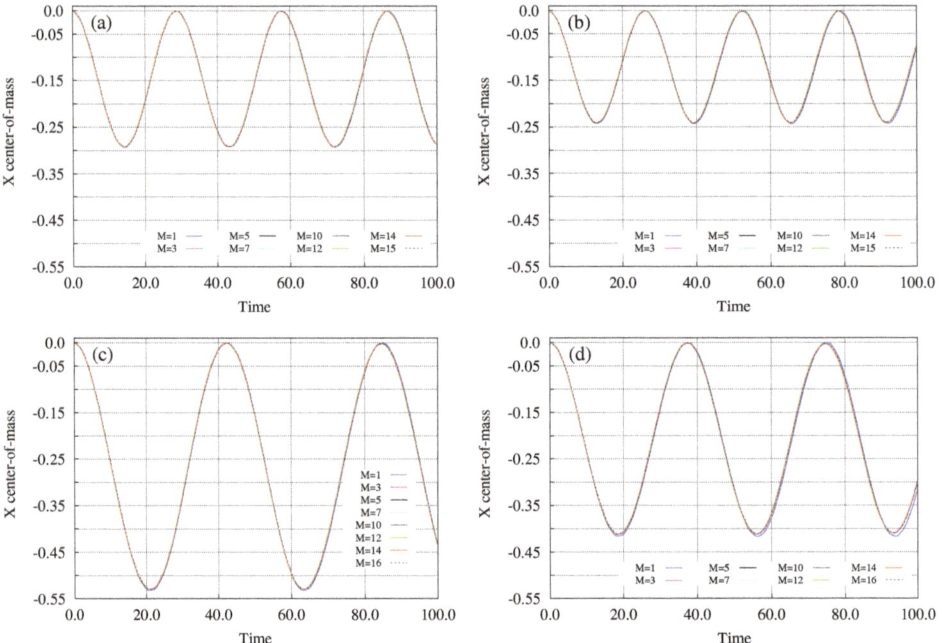

Figure 1. Center-of-mass dynamics following a potential quench. The mean-field ($M = 1$ time-adaptive orbitals) and many-body (using $M = 3$, 5, 7, 10, 12, 14, and 15, 16 time-adaptive orbitals) time-dependent expectation value of the center-of-mass, $\frac{1}{N}\langle\Psi|\hat{X}|\Psi\rangle(t)$, of $N = 10$ bosons in the annuli with barrier heights and interaction strengths: (**a**) $V_0 = 5$, $\lambda_0 = 0.02$; (**b**) $V_0 = 5$, $\lambda_0 = 0.04$; (**c**) $V_0 = 10$, $\lambda_0 = 0.02$; and (**d**) $V_0 = 10$, $\lambda_0 = 0.04$ following a sudden potential tilt by $0.01x$. The corresponding depletions are plotted in Figure 2 and the respective position, momentum, and angular-momentum variances in Figures 3–5. See the text for more details. The quantities shown are dimensionless.

Figure 2 depicts the total number of depleted particles, $N - n_1(t)$, out of $N = 10$ bosons in the tilted annulus. During the dynamics, the depletion is rather small, ranging from less than 0.012 of a particle out of $N = 10$ particles (0.12%) for $V_0 = 5, \lambda_0 = 0.02$ to less than 0.065 of a particle out of $N = 10$ particles (0.65%) for $V_0 = 10, \lambda_0 = 0.04$. Generally, the depletion increases with the annulus radius and interacting strength, implying angular excitations, see [19]. Finally, convergence with M is clearly seen. Now, $M = 3$ orbitals nicely follow and $M = 5$ orbitals accurately describe the depletion dynamics, see Figure 2. The small amount of time-dependent depletion is in line with the accurate description of the center-of-mass dynamics by $M = 1$ time adaptive orbitals, see Figure 1. Let us continue to the variances.

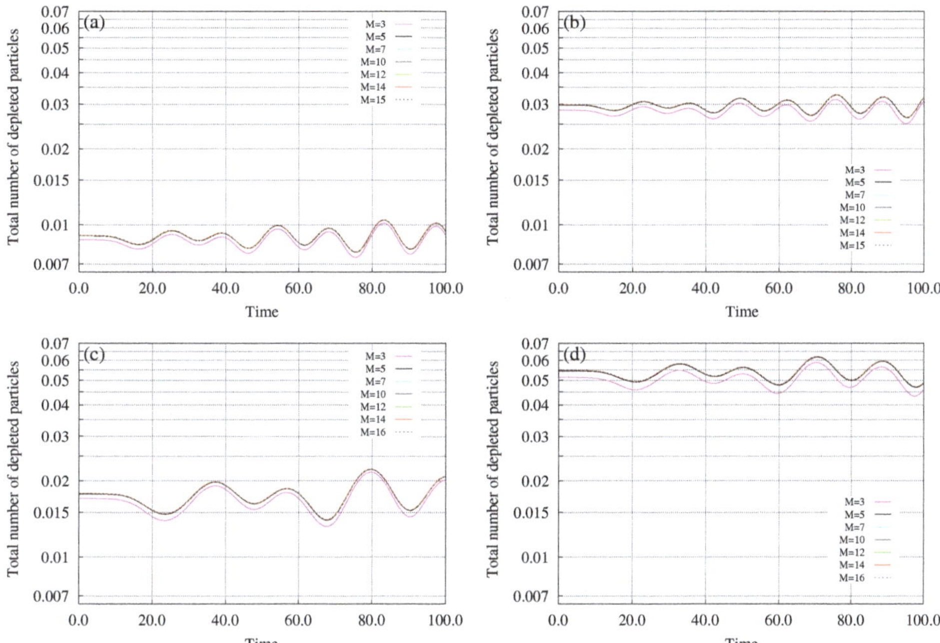

Figure 2. Depletion dynamics following a potential quench. The time-dependent total number of depleted particles, $N - n_1(t)$, of $N = 10$ bosons following a sudden potential tilt by $0.01x$ for annuli with barrier heights and interaction strengths (**a**) $V_0 = 5$, $\lambda_0 = 0.02$; (**b**) $V_0 = 5$, $\lambda_0 = 0.04$; (**c**) $V_0 = 10$, $\lambda_0 = 0.02$; and (**d**) $V_0 = 10$, $\lambda_0 = 0.04$. $M = 3, 5, 7, 10, 12, 14$, and $15, 16$ time-adaptive orbitals are used. The respective position, momentum, and angular-momentum variances are plotted in Figures 3–5. See the text for more details. The quantities shown are dimensionless.

Figure 3 plots the time-dependent many-particle position variance per particle, $\frac{1}{N}\Delta^2_{\hat{X}}(t)$ and $\frac{1}{N}\Delta^2_{\hat{Y}}(t)$, for the two barrier heights and two interaction strengths. There are several features that immediately are seen. First, since rotational symmetry is lifted, the dynamics of respective quantities along the x-axis and y-axis are different [note that at $t = 0$ the variances $\frac{1}{N}\Delta^2_{\hat{X}} = \frac{1}{N}\Delta^2_{\hat{Y}}$ because the initial condition is the ground state of the un-tilted, isotropic annulus]. The mean-field ($M = 1$) and many-body ($M \geq 3$) values are clearly separated from each other, and the former lie about 10–25% above the latter depending on the repulsion strength and barrier height, also see [11,19]. This is despite the small amount of depletion, see Figure 2. Furthermore, the many-body and mean-field variances do not cross each other, see Figure 3, indicating that the dynamics is mild and sufficiently close to the ground state and low-lying manifold of excited states (compare to [18] with interaction-quench dynamics in a single trap).

The mean-field position variance accounts for the geometry of the annulus and shape of the density and weakly depends on the interaction strength. The many-body position variance incorporates the (small amount of) depletion and hence strongly depends on the interaction strength. Both the mean-field and many-body variances oscillate with a relatively small amplitude, albeit with a different frequencies' content, see in this respect [20]. This amplitude slightly decreases with the repulsion strength, which correlates with the dependence of the center-of-mass dynamics on the interaction strength, see Figure 1. Moreover, the amplitude of oscillations of the y-axis variances is smaller than that of the x-axis variances, since the sloshing dynamics is primarily along the x direction. Last but not least is the so-called opposite anisotropy of the (position) variance [18]. During the dynamics, there can occur instances where $\frac{1}{N}\Delta^2_{\hat{X}} > \frac{1}{N}\Delta^2_{\hat{Y}}$ at the many-body level ($M \geq 3$) whereas $\frac{1}{N}\Delta^2_{\hat{X}} < \frac{1}{N}\Delta^2_{\hat{Y}}$ at the mean-field level ($M = 1$) [or, in principle, vice versa, i.e., $\frac{1}{N}\Delta^2_{\hat{X}} < \frac{1}{N}\Delta^2_{\hat{Y}}$ at the many-body level whereas $\frac{1}{N}\Delta^2_{\hat{X}} > \frac{1}{N}\Delta^2_{\hat{Y}}$ at the mean-field level]. Examples for the former can be readily found for $V_0 = 10$, $\lambda_0 = 0.02$, see Figure 3c,g around $t = 70$, and for $V_0 = 10$, $\lambda_0 = 0.04$, see Figure 3d,h around $t = 100$, signifying among others that correlations 'win' over shape. Finally, we see that already $M = 3$ orbitals accurately describe the dynamics of the position variance.

We move to the momentum variance and also make contact with the results of the position variance. Figure 4 displays the many-particle momentum variance per particle, $\frac{1}{N}\Delta^2_{\hat{P}_X}(t)$ and $\frac{1}{N}\Delta^2_{\hat{P}_Y}(t)$, for $V_0 = 5$, $V_0 = 10$ and $\lambda_0 = 0.02$, $\lambda_0 = 0.04$. Just like the results of the position variance, since rotational symmetry is lifted the dynamics of respective quantities along the x-axis and y-axis are different [the initial conditions imply $\frac{1}{N}\Delta^2_{\hat{P}_X} = \frac{1}{N}\Delta^2_{\hat{P}_Y}$ at $t = 0$]. The mean-field ($M = 1$) and many-body ($M \geq 3$) values are, again, separated from each other, but now the former lie below the later, and there is only about 1–4% of a difference depending on the repulsion strength and barrier height, also see [12,19]. Thus, the momentum variance rather weakly depends on the (small amount of) depletion. This is because the matrix elements in (20) are typically smaller with the momentum operator than with the position operator. Yet, despite their small difference, the many-body and mean-field momentum variances do not cross each other, see Figure 4 (contrast with the interaction-quench dynamics in a single trap in [18]).

It is instructive to analyze the momentum-variance dynamics at short times. Whereas $\Delta^2_{\hat{P}_X}(t)$ primarily increases, $\Delta^2_{\hat{P}_Y}(t)$ mainly decreases. This matches the geometry of the sloshing dynamics in the tilted annulus, in which bosons from the 'north' and 'south' poles (on the y-axis) start to move to the left and accumulate in the 'west' pole (on the x-axis), and that the cross section of the rim of an annulus is enlarged when moving away from the center of the annulus. In other words, the dynamics of the momentum variances at short times when moving to the left reflects the relative localization of the bosons in the x direction and the effective broadening of the wavepacket along the y direction. Both the mean-field and many-body variances oscillate with a very small amplitude, note the scale on the y-axis in Figure 4. The high-frequency oscillations mark high-energy radial excitations across the (tight) annulus rim [19]. Like for the position variance, the amplitude of oscillations of the y-axis momentum variances is smaller than that of the x-axis momentum variances. Finally, we see that already $M = 3$ orbitals accurately describe the dynamics of the momentum variance; the difference to the $M > 3$ results is lower than 1%.

We now move to the angular-momentum variance and an interesting inter-connection with the momentum variance. Figure 5 presents the many-particle angular-momentum variance per particle, $\frac{1}{N}\Delta^2_{\hat{L}_Z}(t)$, for the two barrier heights, $V_0 = 5$ and $V_0 = 10$, and the two interaction strengths, $\lambda_0 = 0.02$ and $\lambda_0 = 0.04$. There are several features seen in the dynamics. Since rotational symmetry is lifted, $\frac{1}{N}\Delta^2_{\hat{L}_Z} \neq 0$ expect for the initial conditions at $t = 0$ (the values of the minima for $t > 0$, see below, are close to but not 0). The dynamics of $\frac{1}{N}\Delta^2_{\hat{L}_Z}(t)$ appears to be almost periodic and rather regular, more than that for the respective position and momentum variances, compare to Figures 3 and 4. On the other end, focusing on the dynamics of the center-of-mass in Figure 1, one can clearly observe correlation between the two quantities; Whenever $\frac{1}{N}\langle\Psi|\hat{X}|\Psi\rangle(t)$ has a minimum, i.e., the bosons are maximally localized to the left, $\frac{1}{N}\Delta^2_{\hat{L}_Z}(t)$ has a maximum, and whenever $\frac{1}{N}\langle\Psi|\hat{X}|\Psi\rangle(t)$ has a maximum (which value is about 0), i.e., the bosons are momentarily, approximately equally distributed along the annulus, $\frac{1}{N}\Delta^2_{\hat{L}_Z}(t)$ has a minimum (which value, as mentioned above, is close to 0). Furthermore, the frequencies of the two quantities as well as their relative amplitudes as a function of the barrier height and interaction strength are alike. These observations call for a dedicated analysis.

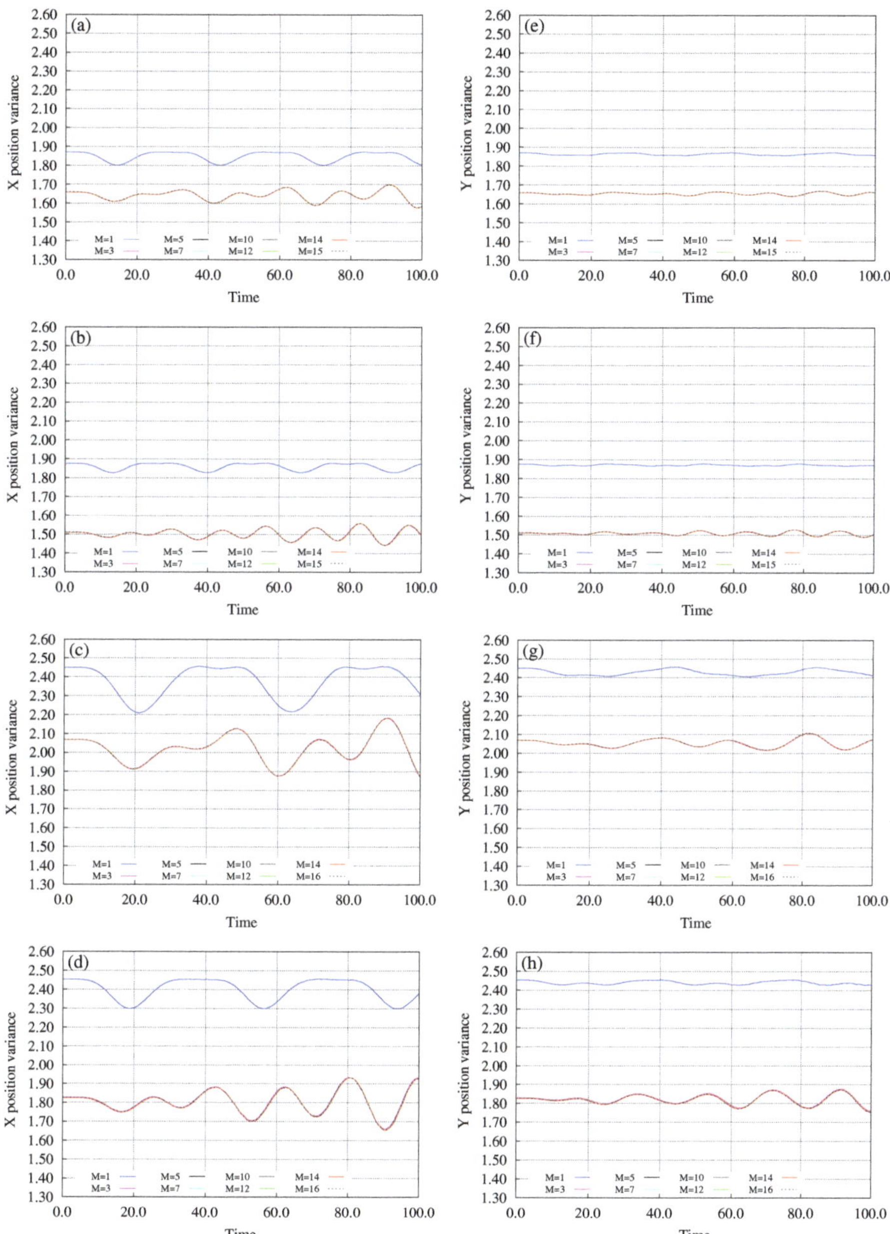

Figure 3. Position variance dynamics following a potential quench. The mean-field ($M = 1$ time-adaptive orbitals) and many-body (using $M = 3, 5, 7, 10, 12, 14$, and $15, 16$ time-adaptive orbitals) time-dependent position variances per particle, $\frac{1}{N}\Delta^2_{\hat{X}}(t)$ [left column, panels (**a–d**)] and $\frac{1}{N}\Delta^2_{\hat{Y}}(t)$ [right column, panels (**e–h**)], of $N = 10$ bosons in the annuli with barrier heights and interaction strengths (**a**,**e**) $V_0 = 5$, $\lambda_0 = 0.02$; (**b**,**f**) $V_0 = 5$, $\lambda_0 = 0.04$; (**c**,**g**) $V_0 = 10$, $\lambda_0 = 0.02$; and (**d**,**h**) $V_0 = 10$, $\lambda_0 = 0.04$ following a sudden potential tilt by $0.01x$. The respective depletions are plotted in Figure 2. See the text for more details. The quantities shown are dimensionless.

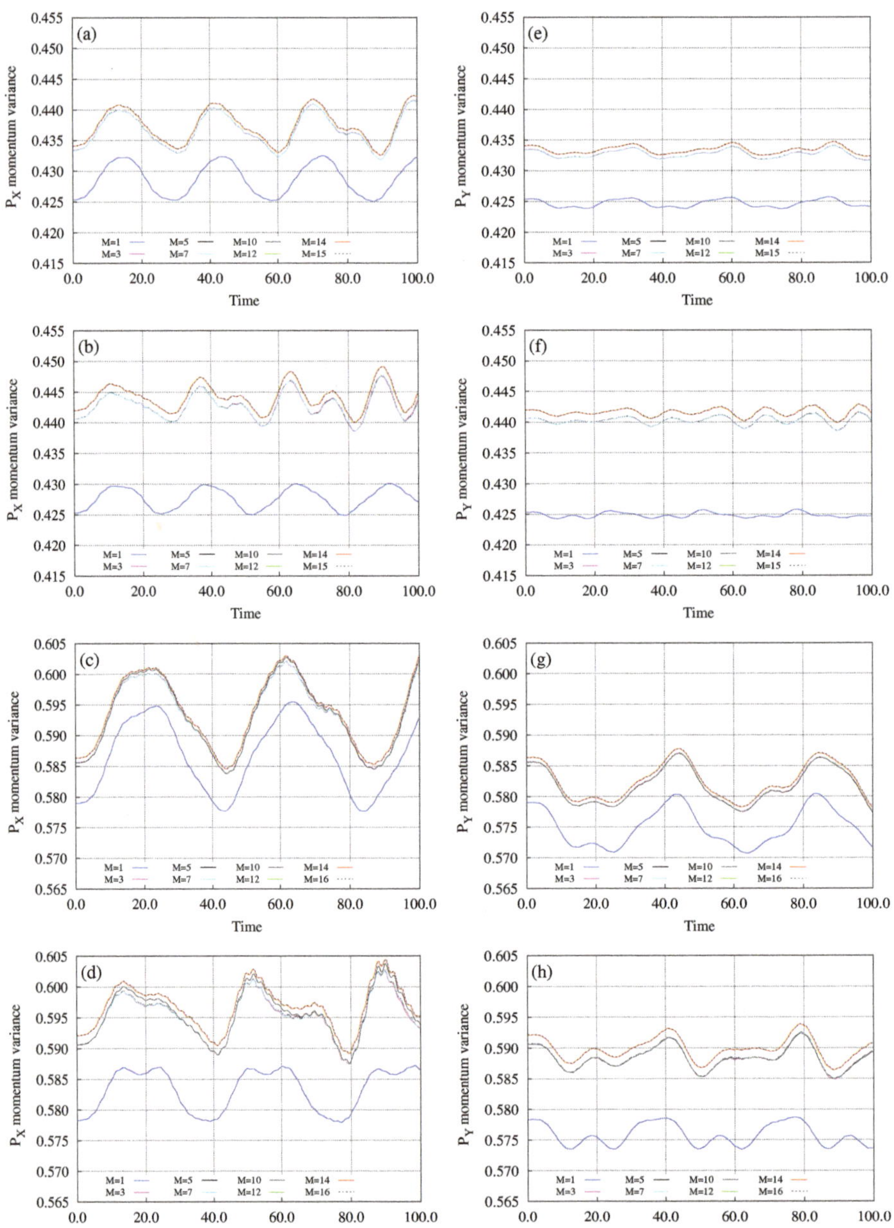

Figure 4. Momentum variance dynamics following a potential quench. The mean-field ($M = 1$ time-adaptive orbitals) and many-body (using $M = 3, 5, 7, 10, 12, 14$, and $15, 16$ time-adaptive orbitals) time-dependent momentum variances per particle, $\frac{1}{N}\Delta^2_{\hat{P}_X}(t)$ [left column, panels (**a–d**)] and $\frac{1}{N}\Delta^2_{\hat{P}_Y}(t)$ [right column, panels (**e–h**)], of $N = 10$ bosons in the annuli with barrier heights and interaction strengths (**a,e**) $V_0 = 5$, $\lambda_0 = 0.02$; (**b,f**) $V_0 = 5$, $\lambda_0 = 0.04$; (**c,g**) $V_0 = 10$, $\lambda_0 = 0.02$; and (**d,h**) $V_0 = 10$, $\lambda_0 = 0.04$ following a sudden potential tilt by $0.01x$. The respective depletions are plotted in Figure 2. See the text for more details. The quantities shown are dimensionless.

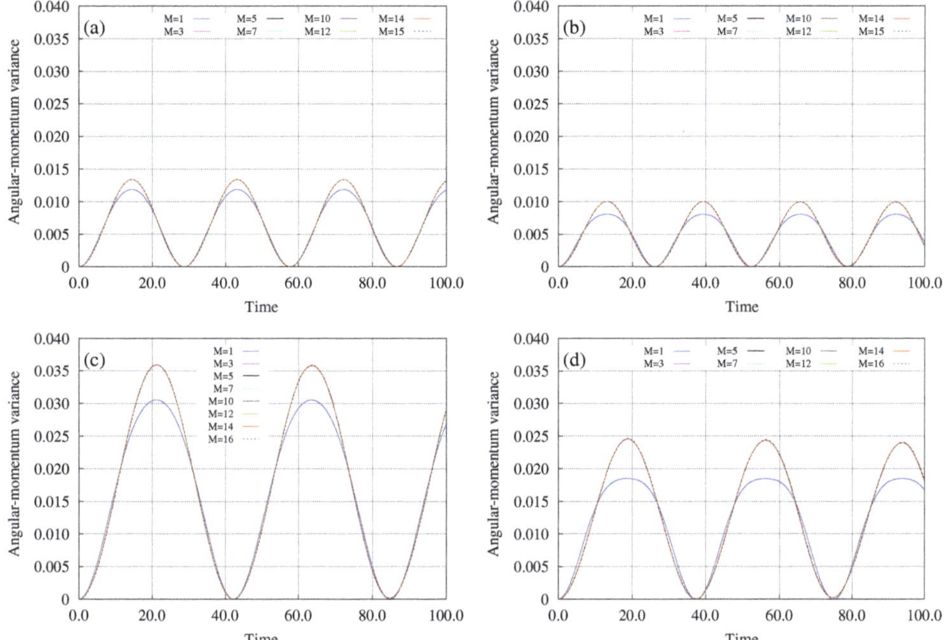

Figure 5. Angular-momentum variance dynamics following a potential quench. The mean-field ($M = 1$ time-adaptive orbitals) and many-body (using $M = 3, 5, 7, 10, 12, 14$, and $15, 16$ time-adaptive orbitals) time-dependent angular-momentum variance per particle, $\frac{1}{N}\Delta^2_{\hat{L}_Z}(t)$, of $N = 10$ bosons in the annuli with barrier heights and interaction strengths (**a**) $V_0 = 5, \lambda_0 = 0.02$; (**b**) $V_0 = 5, \lambda_0 = 0.04$; (**c**) $V_0 = 10, \lambda_0 = 0.02$; and (**d**) $V_0 = 10, \lambda_0 = 0.04$ following a sudden potential tilt by $0.01x$. The respective depletions are plotted in Figure 2. See the text for more details. The quantities shown are dimensionless.

To shed light on the above dynamics of the angular-momentum variance, see Figure 5, we analyze the translational properties of variances in Appendix A. Whereas the position variances and, trivially, the momentum variances, are translationally invariant, this invariance does not hold for the angular-momentum variance. If a wavepacket prepared in the origin has angular-momentum variance $\frac{1}{N}\Delta^2_{\hat{L}_Z}$, then several terms are added when the wavepacket is translated to the point (a,b) in plane, and angular-momentum variance is thereafter computed, see Equation (A3). Now, if this wavepacket is rotationally symmetric, i.e., $\frac{1}{N}\Delta^2_{\hat{L}_Z} = 0$, then several of the terms in (A3) vanish due to spatial symmetry and we are left with the appealing relation, $\frac{1}{N}\Delta^2_{\hat{L}_Z}\Big|_{\Psi(a,b)} = a^2\frac{1}{N}\Delta^2_{\hat{P}_Y}\Big|_{\Psi} + b^2\frac{1}{N}\Delta^2_{\hat{P}_X}\Big|_{\Psi}$ [Equation (A4)], connecting the angular-momentum variance of $\Psi(a,b)$ localized at (a,b) and of Ψ at the origin. The meaning of this relation is that the momentum variances, $\frac{1}{N}\Delta^2_{\hat{P}_X}$ and $\frac{1}{N}\Delta^2_{\hat{P}_Y}$, together with the spatial translations along the y-axis and x-axis, respectively, determine the angular-momentum variance of a translated wavepacket (rotationally symmetric at the origin).

Returning to and combining Figure 5 for the angular-momentum variance, Figure 1 for the center-of-mass dynamics, and Figure 4e–h for $\frac{1}{N}\Delta^2_{\hat{P}_Y}$, we can now discuss and explain their inter-connection. Explicitly, the center-of-mass dynamics is analogous to translating the wavepacket along the x-axis (back and forth to the left), hence, according to Equation (A4), $\frac{1}{N}\Delta^2_{\hat{P}_Y}$ is needed. This is why the dependencies of the frequency and amplitude of oscillations of $\frac{1}{N}\Delta^2_{\hat{L}_Z}(t)$ on the barrier height and interaction strength nicely follow, respectively, those of $\frac{1}{N}\langle\Psi|\hat{X}|\Psi\rangle(t)$, compare

Figures 1 and 5. What is the role of $\frac{1}{N}\Delta^2_{\hat{P}_Y}$ then? The momentum variance helps us understand the deviations between the many-body and mean-field results in Figure 5. We see that the maxima of the many-body $\frac{1}{N}\Delta^2_{\hat{L}_Z}(t)$ $(M > 3)$ are larger than the maxima of the mean-field $\frac{1}{N}\Delta^2_{\hat{L}_Z}(t)$ $(M = 1)$. The difference is about 7–25% (compare to the low depletion, Figure 2), depending on V_0 and λ_0, and follows the respective trend of the many-body and mean-field results for $\frac{1}{N}\Delta^2_{\hat{P}_Y}$, see Figure 4e–h. We note that, although the wavepacket describing the bosons dynamics in the tilted annulus is not a translated, rotationally invariant wavepacket, and the values of deviations (in percents) between the many-body and mean-field results are actually larger for $\Delta^2_{\hat{L}_Z}$ (at the maxima) than for $\Delta^2_{\hat{P}_Y}$, we find the above analytically based analysis to well explain the numerical findings and trends. Last but not least, a close inspection of the many-body and mean-field curves of the angular-momentum variance in Figure 5 shows that there are instances when they cross each other, i.e., one is smaller or larger than the other. This is in contrast with the non-crossing of the many-body and mean-field position and momentum variances, see Figures 3 and 4, respectively. Finally, we find that already $M = 3$ time-adaptive orbitals accurately describe the dynamics of the angular-momentum variance.

Our investigations are nearing their end, what is left to explore is the behavior of the position, momentum, and angular-momentum variances at the particle limit. Which of the above-described detailed findings, plotted in Figures 1–5 for a rather small ($N = 10$ bosons) yet weakly depleted BEC, survive this limit? To answer the question, we concentrate on the system with the higher barrier, $V_0 = 10$, and stronger interaction (for $N = 10$ bosons), $\lambda_0 = 0.04$. We hence fix the interaction parameter $\Lambda = \lambda_0(N-1) = 0.36$, and compute and compare the dynamics for $N = 10$, $N = 100$, and $N = 1000$ bosons using $M = 3$ time-adaptive orbitals. We have seen for $N = 10$ bosons that $M = 3$ time-adaptive orbitals accurately describe the variances. This implies that, keeping the interaction parameter Λ fixed while increasing the number of particles N, using $M = 3$ time-adaptive orbitals for calculating the variances will be (at least) as accurate as for $N = 10$ particles, see in this respect [24]. Before we proceed, a methodological remark. Examining the convergence of properties with the number of particles for $N = 10$, $N = 100$, and $N = 1000$ bosons is (still) far away from infinity, see in this respect [15]. We hence use, interchangeably, the term *en route* to the particle limit. We shall see below that, in effect, the particle limit is practically well achieved for the variances already for $N = 1000$ bosons.

Figure 6 prints the total number of depleted particles, $N - n_1(t)$, for $N = 10$, $N = 100$, and $N = 1000$ bosons for $\Lambda = 0.36$ and $V_0 = 10$ using $M = 3$ time-adaptive orbitals. Convergence of the number of depleted particles with N is nicely seen. Since $N - n_1(t)$ converges to a finite (and small) value with N, the bosons are becoming 100% condensed in the limit of an infinite number of particles, i.e., $\frac{n_1(t)}{N} \to 1$ as $N \to \infty$, at least up to the maximal time of the computation, $t = 100$.

Figure 7 exhibits the position variances per particle, $\frac{1}{N}\Delta^2_{\hat{X}}(t)$ and $\frac{1}{N}\Delta^2_{\hat{Y}}(t)$, momentum variances per particle, $\frac{1}{N}\Delta^2_{\hat{P}_X}(t)$ and $\frac{1}{N}\Delta^2_{\hat{P}_Y}(t)$, angular-momentum variance per particle, $\frac{1}{N}\Delta^2_{\hat{L}_Z}(t)$, and the expectation value of the center-of-mass, $\frac{1}{N}\langle\Psi|\hat{X}|\Psi\rangle(t)$, for $N = 10$, $N = 100$, and $N = 1000$ bosons and for $\Lambda = 0.36$ and $V_0 = 10$ using $M = 3$ time-adaptive orbitals. Once again, convergence of each of the quantities with N is clearly seen. Yet, whereas the center-of-mass dynamics converges to the mean-field dynamics when the number of particles is increased, the variances exhibit many-body dynamics which converges nicely with N, but not to the respective mean-field dynamics. Beyond that, all the above results, for the frequencies, amplitudes, anisotropies, inter-connections, and particularly the differences between the many-body and mean-field position, momentum, and angular-momentum variances persist at the limit of infinite number of particles, despite the bosons becoming 100% condensed. This brings the present analysis to an end.

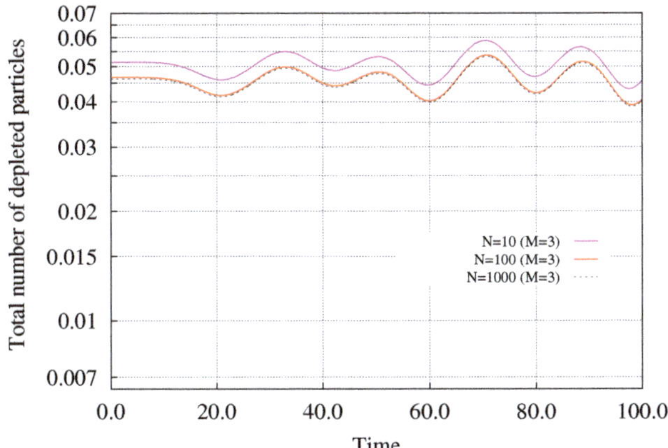

Figure 6. Depletion dynamics following a potential quench *en route* to the particle limit. The time-dependent total number of depleted particles, $N - n_1(t)$, of $N = 10$, $N = 100$, and $N = 1000$ bosons with interaction parameter $\Lambda = \lambda_0(N-1) = 0.36$ for an annulus with barrier height $V_0 = 10$ following a sudden potential tilt by $0.01x$. The number of time-adaptive orbitals is $M = 3$. The respective position, momentum, and angular-momentum variances along with the expectation value of the center-of-mass are plotted in Figure 7. See the text for more details. The quantities shown are dimensionless.

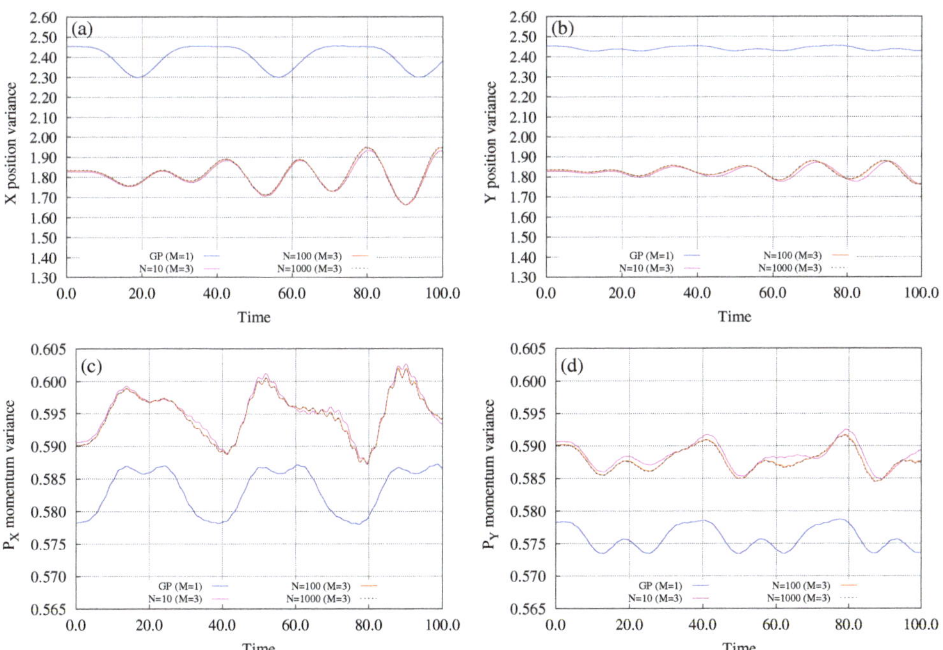

Figure 7. *Cont.*

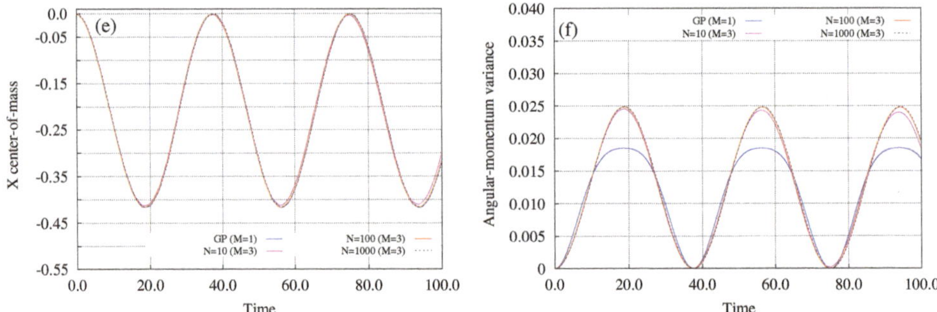

Figure 7. Position, momentum, and angular-momentum variance dynamics following a potential quench *en route* to the particle limit. The mean-field ($M = 1$ time-adaptive orbitals) and many-body (using $M = 3$ time-adaptive orbitals) time-dependent position variances per particle, **(a)** $\frac{1}{N}\Delta^2_{\hat{X}}(t)$ and **(b)** $\frac{1}{N}\Delta^2_{\hat{Y}}(t)$, momentum variances per particle, **(c)** $\frac{1}{N}\Delta^2_{\hat{P}_X}(t)$ and **(d)** $\frac{1}{N}\Delta^2_{\hat{P}_Y}(t)$, and angular-momentum variance per particle, **(f)** $\frac{1}{N}\Delta^2_{\hat{L}_Z}(t)$, of $N = 10$, $N = 100$, and $N = 1000$ bosons with interaction parameter $\Lambda = \lambda_0(N-1) = 0.36$ for an annulus with barrier height $V_0 = 10$ following a sudden potential tilt by $0.01x$. **(e)** The time-dependent expectation value of the center-of-mass, $\frac{1}{N}\langle\Psi|\hat{X}|\Psi\rangle(t)$. The respective depletions are plotted in Figure 6. See the text for more details. The quantities shown are dimensionless.

4. Summary and Outlook

In the present work we studied, analytically and numerically, the position, momentum, and especially the angular-momentum variance of interacting bosons trapped in a two-dimensional anisotropic trap for static and dynamic scenarios. Explicitly, we investigated the ground state of the anisotropic harmonic-interaction model in two spatial dimensions analytically and researched the out-of-equilibrium dynamics of repulsive bosons in tilted two-dimensional annuli numerically accurately by using the MCTDHB method. The differences between the variances at the mean-field level, which are attributed to the shape of the density per particle, and the respective variances at the many-body level, which incorporate a small amount of depletion outside the condensed mode, were used to characterize sometimes large position, momentum, and angular-momentum correlations in the BEC for finite systems and at the limit of an infinite number of particles where the bosons are 100% condensed. Finally, we also explored and utilized inter-connections between the variances, particularly between the angular-momentum and momentum variances, through the analysis of their translational properties.

There are many intriguing directions to follow out of which we list three below. First, variances of BECs in the rotating frame of reference in which high-lying excitations become low-energy excitations and even the ground state. Second, angular-momentum variance of a BEC flowing past an obstacle in which the mean angular-momentum variance vanishes. And third, variances in three-dimensional geometries lacking lower-dimensional analogs, such as a Möbius strip. In all these cases, whether considering a few interacting bosons or a BEC in the particle limit, interesting and exciting results are expected.

Funding: This research was funded by Israel Science Foundation (Grant No. 600/15).

Acknowledgments: This research was supported by the Israel Science Foundation (Grant No. 600/15). We thank Kaspar Sakmann, Anal Bhowmik, Sudip Haldar, and Raphael Beinke for discussions. Computation time on the BwForCluster, the High Performance Computing system Hive of the Faculty of Natural Sciences at University of Haifa, and the Cray XC40 system Hazelhen at the High Performance Computing Center Stuttgart (HLRS) is gratefully acknowledged.

Conflicts of Interest: The authors declare no conflict of interest.

Appendix A. Variances and Translations

Consider the many-particle translation operator in two spatial dimensions $e^{-i(\hat{P}_X a+\hat{P}_Y b)}$, where $\hat{P}_X = \sum_{j=1}^N \hat{p}_{x,j}$ and $\hat{P}_Y = \sum_{j=1}^N \hat{p}_{y,j}$. Its operation on a multi-particle wavefunction Ψ is given by $e^{-i(\hat{P}_X a+\hat{P}_Y b)}\Psi(x_1,y_1,\ldots,x_N,y_N) = \Psi(x_1-a,y_1-b,\ldots,x_N-a,y_N-b) \equiv \Psi(a,b)$. What are the implications on the variances when computed with respect to the translated wavefunction $\Psi(a,b)$?

For the position operator $\hat{X} = \sum_{j=1}^N \hat{x}_j$ (and equivalently for $\hat{Y} = \sum_{j=1}^N \hat{y}_j$) we have $\langle\Psi(a,b)|\hat{X}|\Psi(a,b)\rangle = \langle\Psi|\hat{X}|\Psi\rangle + Na$ and $\langle\Psi(a,b)|\hat{X}^2|\Psi(a,b)\rangle = \langle\Psi|\hat{X}^2|\Psi\rangle + 2Na\langle\Psi|\hat{X}|\Psi\rangle + (Na)^2$, implying that

$$\frac{1}{N}\Delta^2_{\hat{X}}\Big|_{\Psi(a,b)} = \frac{1}{N}\Delta^2_{\hat{X}}\Big|_{\Psi}. \tag{A1}$$

Trivially for the momentum operator $\hat{P}_X$ (and equivalently for $\hat{P}_Y = \sum_{j=1}^N \hat{p}_{y,j}$) one has

$$\frac{1}{N}\Delta^2_{\hat{P}_X}\Big|_{\Psi(a,b)} = \frac{1}{N}\Delta^2_{\hat{P}_X}\Big|_{\Psi}, \tag{A2}$$

i.e., both the position variance and momentum variance are translationally invariant.

For the angular-momentum operator $\hat{L}_Z = \sum_{j=1}^N \left(\hat{x}_j\hat{p}_{y,j} - \hat{y}_j\hat{p}_{x,j}\right)$ the situation is more interesting. From $\langle\Psi(a,b)|\hat{L}_Z|\Psi(a,b)\rangle = \langle\Psi|\hat{L}_Z|\Psi\rangle + a\langle\Psi|\hat{P}_Y|\Psi\rangle - b\langle\Psi|\hat{P}_X|\Psi\rangle$ and $\langle\Psi(a)|\hat{L}_Z^2|\Psi(a)\rangle = \langle\Psi|\hat{L}_Z^2|\Psi\rangle + a^2\langle\Psi|\hat{P}_Y^2|\Psi\rangle + b^2\langle\Psi|\hat{P}_X^2|\Psi\rangle + a\langle\Psi|\hat{L}_Z\hat{P}_Y + \hat{P}_Y\hat{L}_Z|\Psi\rangle - b\langle\Psi|\hat{L}_Z\hat{P}_X + \hat{P}_X\hat{L}_Z|\Psi\rangle - 2ab\langle\Psi|\hat{P}_Y\hat{P}_X|\Psi\rangle$ we have

$$\begin{aligned}\frac{1}{N}\Delta^2_{\hat{L}_Z}\Big|_{\Psi(a,b)} &= \frac{1}{N}\Delta^2_{\hat{L}_Z}\Big|_{\Psi} + a^2\frac{1}{N}\Delta^2_{\hat{P}_Y}\Big|_{\Psi} + b^2\frac{1}{N}\Delta^2_{\hat{P}_X}\Big|_{\Psi} \\ &+a\left(\langle\Psi|\hat{L}_Z\hat{P}_Y + \hat{P}_Y\hat{L}_Z|\Psi\rangle - 2\langle\Psi|\hat{L}_Z|\Psi\rangle\langle\Psi|\hat{P}_Y|\Psi\rangle\right) \\ &-b\left(\langle\Psi|\hat{L}_Z\hat{P}_X + \hat{P}_X\hat{L}_Z|\Psi\rangle - 2\langle\Psi|\hat{L}_Z|\Psi\rangle\langle\Psi|\hat{P}_X|\Psi\rangle\right) \\ &-2ab\left(\langle\Psi|\hat{P}_Y\hat{P}_X|\Psi\rangle - \langle\Psi|\hat{P}_Y|\Psi\rangle\langle\Psi|\hat{P}_X|\Psi\rangle\right).\end{aligned} \tag{A3}$$

Equation (A3) deserves a discussion. In turn, even for the ground state of an interacting many-boson system in a rotationally symmetric [for which $\frac{1}{N}\Delta^2_{\hat{L}_Z} = 0$ holds] but otherwise translated trap, the angular-momentum variance

$$\frac{1}{N}\Delta^2_{\hat{L}_Z}\Big|_{\Psi(a,b)} = a^2\frac{1}{N}\Delta^2_{\hat{P}_Y}\Big|_{\Psi} + b^2\frac{1}{N}\Delta^2_{\hat{P}_X}\Big|_{\Psi} \tag{A4}$$

differs at the many-body level and mean-field level of theory, i.e., when $a,b \neq 0$ and $\lambda_0 \neq 0$. This is, as can be seen in (A4), because of the respective many-body and mean-field momentum variances, $\frac{1}{N}\Delta^2_{\hat{P}_X}$ and $\frac{1}{N}\Delta^2_{\hat{P}_Y}$. The analytical result (A4) is employed to analyze the numerical findings for the time-dependent angular-momentum variance in the main text. Generally in the absence of spatial symmetries, see Equation (A3), more terms contribute to the translated angular-momentum variance.

References

1. Cornell, E.A.; Wieman, C.E. Nobel Lecture: Bose–Einstein condensation in a dilute gas, the first 70 years and some recent experiments. *Rev. Mod. Phys.* **2002**, *74*, 875. [CrossRef]
2. Ketterle, W. Nobel lecture: When atoms behave as waves: Bose–Einstein condensation and the atom laser. *Rev. Mod. Phys.* **2002**, *74*, 1131. [CrossRef]
3. Dalfovo, F.; Giorgini, S.; Pitaevskii, L.P.; Stringari, S. Theory of Bose–Einstein condensation in trapped gases. *Rev. Mod. Phys.* **1999**, *71*, 463. [CrossRef]
4. Leggett, A.J. Bose–Einstein condensation in the alkali gases: Some fundamental concepts. *Rev. Mod. Phys.* **2001**, *73*, 307. [CrossRef]

5. Bloch, I.; Dalibard, J.; Zwerger, W. Many-body physics with ultracold gases. *Rev. Mod. Phys.* **2008**, *80*, 885. [CrossRef]
6. Castin, Y.; Dum, R. Low-temperature Bose–Einstein condensates in time-dependent traps: Beyond the U(1) symmetry breaking approach. *Phys. Rev. A* **1998**, *57*, 3008. [CrossRef]
7. Lieb, E.H.; Seiringer, R.; Yngvason, J. Bosons in a trap: A rigorous derivation of the Gross-Pitaevskii energy functional. *Phys. Rev. A* **2000**, *61*, 043602. [CrossRef]
8. Lieb, E.H.; Seiringer, R. Proof of Bose–Einstein Condensation for Dilute Trapped Gases. *Phys. Rev. Lett.* **2002**, *88*, 170409. [CrossRef]
9. Erdős, L.; Schlein, B.; Yau, H.-T. Rigorous Derivation of the Gross-Pitaevskii Equation. *Phys. Rev. Lett.* **2007**, *98*, 040404. [CrossRef]
10. Erdős, L.; Schlein, B.; Yau, H.-T. Derivation of the cubic non-linear Schrödinger equation from quantum dynamics of many-body systems. *Invent. Math.* **2007**, *167*, 515. [CrossRef]
11. Klaiman, S.; Alon, O.E. Variance as a sensitive probe of correlations. *Phys. Rev. A* **2015**, *91*, 063613. [CrossRef]
12. Klaiman, S.; Streltsov, A.I.; Alon, O.E. Uncertainty product of an out-of-equilibrium many-particle system. *Phys. Rev. A* **2016**, *93*, 023605. [CrossRef]
13. Klaiman, S.; Cederbaum, L.S. Overlap of exact and Gross-Pitaevskii wave functions in Bose–Einstein condensates of dilute gases. *Phys. Rev. A* **2016**, *94*, 063648. [CrossRef]
14. Michelangeli, A.; Olgiati, A. Mean-field quantum dynamics for a mixture of Bose–Einstein condensates. *Anal. Math. Phys.* **2017**, *7*, 377. [CrossRef]
15. Cederbaum, L.S. Exact many-body wave function and properties of trapped bosons in the particle limit. *Phys. Rev. A* **2017**, *96*, 013615. [CrossRef]
16. Alon, O.E. Solvable model of a generic trapped mixture of interacting bosons: reduced density matrices and proof of Bose–Einstein condensation. *J. Phys. A* **2017**, *50*, 295002. [CrossRef]
17. Coleman, A.J.; Yukalov, V.I. *Reduced Density Matrices: Coulson's Challenge*; Lectures Notes in Chemistry; Springer: Berlin, Germany, 2000; Volume 72.
18. Klaiman, S.; Beinke, R.; Cederbaum, L.S.; Streltsov, A.I.; Alon, O.E. Variance of an anisotropic Bose–Einstein condensate. *Chem. Phys.* **2018**, *509*, 45. [CrossRef]
19. Alon, O.E. Condensates in annuli: Dimensionality of the variance. *Mol. Phys.* **2019**, *117*, 2108. [CrossRef]
20. Theisen, M.; Streltsov, A.I. Many-body excitations and deexcitations in trapped ultracold bosonic clouds. *Phys. Rev. A* **2016**, *94*, 053622. [CrossRef]
21. Haldar, S.K.; Alon, O.E. Impact of the range of the interaction on the quantum dynamics of a bosonic Josephson junction. *Chem. Phys.* **2018**, *509*, 72. [CrossRef]
22. Haldar, S.K.; Alon, O.E. Many-body quantum dynamics of an asymmetric bosonic Josephson junction. *New J. Phys.* **2019**, *21*, 103037. [CrossRef]
23. Cosme, J.G.; Weiss, C.; Brand, J. Center-of-mass motion as a sensitive convergence test for variational multimode quantum dynamics. *Phys. Rev. A* **2016**, *94*, 043603. [CrossRef]
24. Alon, O.E.; Cederbaum, L.S. Attractive Bose–Einstein condensates in anharmonic traps: Accurate numerical treatment and the intriguing physics of the variance. *Chem. Phys.* **2018**, *515*, 287. [CrossRef]
25. Sakmann, K.; Schmiedmayer, J. Conserving symmetries in Bose–Einstein condensate dynamics requires many-body theory. *arXiv* **2018**, arXiv:1802.03746v2.
26. Klaiman, S.; Streltsov, A.I.; Alon, O.E. Solvable Model of a Generic Trapped Mixture of Interacting Bosons: Many-Body and Mean-Field Properties. *J. Phys. Conf. Ser.* **2018**, *999*, 012013. [CrossRef]
27. Hall, R.L. Some exact solutions to the translation-invariant N-body problem. *J. Phys. A* **1978**, *11*, 1227. [CrossRef]
28. Hall, R.L. Exact solutions of Schrödinger's equation for translation-invariant harmonic matter. *J. Phys. A* **1978**, *11*, 1235. [CrossRef]
29. Cohen, L.; Lee, C. Exact reduced density matrices for a model problem. *J. Math. Phys.* **1985**, *26*, 3105. [CrossRef]
30. Osadchii, M.S.; Muraktanov, V.V. The System of Harmonically Interacting Particles: An Exact Solution of the Quantum-Mechanical Problem. *Int. J. Quant. Chem.* **1991**, *39*, 173. [CrossRef]
31. Załuska-Kotur, M.A.; Gajda, M.; Orłowski, A.; Mostowski, J. Soluble model of many interacting quantum particles in a trap. *Phys. Rev. A* **2000**, *61*, 033613. [CrossRef]

32. Yan, J. Harmonic Interaction Model and Its Applications in Bose–Einstein Condensation. *J. Stat. Phys.* **2003**, *113*, 623. [CrossRef]
33. Gajda, M. Criterion for Bose–Einstein condensation in a harmonic trap in the case with attractive interactions. *Phys. Rev. A* **2006**, *73*, 023603. [CrossRef]
34. Armstrong, J.R.; Zinner, N.T.; Fedorov, D.V.; Jensen, A.S. Analytic harmonic approach to the *N*-body problem. *J. Phys. B* **2011**, *44*, 055303. [CrossRef]
35. Armstrong, J.R.; Zinner, N.T.; Fedorov, D.V.; Jensen, A.S. Virial expansion coefficients in the harmonic approximation. *Phys. Rev. E* **2012**, *86*, 021115. [CrossRef]
36. Schilling, C. Natural orbitals and occupation numbers for harmonium: Fermions versus bosons. *Phys. Rev. A* **2013**, *88*, 042105. [CrossRef]
37. Benavides-Riveros, C.L.; Toranzo, I.V.; Dehesa, J.S. Entanglement in N-harmonium: Bosons and fermions. *J. Phys. B* **2014**, *47*, 195503. [CrossRef]
38. Bouvrie, P.A.; Majtey, A.P.; Tichy, M.C.; Dehesa, J.S.; Plastino, A.R. Entanglement and the Born-Oppenheimer approximation in an exactly solvable quantum many-body system. *Eur. Phys. J. D* **2014**, *68*, 346. [CrossRef]
39. Armstrong, J.R.; Volosniev, A.G.; Fedorov, D.V.; Jensen, A.S.; Zinner, N.T. Analytic solutions of topologically disjoint systems. *J. Phys. A* **2015**, *48*, 085301. [CrossRef]
40. Schilling, C.; Schilling, R. Number-parity effect for confined fermions in one dimension. *Phys. Rev.* **2016**, *93*, 021601. [CrossRef]
41. Klaiman, S.; Streltsov, A.I.; Alon, O.E. Solvable model of a trapped mixture of Bose–Einstein condensates. *Chem. Phys.* **2017**, *482*, 362. [CrossRef]
42. Sakmann, K.; Streltsov, A.I.; Alon, O.E.; Cederbaum, L.S. Exact ground state of finite Bose–Einstein condensates on a ring. *Phys. Rev. A* **2005**, *72*, 033613. [CrossRef]
43. Gupta, S.; Murch, K.W.; Moore, K.L.; Purdy, T.P.; Stamper-Kurn, D.M. Bose–Einstein Condensation in a Circular Waveguide. *Phys. Rev. Lett.* **2005**, *95*, 143201. [CrossRef] [PubMed]
44. Cozzini, M.; Jackson, B.; Stringari, S. Vortex signatures in annular Bose–Einstein condensates. *Phys. Rev. A* **2006**, *73*, 013603. [CrossRef]
45. Bao, C.G. Oscillation bands of Bose–Einstein condensates on a ring: Beyond the mean-field theory. *Phys. Rev. A* **2007**, *75*, 063626. [CrossRef]
46. Smyrnakis, J.; Bargi, S.; Kavoulakis, G.M.; Magiropoulos, M.; Kärkkäinen, K.; Reimann, S.M. Mixtures of Bose Gases Confined in a Ring Potential. *Phys. Rev. Lett.* **2009**, *103*, 100404. [CrossRef] [PubMed]
47. Halkyard, P.L.; Jones, M.P.A.; Gardiner, S.A. Rotational response of two-component Bose–Einstein condensates in ring traps. *Phys. Rev. A* **2010**, *81*, 061602. [CrossRef]
48. Mathey, L.; Ramanathan, A.; Wright, K.C.; Muniz, S.R.; Phillips, W.D.; Clark, C.W. Phase fluctuations in anisotropic Bose–Einstein condensates: From cigars to rings. *Phys. Rev. A* **2010**, *82*, 033607. [CrossRef]
49. Sherlock, B.E.; Gildemeister, M.; Owen, E.; Nugent, E.; Foot, C.J. Time-averaged adiabatic ring potential for ultracold atoms. *Phys. Rev. A* **2011**, *83*, 043408. [CrossRef]
50. Zöllner, S.; Bruun, G.M.; Pethick, C.J.; Reimann, S.M. Bosonic and Fermionic Dipoles on a Ring. *Phys. Rev. Lett.* **2011**, *107*, 035301. [CrossRef]
51. Adhikari, S.K. Dipolar Bose–Einstein condensate in a ring or in a shell. *Phys. Rev. A* **2012**, *85*, 053631. [CrossRef]
52. Dubessy, R.; Liennard, T.; Pedri, P.; Perrin, H. Critical rotation of an annular superfluid Bose–Einstein condensate. *Phys. Rev. A* **2012**, *86*, 011602. [CrossRef]
53. Woo, S.J.; Son, Y.-W. Vortex dynamics in an annular Bose–Einstein condensate. *Phys. Rev. A* **2012**, *86*, 011604. [CrossRef]
54. Moulder, S.; Beattie, S.; Smith, R.P.; Tammuz, N.; Hadzibabic, Z. Quantized supercurrent decay in an annular Bose–Einstein condensate. *Phys. Rev. A* **2012**, *86*, 013629. [CrossRef]
55. Toikka, L.A.; Suominen, K.-A. Snake instability of ring dark solitons in toroidally trapped Bose–Einstein condensates. *Phys. Rev. A* **2013**, *87*, 043601. [CrossRef]
56. Eckel, S.; Lee, J.G.; Jendrzejewski, F.; Murray, N.; Clark, C.W.; Lobb, C.J.; Phillips, W.D.; Edwards, M.; Campbell, G.K. Hysteresis in a quantized superfluid 'atomtronic' circuit. *Nature* **2014**, *506*, 200. [CrossRef]
57. Mateo, A.M.; Gallemí, A.; Guilleumas, M.; Mayol, R. Persistent currents supported by solitary waves in toroidal Bose–Einstein condensates. *Phys. Rev. A* **2015**, *91*, 063625. [CrossRef]

58. Kolář, M.; Opatrný, T.; Das, K.K. Criticality and spin squeezing in the rotational dynamics of a Bose–Einstein condensate on a ring lattice. *Phys. Rev. A* **2015**, *92*, 043630. [CrossRef]
59. Roy, A.; Angom, D. Geometry-induced modification of fluctuation spectrum in quasi-two-dimensional condensates. *New J. Phys.* **2016**, *18*, 083007. [CrossRef]
60. Roussou, A.; Smyrnakis, J.; Magiropoulos, M.; Efremidis, N.K.; Kavoulakis, G.M. Rotating Bose–Einstein condensates with a finite number of atoms confined in a ring potential: Spontaneous symmetry breaking beyond the mean-field approximation. *Phys. Rev. A* **2017**, *95*, 033606. [CrossRef]
61. Wang, J.-G.; Xu, L.-L.; Yang, S.-J. Ground-state phases of the spin-orbit-coupled spin-1 Bose gas in a toroidal trap. *Phys. Rev. A* **2017**, *96*, 033629. [CrossRef]
62. Guenther, N.-E.; Massignan, P.; Fetter, A.L. Quantized superfluid vortex dynamics on cylindrical surfaces and planar annuli. *Phys. Rev. A* **2017**, *96*, 063608. [CrossRef]
63. Roy, A.; Angom, D. Ramifications of topology and thermal fluctuations in quasi-2D condensates. *J. Phys. B* **2017**, *50*, 225301. [CrossRef]
64. Padavić, K.; Sun, K.; Lannert, C.; Vishveshwara, S. Physics of hollow Bose–Einstein condensates. *Europhys. Lett.* **2017**, *120*, 20004. [CrossRef]
65. Eckel, S.; Kumar, A.; Jacobson, T.; Spielman, I.B.; Campbell, G.K. A Rapidly Expanding Bose–Einstein Condensate: An Expanding Universe in the Lab. *Phys. Rev. X* **2018**, *8*, 021021. [CrossRef] [PubMed]
66. Sun, K.; Padavić, K.; Yang, F.; Vishveshwara, S.; Lannert, C. Static and dynamic properties of shell-shaped condensates. *Phys. Rev. A* **2018**, *98*, 013609. [CrossRef]
67. Streltsov, A.I.; Alon, O.E.; Cederbaum, L.S. Role of Excited States in the Splitting of a Trapped Interacting Bose–Einstein Condensate by a Time-Dependent Barrier. *Phys. Rev. Lett.* **2007**, *99*, 030402. [CrossRef] [PubMed]
68. Alon, O.E.; Streltsov, A.I.; Cederbaum, L.S. Multiconfigurational time-dependent Hartree method for bosons: Many-body dynamics of bosonic systems. *Phys. Rev. A* **2008**, *77*, 033613. [CrossRef]
69. Lode, A.U.J.; Lévêque, C.; Madsen, L.B.; Streltsov, A.I.; Alon, O.E. Multiconfigurational time-dependent Hartree approaches for indistinguishable particles. *arXiv* **2019**, arXiv:1908.03578v1.
70. Alon, O.E.; Streltsov, A.I.; Cederbaum, L.S. Multiconfigurational time-dependent Hartree method for mixtures consisting of two types of identical particles. *Phys. Rev. A.* **2007**, *76*, 062501. [CrossRef]
71. Sakmann, K.; Streltsov, A.I.; Alon, O.E.; Cederbaum, L.S. Exact Quantum Dynamics of a Bosonic Josephson Junction. *Phys. Rev. Lett.* **2009**, *103*, 220601. [CrossRef]
72. Grond, J.; Schmiedmayer, J.; Hohenester, U. Optimizing number squeezing when splitting a mesoscopic condensate. *Phys. Rev. A* **2009**, *79*, 021603. [CrossRef]
73. Lode, A.U.J.; Sakmann, K.; Alon, O.E.; Cederbaum, L.S.; Streltsov, A.I. Numerically exact quantum dynamics of bosons with time-dependent interactions of harmonic type. *Phys. Rev. A* **2012**, *86*, 063606. [CrossRef]
74. Krönke, S.; Cao, L.; Vendrell, O.; Schmelcher, P. Non-equilibrium quantum dynamics of ultra-cold atomic mixtures: the multi-layer multi-configuration time-dependent Hartree method for bosons. *New J. Phys.* **2013**, *15*, 063018. [CrossRef]
75. Cao, L.; Krönke, S.; Vendrell, O.; Schmelcher, P. The multi-layer multi-configuration time-dependent Hartree method for bosons: Theory, implementation, and applications. *J. Chem. Phys.* **2013**, *139*, 134103. [CrossRef]
76. Streltsov, A.I. Quantum systems of ultracold bosons with customized interparticle interactions. *Phys. Rev. A* **2012**, *88*, 041602. [CrossRef]
77. Streltsova, O.I.; Alon, O.E.; Cederbaum, L.S.; Streltsov, A.I. Generic regimes of quantum many-body dynamics of trapped bosonic systems with strong repulsive interactions. *Phys. Rev. A* **2014**, *89*, 061602. [CrossRef]
78. Fischer, U.R.; Lode, A.U.J.; Chatterjee, B. Condensate fragmentation as a sensitive measure of the quantum many-body behavior of bosons with long-range interactions. *Phys. Rev. A* 2015, *91*, 063621. [CrossRef]
79. Fasshauer, E.; Lode, A.U.J. Multiconfigurational time-dependent Hartree method for fermions: Implementation, exactness, and few-fermion tunneling to open space. *Phys. Rev. A* **2016**, *93*, 033635. [CrossRef]
80. Lode, A.U.J. Multiconfigurational time-dependent Hartree method for bosons with internal degrees of freedom: Theory and composite fragmentation of multicomponent Bose–Einstein condensates. *Phys. Rev. A* **2016**, *93*, 063601. [CrossRef]

81. Sakmann, K.; Kasevich, M. Single-shot simulations of dynamic quantum many-body systems. *Nat. Phys.* **2016**, *12*, 451. [CrossRef]
82. Cao, L.; Bolsinger, V.; Mistakidis, S.I.; Koutentakis, G.M.; Krönke, S.; Schurer, J.M.; Schmelcher, P. A unified ab initio approach to the correlated quantum dynamics of ultracold fermionic and bosonic mixtures. *J. Chem. Phys.* **2017**, *147*, 044106. [CrossRef] [PubMed]
83. Bolsinger, V.J.; Krönke, S.; Schmelcher, P. Beyond mean-field dynamics of ultra-cold bosonic atoms in higher dimensions: facing the challenges with a multi-configurational approach. *J. Phys. B* **2017**, *50*, 034003. [CrossRef]
84. Lévêque, C.; Madsen, L.B. Time-dependent restricted-active-space self-consistent-field theory for bosonic many-body systems. *New J. Phys.* **2017**, *19*, 043007. [CrossRef]
85. Weiner, S.E.; Tsatsos, M.C.; Cederbaum, L.S.; Lode, A.U.J. Phantom vortices: Hidden angular momentum in ultracold dilute Bose–Einstein condensates. *Sci Rep.* **2017**, *7*, 40122. [CrossRef]
86. Lode, A.U.J.; Bruder, C. Fragmented Superradiance of a Bose–Einstein Condensate in an Optical Cavity. *Phys. Rev. Lett.* **2017**, *118*, 013603. [CrossRef]
87. Bolsinger, V.J.; Krönke, S.; Schmelcher, P. Ultracold bosonic scattering dynamics off a repulsive barrier: Coherence loss at the dimensional crossover. *Phys. Rev. A* **2017**, *96*, 013618. [CrossRef]
88. Katsimiga, G.C.; Mistakidis, S.I.; Koutentakis, G.M.; Kevrekidis, P.G.; Schmelcher, P. Many-body quantum dynamics in the decay of bent dark solitons of Bose–Einstein condensates. *New J. Phys.* **2017**, *19*, 123012. [CrossRef]
89. Schurer, J.M.; Negretti, A.; Schmelcher, P. Unraveling the Structure of Ultracold Mesoscopic Collinear Molecular Ions. *Phys. Rev. Lett.* **2017**, *119*, 063001. [CrossRef]
90. Chen, J.; Schurer, J.M.; Schmelcher, P. Entanglement Induced Interactions in Binary Mixtures. *Phys. Rev. Lett.* **2018**, *121*, 043401. [CrossRef]
91. Lévêque, C.; Madsen, L.B. Multispecies time-dependent restricted-active-space self-consistent-field-theory for ultracold atomic and molecular gases. *J. Phys. B* **2018**, *51*, 155302. [CrossRef]
92. Roy, R.; Gammal, A.; Tsatsos, M.C.; Chatterjee, B.; Chakrabarti, B.; Lode, A.U.J. Phases, many-body entropy measures, and coherence of interacting bosons in optical lattices. *Phys. Rev. A* **2018**, *97*, 043625. [CrossRef]
93. Elsayed, T.A.; Streltsov, A.I. Probing quantum states with momentum boosts. *Phys. Rev. A* **2018**, *98*, 013618. [CrossRef]
94. Nguyen, J.H.V.; Tsatsos, M.C.; Luo, D.; Lode, A.U.J.; Telles, G.D.; Bagnato, V.S.; Hulet, R.G. Parametric Excitation of a Bose–Einstein Condensate: From Faraday Waves to Granulation. *Phys. Rev. X* **2019**, *9*, 011052. [CrossRef]
95. Marchukov, O.V.; Fischer, U.R. Self-consistent determination of the many-body state of ultracold bosonic atoms in a one-dimensional harmonic trap. *Ann. Phys.* **2019**, *405*, 274. [CrossRef]
96. Streltsov, A.I.; Streltsova, O.I. MCTDHB-Lab, Version 1.5. 2015. Available online: http://www.mctdhb-lab.com (accessed on 29 September 2019).
97. Streltsov, A.I.; Cederbaum, L.S.; Alon, O.E.; Sakmann, K.; Lode, A.U.J.; Grond, J.; Streltsova, O.I.; Klaiman, S.; Beinke, R. The Multiconfigurational Time-Dependent Hartree for Bosons Package, Version 3.x. Available online: http://mctdhb.org (accessed on 29 September 2019).
98. Streltsov, A.I.; Alon, O.E.; Cederbaum, L.S. General variational many-body theory with complete self-consistency for trapped bosonic systems. *Phys. Rev. A* **2006**, *73*, 063626. [CrossRef]
99. Meyer, H.-D.; Manthe, U.; Cederbaum, L.S. The multi-configurational time-dependent Hartree approach. *Chem. Phys. Lett.* **1990**, *165*, 73. [CrossRef]
100. Manthe, U.; Meyer, H.-D.; Cederbaum, L.S. Wave-packet dynamics within the multiconfiguration Hartree framework: General aspects and application to NOCl. *J. Chem. Phys.* **1992**, *97*, 3199. [CrossRef]
101. Beck, M.H.; Jäckle, A.; Worth, G.A.; Meyer, H.D. The multiconfiguration time-dependent Hartree (MCTDH) method: A highly efficient algorithm for propagating wavepackets. *Phys. Rep.* **2000**, *324*, 1. [CrossRef]
102. Meyer, H.-D.; Gatti, F.; Worth, G.A. (Eds.) *Multidimensional Quantum Dynamics: MCTDH Theory and Applications*; Wiley-VCH: Weinheim, Germany, 2009.
103. Wang, H.; Thoss, M. Multilayer formulation of the multiconfiguration time-dependent Hartree theory. *J. Chem. Phys.* **2003**, *119*, 1289. [CrossRef]

104. Manthe, U. A multilayer multiconfigurational time-dependent Hartree approach for quantum dynamics on general potential energy surfaces. *J. Chem. Phys.* **2008**, *128*, 164116. [CrossRef]
105. Vendrell, O.; Meyer, H.-D. Multilayer multiconfiguration time-dependent Hartree method: Implementation and applications to a Henon-Heiles Hamiltonian and to pyrazine. *J. Chem. Phys.* **2011**, *134*, 044135. [CrossRef] [PubMed]

symmetry

MDPI

Article

Nonlinear Dynamics of Wave Packets in Tunnel-Coupled Harmonic-Oscillator Traps

Nir Hacker [1] and Boris A. Malomed [1,2,*]

1 Department of Physical Electronics, School of Electrical Engineering, Faculty of Engineering, and Center for Light-Matter interaction, Tel Aviv University, Tel Aviv 69978, Israel; nirhack@gmail.com
2 Instituto de Alta Investigación, Universidad de Tarapacá, Casilla 7D, 1000000 Arica, Chile
* Correspondence: malomed@tauex.tau.ac.il

Abstract: We consider a two-component linearly coupled system with the intrinsic cubic nonlinearity and the harmonic-oscillator (HO) confining potential. The system models binary settings in BEC and optics. In the symmetric system, with the HO trap acting in both components, we consider Josephson oscillations (JO) initiated by an input in the form of the HO's ground state (GS) or dipole mode (DM), placed in one component. With the increase of the strength of the self-focusing nonlinearity, spontaneous symmetry breaking (SSB) between the components takes place in the dynamical JO state. Under still stronger nonlinearity, the regular JO initiated by the GS input carries over into a chaotic dynamical state. For the DM input, the chaotization happens at smaller powers than for the GS, which is followed by SSB at a slightly stronger nonlinearity. In the system with the defocusing nonlinearity, SSB does not take place, and dynamical chaos occurs in a small area of the parameter space. In the asymmetric half-trapped system, with the HO potential applied to a single component, we first focus on the spectrum of confined binary modes in the linearized system. The spectrum is found analytically in the limits of weak and strong inter-component coupling, and numerically in the general case. Under the action of the coupling, the existence region of the confined modes shrinks for GSs and expands for DMs. In the full nonlinear system, the existence region for confined modes is identified in the numerical form. They are constructed too by means of the Thomas–Fermi approximation, in the case of the defocusing nonlinearity. Lastly, particular (non-generic) exact analytical solutions for confined modes, including vortices, in one- and two-dimensional asymmetric linearized systems are found. They represent bound states in the continuum.

Keywords: Bose-Einstein condensates; Josephson oscillations; spontaneous symmetry breaking; Thomas-Fermi approximation; dynamical chaos; ground states; perturbation theory

Citation: Hacker, N.; Malomed, B.A. Nonlinear Dynamics of Wave Packets in Tunnel-Coupled Harmonic-Oscillator Traps. *Symmetry* **2021**, *13*, 372. https://doi.org/10.3390/sym13030372

Academic Editor: V. I. Yukalov and Rashid G. Nazmitdinov

Received: 06 February 2021
Accepted: 21 February 2021
Published: 25 February 2021

1. Introduction

The combination of a harmonic-oscillator (HO) trapping potential and cubic nonlinearity is a ubiquitous setting which occurs in diverse microscopic, mesoscopic, and macroscopic physical settings. A well-known realization is offered by Bose-Einstein condensates (BECs) with collisional nonlinearity [1–3], loaded in a magnetic or optical trap—see, e.g., Refs. [4–14]. A similar combination of the effective confinement, approximated by the parabolic profile of the local refractive index, and the Kerr term is relevant as a model of optical waveguides [15–18]. Models of the same type appear in other physical systems too, such as networks of Josephson oscillators [19].

The character of states created by the interplay of the intrinsic nonlinearity and externally applied trapping potential strongly depends on the sign of the nonlinearity. In the case of the self-attraction (or self-phase-modulation, SPM, in terms of optics [20]), localized modes, similar to solitons, arise spontaneously. On the other hand, self-repulsion tends to create flattened configurations, which, in turn, may support dark solitons in various static and dynamical states [6–14]. A specific situation occurs in a system combining repulsion and a weak parabolic potential with an additional spatially periodic one (it represents an optical lattice in BEC [21], or a photonic crystal in optics and plasmonics [22–24]): the interplay of the periodic potential with the self-repulsion gives rise to bright gap solitons [21,25], whose effective mass is negative. For this reason, gap solitons are expelled by the HO potential, but are trapped by the inverted one that would expel modes with positive masses [26].

Another noteworthy feature of the dynamics of one-dimensional (1D) nonlinear fields trapped in confining potentials is the degree of its *nonintegrability*. The generic model for such settings is provided by the nonlinear Schrödinger equation (NLSE, alias the Gross–Pitaevskii equation, in terms of BEC [1–3]) with the cubic term, whose integrability in the 1D space [27] is broken by the presence of the HO potential. Nevertheless, systematic numerical simulations, performed for the repulsive sign of the cubic nonlinearity, have demonstrated that the long-time evolution governed by NLSE with the HO potential term does not lead to establishment of spatiotemporal chaos ("turbulence"), which would be expected in the case of generic nonintegrability [28,29]. Instead, the setup demonstrates quasi-periodic evolution, represented by a quasi-discrete power spectrum, in terms of a multi-mode truncation (Galerkin approximation) [14]. This observation is specific for the harmonic (quadratic) confining potential, while anharmonic ones quickly lead to the onset of clearly observed chaos [30,31]. Thermalization of the model with the HO potential was recently explored, in the framework of a stochastically driven dissipative Gross–Pitaevskii equation, in Ref. [32].

The interplay of the cubic nonlinearity and trapping potentials also occurs in two-component systems, which represent, in particular, binary BEC [12,33–35]. Here, we aim to consider the system with linear coupling between the components. In optics, if two modes carrying orthogonal polarizations of light propagate in the same waveguide, linear mixing between them is induced by a twist of the guiding structure, see, e.g., Ref. [36]. In BEC, different atomic states which correspond to the interacting components can be linearly coupled by a resonant electromagnetic wave [37–40]. Another realization of linearly mixed systems is offered by dual-core waveguides, coupled by tunneling of the field across a barrier separating the cores. The dual-core schemes are equally relevant to optics and BEC [41]. In particular, experiments with temporal solitons in dual-core optical fibers were reported in recent work [42].

The coupling between the components enhances the complexity of the system and makes it possible to find new static and dynamical states in it. In particular, the symmetric system combining the attractive cubic terms of the SPM type and (optionally) HO potential acting in each component, with linear coupling between them, gives rise to spontaneous symmetry breaking (SSB) of two-component states [35,42]. In the case of repulsive SPM acting in each component and nonlinear repulsion between them (cross-phase modulation, XPM), it was found [33] that the linear mixing shifts the miscibility-immiscibility transition [43] in the trapped condensate. Furthermore, effects of nonintegrability may be stronger in the linearly coupled system, because the linear coupling makes the system of one-dimensional NLSEs nonintegrable, even in the absence of the HO potential [44], although the integrability is kept by the system with the linear coupling added to the SPM and XPM terms with equal strengths (the Manakov's nonlinearity) [45].

The objective of the present work is two-fold. First, in Section 2 we aim to analyze the onset of chaos, as well as SSB, in the symmetric linearly coupled system, with both attractive and repulsive signs of the SPM terms, starting from an input in the form of the ground state (GS), or the first excited state (the dipole mode (DM), represented by a spatially odd wave function) of the HO, which is initially launched in one core (component), while the other one is empty. This type of the input is, in particular, experimentally relevant in optics [42]. As a nonchaotic dynamical regime in this case, one may expect Josephson oscillations (JO) of the optical field [42,44,46–49], or of the BEC wave function [50–56], between the cores. As a measure of the transition to dynamical chaos in the system, we use a relative spread of the power spectrum of oscillations produced by simulations of the coupled NLSEs. Naturally, the chaotization sets in above a certain threshold value of the nonlinearity strength, and the chaos is much weaker in the case of the repulsive nonlinearity. In the dynamical state initiated by the GS, the SSB takes place prior to the onset of the dynamical chaos, while the DM input undergoes the chaotization occurs first, followed by the SSB at a slightly stronger nonlinearity.

The second objective, which is presented below in Section 3, is to construct stationary GS and DM in the asymmetric (*half-trapped*) linearly coupled system, with the HO potential applied to one component only. The latter system can be realized in the experiment, applying, for instance, the trapping potential only to one core of the double waveguide for matter waves in BEC (e.g., by focusing laser beams, which induce the trapping, on the single core). In optics, a similar setup may be built as a coupler with two widely different cores, narrow and broad ones, with the narrow core emulating the component carrying a tightly confining potential, cf. Ref. [57]. A remarkable peculiarity of such a system is that the linear coupling mixes completely different types of the asymptotic behavior at $|x| \to \infty$: the trapped component is always confined, in the form of a Gaussian, by the HO potential, while the untrapped one is free to escape. In addition, the asymmetry between the coupled cores makes it necessary to take into regard a difference in the chemical potentials or propagation constants (in terms of the BEC and optics, respectively) between them, which is represented by parameter ω in Equation (10), see below. Some results for the half-trapped system are obtained in an analytical form, in the weak- and strong-coupling limits, as well as by means of the Thomas–Fermi approximation (TFA), and full results are produced numerically. In addition, particular solutions for localized states of the linear half-trapped system (including vortex states in its 2D version) are found in an exact form. The exact solutions belong to the class of *bound states in the continuum* [58–60], alias *embedded* [61] ones, which are spatially confined modes existing, as exceptional states, with the carrier frequency falling in the continuous spectrum. The paper is concluded by Section 4.

2. The Symmetric System

2.1. The Coupled Equations

As outlined above, we consider systems modeled by a pair of coupled NLSEs for complex amplitudes $u(x,z)$ and $v(x,z)$ of two interacting waves. In the normalized form, the equations are written in terms of the spatial-domain propagation in an optical waveguide with propagation distance z and transverse coordinate x:

$$\begin{aligned} i\frac{\partial u}{\partial z}+\frac{1}{2}\frac{\partial^2 u}{\partial x^2}+\lambda v-\frac{\Omega^2}{2}x^2 u+\sigma\left(|u|^2+g|v|^2\right)u &= 0, \\ i\frac{\partial v}{\partial z}+\frac{1}{2}\frac{\partial^2 v}{\partial x^2}+\lambda u-\frac{\Omega^2}{2}x^2 v+\sigma\left(|v|^2+g|u|^2\right)v &= 0. \end{aligned} \tag{1}$$

Here, coefficients $\sigma > 0$ or $\sigma < 0$ represents the strength of the focusing or defocusing SPM in each core, while λ and σg represent the linear mixing and XPM interaction, respectively. By means of rescaling, the strength of the HO trapping potential is set to be $\Omega = 1$ (unless it is zero in one core). In other words, x is measured in units of the respective HO length. This implies that the unit of the transverse coordinate in the optical waveguides takes typical values in the range of 10–30 μm, hence the respective unit of the propagation distance (the Rayleigh/diffraction distance corresponding to the OH length) is estimated to be between 1 mm and 1 cm, for the carrier wavelength $\sim$ 1μm. In the matter-wave realization of the system, typical units of x and time (replacing z in Equation (1) are $\sim$10 μm and 10 ms, respectively.

The system conserves two dynamical invariants., *viz.*, the total norm (or power, in terms of optics),

$$P=\int_{-\infty}^{+\infty}\left[|u(x)|^2+|v(x)|^2\right]dx, \tag{2}$$

and the Hamiltonian (in which $\Omega = 1$ is set),

$$H = \int_{-\infty}^{+\infty} \left[\frac{1}{2} \left(\left| \frac{\partial^2 u}{\partial x^2} \right|^2 + \left| \frac{\partial^2 v}{\partial x^2} \right|^2 \right) - \frac{\sigma}{2} \left(|u|^4 + |v|^4 \right) \right.$$
$$\left. -\sigma g |u|^2 |v|^2 + \frac{1}{2} x^2 \left(|u|^2 + |v|^2 \right) - \lambda (uv^* + u^* v) \right] dx, \tag{3}$$

where $*$ stands for the complex conjugate. The remaining scaling invariance of Equation (1) makes it possible to either set $|\sigma| = 1$, or keep nonlinearity coefficient σ as a free parameter, but fix $P \equiv 1$. All simulations performed in this work comply with the conservation of P and H, up to the accuracy of the numerical codes.

It is relevant to mention that the present two-component system resembles nonlinear models with a double-well potential, in the case when the wave functions in two wells are linearly coupled by tunneling across the potential barrier, see, e.g., Refs. [62,63]. Nevertheless, the exact form of the system and its solutions are different.

In the limit of a small amplitude A_0 of the input, linearized Equation (1) with $\Omega = 1$ admit exact solutions for inter-core JO of the ground and dipole states (exact solutions for higher-order states can be readily found too):

$$\begin{aligned} u_{\mathrm{JO}}^{(\mathrm{GS})}(x,t) &= A_0 \exp\left(-\frac{1}{2} x^2 - \frac{1}{2} iz \right) \cos(\lambda z), \\ v_{\mathrm{JO}}^{(\mathrm{GS})}(x,t) &= iA_0 \exp\left(-\frac{1}{2} x^2 - \frac{1}{2} iz \right) \sin(\lambda z), \end{aligned} \tag{4}$$

$$\begin{aligned} u_{\mathrm{JO}}^{(\mathrm{DM})}(x,t) &= A_0 x \exp\left(-\frac{1}{2} x^2 - \frac{3}{2} iz \right) \cos(\lambda z), \\ v_{\mathrm{JO}}^{(\mathrm{DM})}(x,t) &= iA_0 x \exp\left(-\frac{1}{2} x^2 - \frac{3}{2} iz \right) \sin(\lambda z). \end{aligned} \tag{5}$$

Expressions given by Equations (4) and (5) at $z = 0$ are used below as inputs in simulations of the full nonlinear Equation (1). The simulations presented in this section were performed in domain $|x| < 10$. This size, tantamount to 20 HO lengths, is sufficient to display all details of the solutions. Standard numerical methods were used, *viz.*, the split-step fast-Fourier-transform scheme for simulations of the evolution governed by Equations (1) and (10), and the relaxation algorithm for finding solutions to stationary equations, such as Equations (12) and (13), see below.

2.2. The Transition from Regular to Chaotic Dynamics

Increase of amplitude A_0 of the input leads to nonlinear deformation of the oscillations, and eventually to the onset of dynamical chaos. A typical example of an essentially nonlinear but still regular JO dynamical regime, produced by numerical simulations of Equation (1) with $g = 0$ (no XPM interaction), in interval $0 < z < Z$, with the DM input, taken as per Equation (5) at $z = 0$, is presented in Figure 1. In particular, the left bottom panel of the figure displays oscillations of peak intensities of the fields,

$$\left\{ U_{\max}^2(z), V_{\max}^2(z) \right\} \equiv \max_x \left\{ |u(x,z)|^2, |v(x,z)|^2 \right\}, \tag{6}$$

and, as a characteristic of the dynamics, in the right bottom panel we plot power spectra, $|P(\kappa)|^2, |Q(\kappa)|^2$, produced by the Fourier transform of the peak intensities:

$$\{P(\kappa), Q(\kappa)\} = \int_0^Z \left\{ U_{\max}^2(z), V_{\max}^2(z) \right\} \exp(-i\kappa z) dz, \tag{7}$$

where κ is a real propagation constant. Very slow decay of the peak intensities, observed in the former panel, is a manifestation of the system's nonintegrability (in this connection, we again stress that the total norm is conserved during the simulations).

The regularity of the dynamical regime displayed in Figure 1 is clearly demonstrated by its spectral structure, which exhibits a single narrow peak at $\kappa_{\mathrm{peak}} \approx 2$. The peak's width, $\Delta\kappa \simeq 0.1$, which corresponds to the relative width, $\Delta\kappa/\kappa_{\mathrm{peak}} \approx 0.05$, is comparable to the spread of the Fourier transform, corresponding to $Z = 100$ in Figure 1. It can be estimated as $\delta\kappa = 2\pi/Z \approx 0.06$. Note the overall symmetry between the two components in Figure 1 (in particular, their spectra are identical in the bottom panel).

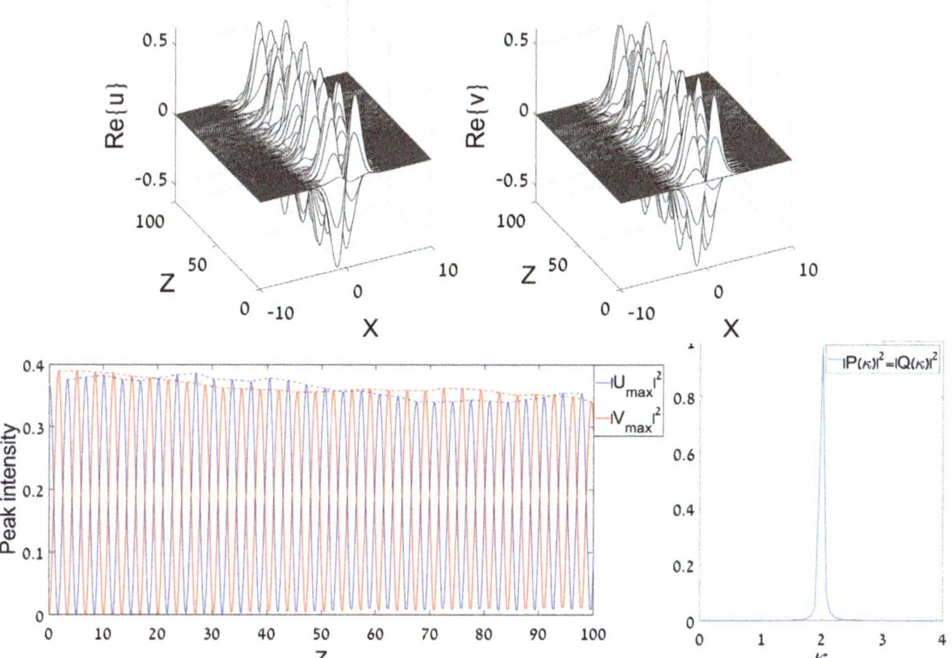

Figure 1. A typical example of a regular Josephson dynamical regime, initiated by the DM (dipole mode) input launched in the u component (as given by Equation (5) with $z = 0$ and $A_0 = 1$). The solution is produced by simulations of Equation (1) with $\lambda = \sigma = \Omega = 1$, $g = 0$. Plots in the top row display the evolution of components $u(x,z)$ and $v(x,z)$. Left bottom: The evolution of the peak intensities of both components, $U^2_{\max}(z) \equiv \max_x\left\{|u(x,z)|^2\right\}$ and $V^2_{\max}(z) \equiv \max_x\left\{\left[|v(x,z)|^2\right]\right\}$. Right bottom: The power spectrum of oscillations of the two components, defined as per Equation (7). The spectra are virtually identical for both components.

The simulations with the same input, but larger values of A_0, give rise to chaotic ("turbulent") dynamical states with a broad dynamical spectrum, see a typical example in Figure 2. Note that both the regular and chaotic dynamical pictures displayed in Figures 1 and 2 extend over the distance estimated to be $\sim$10 Rayleigh (diffraction) lengths corresponding to the width of the DM input. This estimate is sufficient to make conclusions about the character of the dynamics.

Systematic simulations with the GS input, provided by Equation (4) at $z = 0$, produce similar results (not shown here in detail). In particular, as well as in the case of the DM input, amplitudes $A_0 = 1$ and 4 initiate, severally, quasi-regular and chaotic evolution.

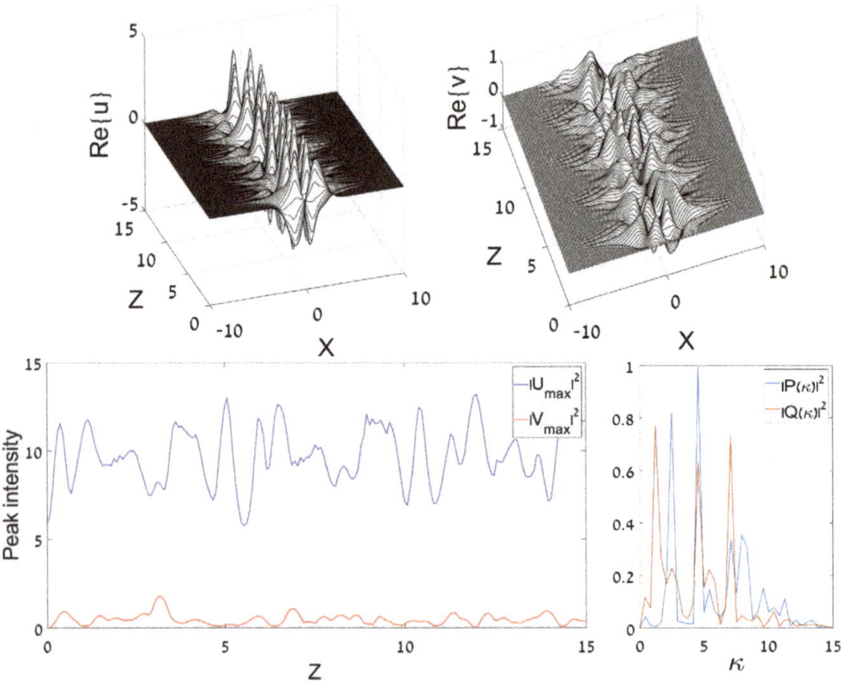

Figure 2. The same as in Figure 1, but for a typical example of a chaotic dynamical regime, initiated by the DM input (5) with a larger amplitude, $A_0 = 4$. Note different scales on vertical axes in two plots in the left panel.

The results for the transition from regular JO to chaotic dynamics, initiated by the GS and DM inputs, taken as per Equations (4) and (5) at $z = 0$, are summarized in charts plotted in the plane of $\left(\lambda, A_0^2\right)$ in Figure 3. They display heatmaps of values of the parameter which quantifies the sharpness of the central peak in the spectrum of the dynamical state:

$$\text{Sharpness} \equiv \frac{\int_{\text{FWHM}} |P(\kappa)|^2 d\kappa}{\int_0^\infty |P(\kappa)|^2 d\kappa}, \tag{8}$$

where the integration in the numerator is performed over the section of the central spectral peak selected according to the standard definition of the full width at half-maximum: $|P(\kappa_{\text{FWHM}})|^2 = (1/2)(P(\kappa))_{\text{max}}$. Values of Sharpness close to 1 imply the domination of a single sharp peak, such as one in Figure 1, which corresponds to a regular dynamical regime, while decrease of this parameter indicates a transition to a broad spectrum, which is a telltale of the onset of chaotic dynamics—see, e.g., Figure 2.

Figure 3 clearly demonstrates decay of the central peak's sharpness, i.e., transition to dynamical chaos, with the increase of the input's intensity, A_0^2, in the case of the attractive SPM, $\sigma = +1$. Such a trend is not straightforward in the opposite case of the self-repulsion, $\sigma = -1$. In particular, the chaotization is not observed at all in the latter case at small values of λ. This conclusion agrees with findings reported in work [14] for the single-component NLSE (which corresponds to $\lambda = 0$), with the HO potential and $\sigma = -1$. Computations of the spectrum, reported in that work, demonstrate that no transition to dynamical chaos takes place at all values of parameters. The fact that an area of weak chaotization is, nevertheless, observed in the right panels of Figure 3 is explained by the above-mentioned circumstance, that the addition of the linear coupling to a pair of NLSEs destroys their integrability (in free space). On the other hand, the increase of the sharpness with the increase of λ, observed at relatively small values of A_0^2 (which is observed in an especially salient form in the left bottom panel of Figure 3) also has a simple explanation: the increase of λ makes the linear terms in the system dominating over nonlinear ones, thus tending to maintain a quasi-linear behavior.

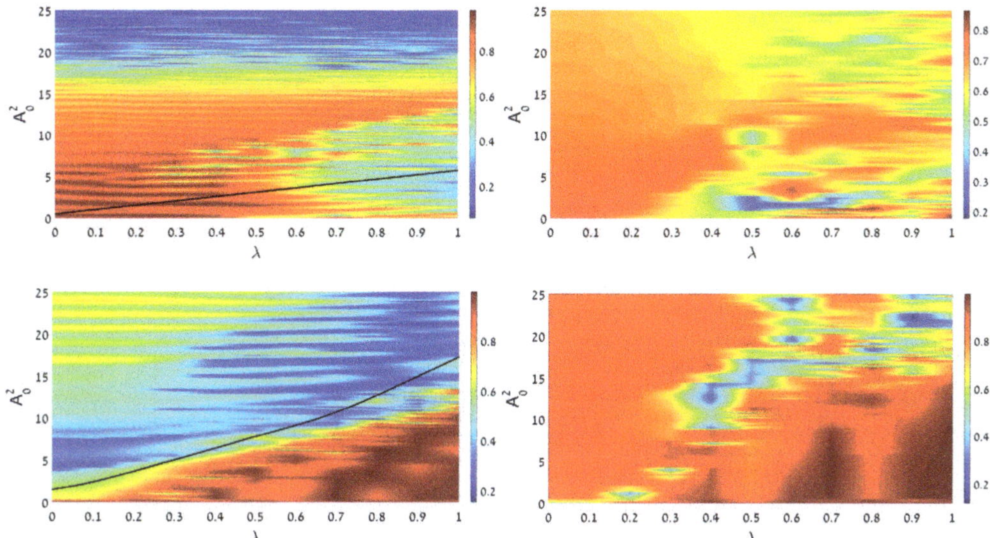

Figure 3. Heatmaps of values of sharpness (8) of the central spectral peak quantifying proximity of the system's dynamics to the regular regime. The maps are plotted in the plane of the linear-coupling strength, λ, and intensity of the input, A_0^2, which is launched in one component. (**Top left**): The GS input, given by Equation (4) at $z = 0$, in the case of the self-attraction ($\sigma = +1$). (**Top right**): The same, but in the case of self-repulsion ($\sigma = -1$). (**Bottom left**): The same as in the top left panel, but produced by the DM input, given by Equation (5) at $z = 0$. (**Bottom right**): The same as in the bottom left panel, but in the case of self-repulsion ($\sigma = -1$). In all cases, $g = 0$ is set in Equation (1) (no XPM interaction between the components). Black curves cutting the left panels in their lower areas designate the onset of SSB (spontaneous symmetry breaking), signalized by appearance of $\theta \neq 0$, see Equation (9).

Lastly, the comparison of the left and right panels in Figure 3 suggests that the chaotization sets in faster in dynamical regimes initiated by the DM input, in comparison to their counterparts originating from the GS, for the same values of the input's intensity, A_0^2. The difference between the GS and DM dynamical regimes is salient for relatively small values of λ. It may be explained by the fact that attractive SPM naturally tends to form a stable bright soliton from the GS input, which then maintains regular motion in the HO potential [64]. On the other hand, spatially odd bright solitons do not exist in free space, which impedes transformation of the DM input into a regular dynamical state.

As said above, the heatmaps are displayed in Figure 3 for $g = 0$, i.e., in the absence of the XPM coupling between the components, which is the case for dual-core couplers. On the other hand, in the case of the Manakov's nonlinearity, i.e., $g = 1$, the above-mentioned integrability of such a system of NLSEs with the linear coupling [45] (but without the trapping potential) suggests that the full system will be closer to integrability and farther from the onset of chaos. Indeed, numerical results collected from simulations of Equation (1) with $\sigma = g = +1$ (not shown here in detail) demonstrate a much smaller chaotic area in the $\left(\lambda, A_0^2\right)$ plane. In particular, the GS input generates "turbulent" behavior only at $A_0^2 \gtrsim 200$, being limited to $\lambda \lesssim 0.15$, cf. the top left panel in Figure 3.

2.3. Spontaneous Symmetry Breaking (SSB) between the Coupled Components

A noteworthy feature of the dynamical state presented in Figure 2 is breaking of the symmetry between fields u and v (while the patterns initiated by the GS and DM inputs keep their parities, i.e., spatial symmetry (evenness) and antisymmetry (oddness), respectively). This is a manifestation of the general effect, which is well known, in diverse forms, in linearly coupled dual-core systems with intrinsic attractive SPM [41]. In particular, SSB of stationary states in systems with the HO trapping potential acting in both cores was addressed in Refs. [35,56], while Figure 2 demonstrates the symmetry breaking in the *dynamical* JO state.

The SSB effect may be quantified, as usual [44], by asymmetry of the dynamical states, which we define as

$$\Theta \equiv \frac{\int_0^\infty |P(\kappa)|^2 d\kappa - \int_0^\infty |Q(\kappa)|^2 d\kappa}{\int_0^\infty |P(\kappa)|^2 d\kappa + \int_0^\infty |Q(\kappa)|^2 d\kappa}. \quad (9)$$

The SSB occurs as a transition from $\Theta = 0$ to $\Theta \neq 0$ with the increase of A_0^2 at some critical point, which is a generic property of stationary states in dual-core systems with intrinsic attractive nonlinearity [41,44], while here we consider it in the dynamical setting. On the other hand, in the case of the repulsive nonlinearity, $\sigma = -1$ in Equation (1), SSB of stationary states takes place in this system only at $g > 1$ [56], while we here focus on the most relevant case of $g = 0$. Accordingly, the present system with $\sigma = -1$ does not feature SSB.

For $\sigma = +1$, the SSB boundaries in the parameter planes of the GS and DM solutions are shown by bold black lines in the top and bottom left panels of Figure 3, respectively. The SSB bifurcation of the dynamical states under consideration is of the *supercritical*, alias *forward*, type [65], in terms of dependence $\Theta(A_0^2)$, as shown in Figure 4 for the dynamical states initiated by the GS and DM inputs. It is observed that, naturally, the critical value of A_0^2 increases with λ, as the symmetry is maintained by the linear coupling, hence stronger coupling needs stronger nonlinearity to break the symmetry. Note also that the transition from $\Theta = 0$ to $\Theta \approx 1$ (a strongly asymmetric state) is steeper at larger λ.

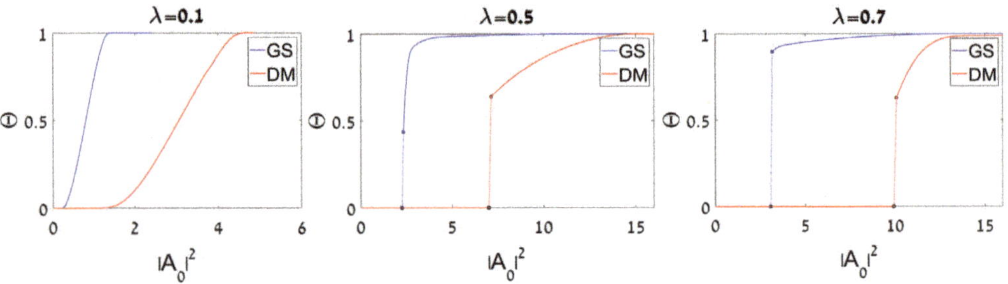

Figure 4. Plots of the SSB (spontaneous symmetry breaking) in the dynamical states initiated by the GS (ground state) and DM (dipole mode) inputs: the asymmetry parameter, defined as per Equation (9), is plotted versus the intensity of the input, A_0^2, for three different fixed values of the linear-coupling constant, λ, as indicated in the figure.

It is worth noting that, as clearly shown by the SSB boundary (the black line) in the top left panel of Figure 3, the SSB of the GS mode happens prior to the transition to the dynamical chaos. This conclusion agrees with known results showing that SSB of stationary modes does not, normally, lead to chaotization of the system's dynamics [41,44]. On the other hand, the bottom left panel in Figure 3 demonstrates a different situation for the DM states, which exhibit the chaotization prior to the SSB, although the separation between these transitions is small. This conclusion agrees with the fact that, as clearly seen in Figure 4, the SSB in DM states occurs at values of the input's amplitude essentially higher than those which determine the SSB threshold of the GS solutions.

3. The Half-Trapped System

The asymmetric system of linearly coupled NLSEs, with the HO potential included in one equation only, is written as

$$
\begin{aligned}
iu_z + \frac{1}{2}u_{xx} + \lambda v - \frac{1}{2}x^2 u + \sigma\left(|u|^2 + g|v|^2\right)u &= -\omega u, \\
iv_z + \frac{1}{2}v_{xx} + \lambda u + \sigma\left(|v|^2 + g|u|^2\right)v &= 0,
\end{aligned} \tag{10}
$$

cf. Equation (1). Here, as said above, $\Omega = 1$ is set in the first equation, and the propagation-constant mismatch, ω (in terms of BEC, it represents a difference in the chemical potentials between the two wave functions) is a common feature of asymmetric systems. Stationary solutions to Equation (10) are looked for as

$$
\{u, v\} = \{U(x), V(x)\} \exp(-i\mu z), \tag{11}
$$

with real propagation constant $-\mu$ (in BEC, with z replaced by t, μ is the chemical potential), and real functions $U(x)$ and $V(x)$ satisfying equations

$$
(\mu + \omega)U + \frac{1}{2}\frac{d^2 U}{dx^2} + \lambda V - \frac{1}{2}x^2 U + \sigma\left(U^2 + gV^2\right)U = 0, \tag{12}
$$

$$
\mu V + \frac{1}{2}\frac{d^2 V}{dx^2} + \lambda U + \sigma\left(V^2 + gU^2\right)V = 0. \tag{13}
$$

Most results are produced below disregarding the XPM coupling between the components ($g = 0$). Nevertheless, the XPM terms are included when collecting numerical results for the threshold of the existence of bound states.

3.1. The Linearized System: Analytical and Numerical Results

3.1.1. Emission of Radiation in the Untrapped Component

In the linear limit, $\sigma = 0$, two decoupled equations in system (10) with $\lambda = 0$ produce completely different results: all excitations of component u stay confined in the HO trap, while the v component with any $\mu > -\omega$ freely expands. In particular, in the limit of $\lambda = 0$, obvious bound-state GS and DM solutions to the linearized version of Equations (12) and (13), with zero v component, are

$$
U_{\mathrm{GS}}^{(0)}(x) = \frac{1}{\pi^{1/4}} \exp\left(-\frac{x^2}{2}\right), V_{\mathrm{GS}}^{(0)} = 0, \tag{14}
$$

$$
U_{\mathrm{DM}}^{(0)}(x) = \frac{\sqrt{2}}{\pi^{1/4}} x \exp\left(-\frac{x^2}{2}\right), V_{\mathrm{DM}}^{(0)} = 0, \tag{15}
$$

where the pre-exponential constants are determined by the normalization condition,

$$
\int_{-\infty}^{+\infty} \left[U^2(x) + V^2(x)\right] dx = 1, \tag{16}
$$

which we adopt in this section. The eigenvalues corresponding to eigenmodes (14) and (15) are

$$
\mu_{\mathrm{GS}}^{(0)} = 1/2 - \omega; \ \mu_{\mathrm{DM}}^{(0)} = 3/2 - \omega. \tag{17}
$$

Proceeding to dynamical states, in the lowest approximation with respect to small λ the evolution of the v field is driven by the respective linearized equation in system (10),

$$
iv_z + \frac{1}{2}v_{xx} = -\lambda U_{\mathrm{GS,DM}}^{(0)}(x) \exp\left(-i\mu_{\mathrm{GS,DM}}^{(0)} z\right). \tag{18}
$$

Obviously, Equation (18) gives rise to emission of propagating waves ("radiation"), in the form of $v \sim \lambda \exp\left(ikx - i\mu^{(0)}_{\mathrm{GS,DM}} z\right)$, at resonant wavenumbers $k = \pm\sqrt{2\mu^{(0)}_{\mathrm{GS,DM}}}$, provided that $\mu^{(0)}_{\mathrm{GS,DM}}$ is positive, i.e.,

$$v_{\mathrm{rad}} \sim \lambda \exp\left(\pm i\sqrt{2\mu^{(0)}_{\mathrm{GS,DM}}}\left(x - V_{\mathrm{ph}} z\right)\right), \tag{19}$$

where the phase velocity is

$$V_{\mathrm{ph}} = \frac{k}{2} \equiv \pm\frac{1}{2}\sqrt{2\mu^{(0)}_{\mathrm{GS,DM}}} \tag{20}$$

(in terms of the spatial-domain propagation in the optical waveguide, it is actually the beam's slope). The expansion of the area in the (x, z) plane occupied by the radiation field is bounded by the group velocity, $|V_{\mathrm{gr}}| = |k| \equiv 2V_{\mathrm{ph}}$.

An illustration of this dynamics is presented in Figure 5. Straight red lines designate the wave-propagation directions, which exactly agree with the phase velocity predicted by Equation (20), and the expansion of the area occupied by the radiation complies with the prediction based on the group velocity.

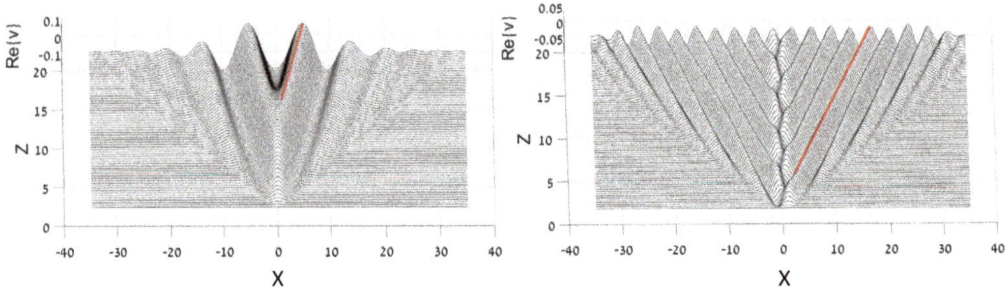

Figure 5. Simulations of the evolution of the linearized half-trapped system (10), displayed in the untrapped component by plotting $\mathrm{Re}(v(x,z))$. (**Left**): Emission of radiation generated by the GS (ground state) populating the trapped component, $u(x, z)$ (see Equation (14)), with $\omega = 0.25$ in Equation (10). (**Right**): The same, but for the radiation generated by the DM (dipole mode) in the trapped component (see Equation (15)), with $\omega = 0$. In both cases, the linear-coupling constant is $\lambda = 0.05$.

The emission of radiation into the v core gives rise to a gradual decay of the amplitude in the u core, due to the conservation of the total norm, see Equation (16). An example of the decay is displayed in Figure 6, for a small initial amplitude of the GS input, $A_0 = 0.1$ (which corresponds to the quasi-linear dynamical regime), and a relatively large coupling constant, $\lambda = 1$, which makes the transfer of the norm (power) from u to v faster.

On the other hand, the same input with large A_0 makes the u-v coupling a weak effect, in comparison with the dominant nonlinearity, hence the input mode in the u core seems quite stable, as shown in Figure 7.

3.1.2. The Shift of the GS and DM Existence Thresholds at Small Values of Coupling Constant λ

At $\mu^{(0)}_{\mathrm{GS,DM}} < 0$ the radiation is not generated by Equation (19), hence two-component bound states may exist in this case. Our next objective is to find, for the coupled system with $\lambda > 0$, threshold values $(\omega_{\mathrm{GS,DM}})_{\mathrm{thr}}$ of the mismatch parameter ω in Equations (12) and (13), such that the bound states of the GS and DM types exist at

$$\omega > (\omega_{\mathrm{GS,DM}})_{\mathrm{thr}}, \tag{21}$$

respectively. Obviously, $(\omega_{\mathrm{GS}})_{\mathrm{thr}} = 1/2$ and $(\omega_{\mathrm{DM}})_{\mathrm{thr}} = 3/2$ in the limit of $\lambda = 0$, see Equation (17).

First, we aim to find lowest-order corrections to the GS and DM eigenvalues (17) for small λ. Then, a shift of the respective thresholds can be identified by setting $\mu_{\mathrm{GS,DM}} = 0$. In the limit of $\lambda = 0$, the GS and DM wave functions are taken as per Equations (14) and (15), respectively. With small λ, the first-order solution for $V(x)$, *viz.*, $V(x) \equiv \lambda V^{(1)}_{\mathrm{GS,DM}}(x)$, must be found from the inhomogeneous equation, which follows from Equation (13), in which $\mu = 0$ is set:

$$\frac{d^2}{dx^2} V^{(1)}_{\mathrm{GS,DM}} = -2U^{(0)}_{\mathrm{GS,DM}}(x). \tag{22}$$

Straightforward integration of Equation (22), with expressions (14) and (15) substituted on the right-hand side, yields

$$V^{(1)}_{\mathrm{GS}}(x) = -\sqrt{2}\pi^{1/4}\left[\frac{x}{\sqrt{2}}\mathrm{erf}\left(\frac{x}{\sqrt{2}}\right) + \sqrt{\frac{2}{\pi}}\exp\left(-\frac{x^2}{2}\right)\right], \tag{23}$$

$$V^{(1)}_{\mathrm{DM}}(x) = 2\pi^{1/4}\mathrm{erf}\left(\frac{x}{\sqrt{2}}\right), \tag{24}$$

where $\mathrm{erf}(x)$ is the standard error function, which is an odd function of x.

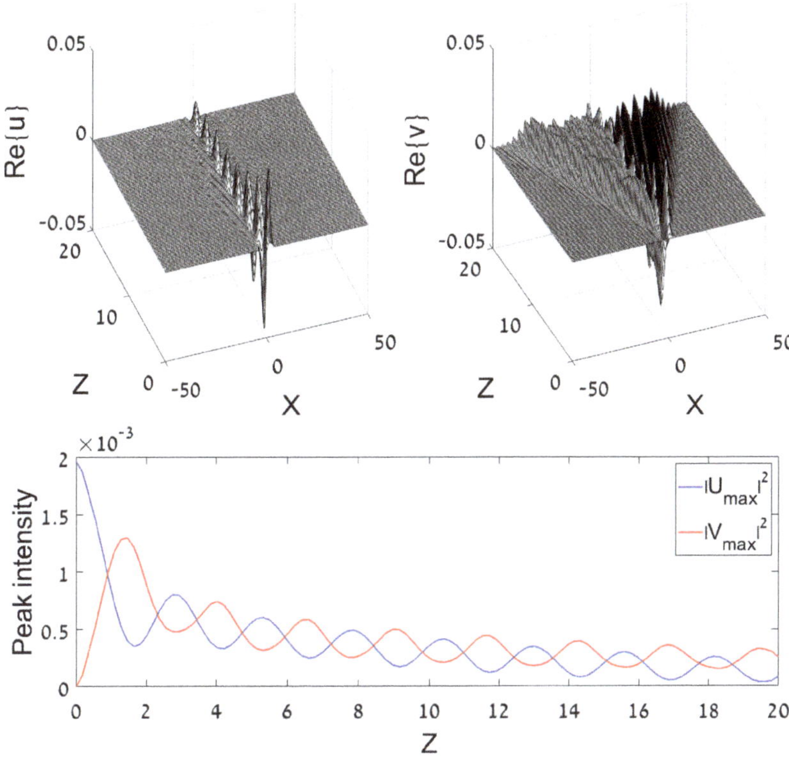

Figure 6. (**Left**): The evolution of fields u and v in the half-trapped system, as produced by simulations of Equation (10) with $\omega = 0$ and coupling constant $\lambda = 1$, initiated by the DM input in the u core, taken as per Equation (14) with $A_0 = 0.1$. Here and in Figure 7, spurious left-right asymmetry of the radiation field in the v component is an illusion produced by plotting. (**Right**): The evolution of the peak intensities of both components, $\max_x\left\{|u(x,t)|^2\right\}$ and $\max_x\left\{|v(x,t)|^2\right\}$.

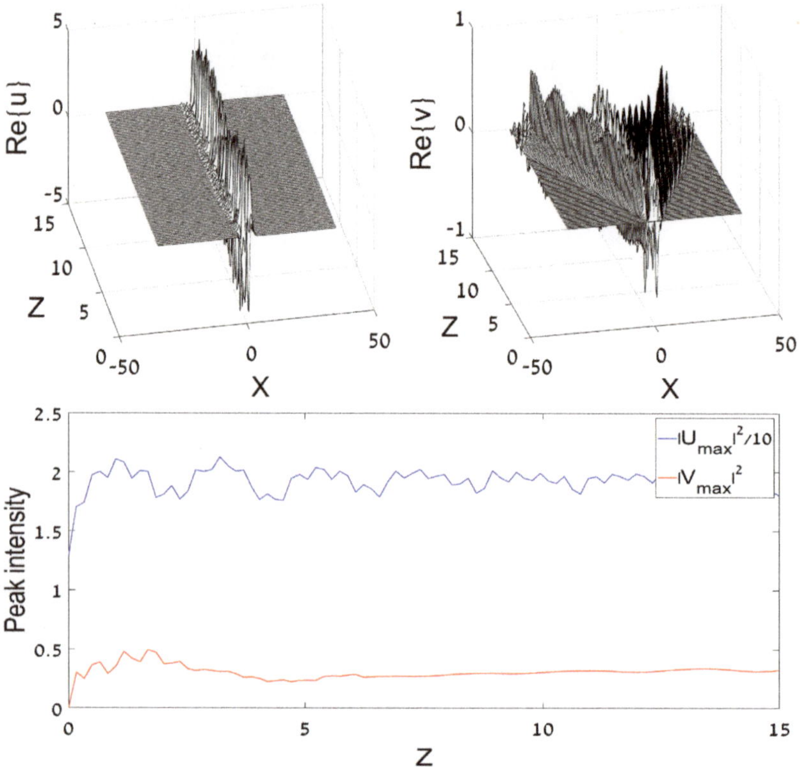

Figure 7. The same as in Figure 6, but for a large amplitude of the DM input in the u component, $A_0 = 8$.

Next, the small perturbation in Equation (12), represented by term λV, produces a small shift $\delta\mu$ of the eigenvalue, as a feedback from component V. According to the standard rule of quantum mechanics, which deals with the linear Schrödinger equation [66], in the first approximation of the perturbation theory, when $V(x)$ is replaced by expression (23) or (24), the result is

$$\delta\mu_{\mathrm{GS,DM}} = -\lambda^2 \int_{-\infty}^{+\infty} V^{(1)}_{\mathrm{GS,DM}}(x) U^{(0)}_{\mathrm{GS,DM}}(x) dx \equiv -I_{\mathrm{GS,DM}}\lambda^2, \tag{25}$$

with coefficients

$$I_{\mathrm{GS}} = -3.414, I_{\mathrm{DM}} = 4, \tag{26}$$

where the former and latter ones are computed, respectively, in a numerical form and analytically (note *opposite signs* of these coefficients). Then, the accordingly shifted threshold values sought for are

$$(\omega_{\mathrm{GS}})_{\mathrm{thr}} = 1/2 - I_{\mathrm{GS}}\lambda^2,\ (\omega_{\mathrm{DM}})_{\mathrm{thr}} = 3/2 - I_{\mathrm{DM}}\lambda^2. \tag{27}$$

The analytical predictions are compared to numerical results, obtained from a solution of the linearized variant of Equations (12) and (13), in Figure 8. Note that the linear u-v coupling *facilitates* the formation of the DM bound state, by lowering $(\omega_{\mathrm{DM}})_{\mathrm{thr}}$, but *impedes* to form the GS, by making the respective threshold, $(\omega_{\mathrm{DM}})_{\mathrm{thr}}$, higher. It is seen that for the DM the analytical approximation is essentially more accurate than for the GS.

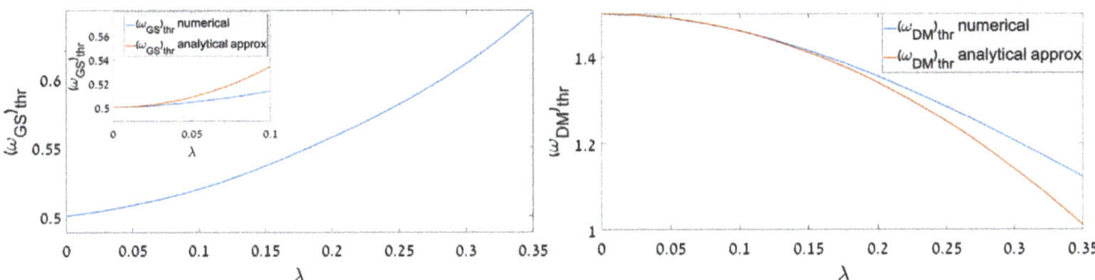

Figure 8. The analytically predicted and numerically found threshold values of the mismatch parameter, ω, above which the GS and DM solutions (the left and right panels, respectively) are produced, for the half-trapped system, by the linearized version of Equations (12) and (13), vs. coupling constant λ. The analytical results are produced by Equation (27). For the GS, they are shown (in the inset) only for relatively small values of λ, as in this case the analytical approximation is inaccurate at larger λ.

An example of a bound-state solution of linearized Equations (12) and (13) of the DM type, numerically found at $\lambda = 0.225$ and $\omega = 1.4$, i.e., *below* the threshold value $(\omega_{\mathrm{DM}})_{\mathrm{thr}} = 1.5$, corresponding to the limit of $\lambda \to 0$, is displayed in Figure 9. The existence of the DM at this point agrees with the right panel of Figure 8.

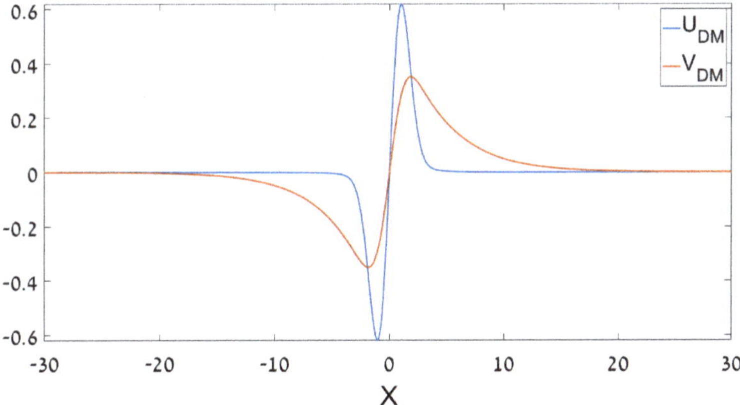

Figure 9. A bound state of the DM (dipole mode) type in the half-trapped system, found as a numerical solution of Equations (12) and (13) with $\omega = 1.4$, $\lambda = 0.225$, and $\sigma = 0$ (the linearized version). The eigenvalue corresponding to this solution is $\mu \approx -0.036$.

3.1.3. The Analysis for Large Values of λ

In the opposite limit of large coupling constant λ, an analytical approximation for the discrete eigenvalues can be developed too. In this case, $|\mu|$ may also be large, $\sim\lambda$. In the zero-order approximation, one may neglect the derivative term in Equation (13), to obtain $V \approx -(\lambda/\mu)U$. Then, substituting this relation back into the originally neglected derivative term, one obtains a necessary correction to this relation:

$$V \approx -\frac{\lambda}{\mu}U + \frac{\lambda}{2\mu^2}\frac{d^2U}{dx^2}. \tag{28}$$

The subsequent substitution of this expression in the linearized Equation (12) leads to an equation for U which is tantamount to the usual stationary linear Schrödinger equation with the HO potential:

$$\left(\mu - \frac{\lambda^2}{\mu} + \omega\right)U + \frac{1}{2}\left(1 + \frac{\lambda^2}{\mu^2}\right)\frac{d^2U}{dx^2} - \frac{1}{2}x^2U = 0. \tag{29}$$

Then, the standard solution for the quantum-mechanical HO yields an equation which determines the spectrum of the eigenvalues:

$$\mu - \frac{\lambda^2}{\mu} + \omega = \left(\frac{1}{2} + n\right)\sqrt{1 + \frac{\lambda^2}{\mu^2}}, \tag{30}$$

where $n = 0, 1, 2, \ldots$ is the quantum number. Taking into regard that λ is now a large parameter, Equation (30) produces a final result for the spectrum,

$$\mu \approx -\lambda - \frac{\omega}{2} + \frac{1}{\sqrt{2}}\left(\frac{1}{2} + n\right). \tag{31}$$

The spectrum remains equidistant in the current approximation, while further corrections $\sim 1/\lambda$ give rise to terms $\sim (1/2 + n)^2$, which break this property. As concerns the existence threshold for the bound states, Equation (31) predicts $\omega_{\text{thr}} \approx -2\lambda$. The coefficient in this relation is not accurate, as the derivation is not valid for small $|\mu|$, but the implication is that, for large λ, ω_{thr} drops to negative values with a large modulus, $\sim -\lambda$.

The prediction of the GS and DM eigenvalues, given by Equation (31) with $n = 0$ and $n = 1$, respectively, is compared to numerically found counterparts in Figure 10, which shows proximity between the analytical and numerical results. The plots do not terminate in the displayed domain, i.e., they do not reach the existence boundary.

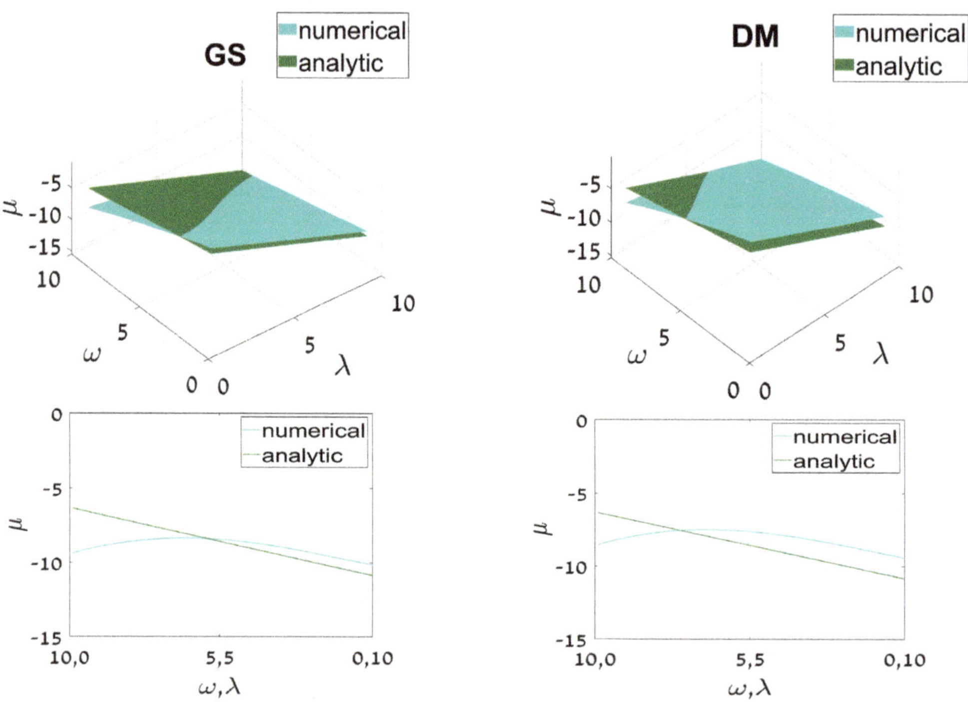

Figure 10. (**The top row**): left and right panels display the analytically predicted eigenvalues given by Equation (31) with $n = 0$ and $n = 1$, for the ground state (GS) and dipole mode (DM), respectively, of the half-trapped system, and their counterparts produced by the numerical solution of linearized coupled Equations (12) and (13), as functions of the linear-coupling constant, λ, and mismatch parameter, ω. (**The bottom row**): cross sections of the respective top panels along the diagonal connecting points $(\lambda, \omega) = (0.10)$ and $(10, 0)$. The results shown in these plots are relevant for relatively large values of λ.

3.1.4. Exact Solutions for One- and Two-Dimensional *Bound States in the Continuum* (BIC) in the Linear System

A remarkable property of the coupled half-trapped system, represented by the linearized version of Equations (10), (12) and (13), is that it admits particular spatially confined solutions in an exact analytical form. These are exceptional solutions, which, for an arbitrary value of the linear-coupling constant, λ, exist at a single, specially selected, value of the mismatch parameter, ω, and with a single value of the eigenvalue, μ. First, it is possible to find an exact mode which is a fundamental one (GS) in the V component, and a *second-order mode* in U:

$$\begin{aligned} U(x) &= U_0\left[\left(\lambda^2-\frac{1}{2}\right)+x^2\right]\exp\left(-\frac{x^2}{2}\right), \\ V(x) &= -2\lambda U_0\exp\left(-\frac{x^2}{2}\right), \\ U_0^2 &= \pi^{-1/2}\left(\lambda^4+4\lambda^2+1/2\right)^{-1}, \\ \omega &= \frac{9}{4}-\frac{\lambda^2}{2},\ \mu=\frac{1}{2}\left(\lambda^2+\frac{1}{2}\right), \end{aligned} \tag{32}$$

with amplitude U_0 defined by condition $P=1$, see Equation (16). We stress that, as seen in Equation (32), this exact solution may only have $\mu>0$, i.e., it is BIC (a *bound state in the continuous spectrum* [58–60]), alias an *embedded mode* [61]. It is worthy to note that this BIC mode and additional ones, presented below, are found in the coupled system, with one component trapped in the HO potential.

In addition to the above spatially even solution, an odd one of the BIC type is available too. It is composed of a DM in the V component and a *third-order mode* in U. In the normalized form, i.e., with $P=1$ (as per Equation (16)), the solution is

$$\begin{aligned} U(x) &= U_0x\left[\left(\lambda^2-\frac{3}{2}\right)+x^2\right]\exp\left(-\frac{x^2}{2}\right), \\ V(x) &= -2\lambda U_0x\exp\left(-\frac{x^2}{2}\right), \\ U_0^2 &= 2\pi^{-1/2}\left(\lambda^4+4\lambda^2+3/2\right)^{-1}, \\ \omega &= \frac{11}{4}-\frac{\lambda^2}{2},\ \mu=\frac{1}{2}\left(\lambda^2+\frac{3}{2}\right), \end{aligned} \tag{33}$$

which also exists at a single value of ω, and with a single eigenvalue, $\mu>0$. Note that both exact solutions, given by Equations (32) and (33), may exists at positive and negative values of ω, as well as at $\omega=0$ (in Equations (32) and (33), $\omega=0$ at $\lambda^2=9/2$ and $\lambda^2=11/2$, respectively).

These exact solutions for BIC states in the two-component systems are somewhat similar to those found in Ref. [67], which addressed a system of spin-orbit-coupled linear Gross–Pitaevskii equations for a binary BEC. In that work, exact solutions were produced for a specially designed form of the trapping potential.

The exact solutions of the linearized system may be tried as inputs in simulations of the full nonlinear system based on Equation (10), with ω selected as per the solutions. The simulations, performed with moderate values of the initial amplitude, produce a robust state with steady internal oscillations, as shown in Figure 11 for the attractive ($\sigma=1$) and repulsive ($\sigma=-1$) SPM nonlinearity. In addition, the simulations demonstrate weak emission of radiation in the untrapped (v) component. In the case of the self-attraction (the left panel in Figure 7), the radiation is almost invisible. The self-repulsion in the v component, naturally, enhances the emission, which becomes visible in the right panel of the figure. Still larger amplitudes of the input lead to irregular oscillations and conspicuous emission of radiation (not shown here in detail).

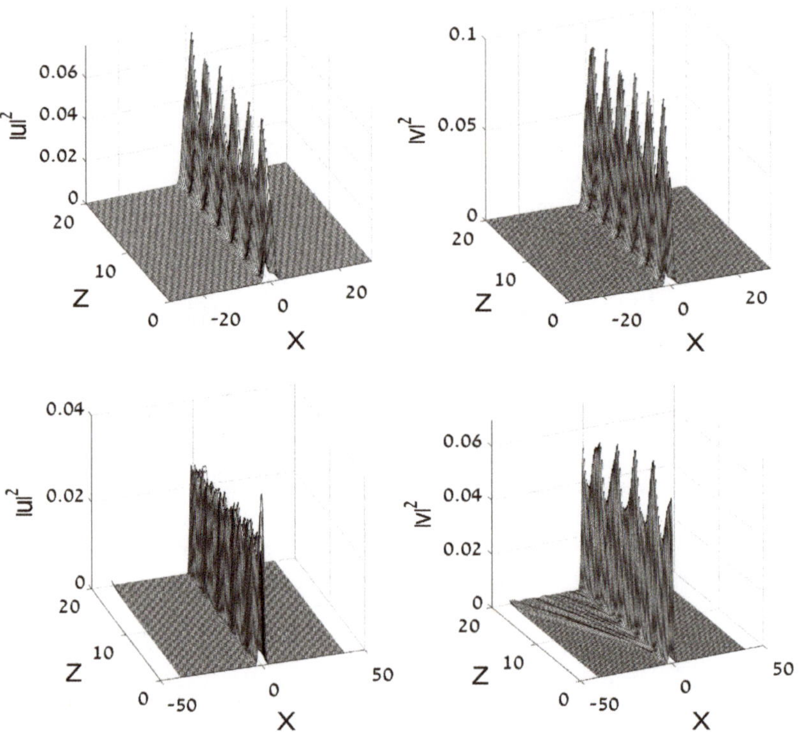

Figure 11. The evolution initiated by the exceptional (BIC) exact solution (32) of the system of linearized Equations (12) and (13), as produced by simulations of the full nonlinear half-trapped system (10), with the attractive SPM, $\sigma = 1$ in the top row, and repulsive $\sigma = -1$ in the bottom one. Other parameters are $g = 0$, $\lambda = 7/2$ and $\omega = 1/2$, which are related as per Equation (32).

It may happen that the linearized system of Equations (12) and (13) produces isolated BIC/embedded modes in a numerical form at other values of parameters. The present work does not aim to carry our comprehensive search for such solutions. On the other hand, it is relevant to mention that a straightforward two-dimensional (2D) extension of the present system readily produces exceptional exact solutions for BIC/embedded modes of both fundamental (GS, alias zero-vorticity) and vortex types.

The 2D extension of the linearized form of Equation (10) is

$$\begin{aligned} iu_z + \frac{1}{2}(u_{xx} + u_{yy}) + \lambda v - \frac{1}{2}\left(x^2 + y^2\right)u &= -\omega u, \\ iv_z + \frac{1}{2}(v_{xx} + v_{yy}) + \lambda u &= 0. \end{aligned} \tag{34}$$

Although the realization of this model in optics in not straightforward, it may be implemented for matter waves in a dual-core "pancake-shaped" holder of BEC [35,68]. In polar coordinates (r,θ), particular exact solutions of linear Equation (34), with all integer values of the vorticity, $S = 0,1,2,\ldots$, are found as

$$u = U_0 r^S \left[\left(\lambda^2 - 1 - S\right) + r^2\right] \exp\left(-i\mu z + iS\theta - \frac{r^2}{2}\right),$$
$$v = -2\lambda U_0 r^S \exp\left(-i\mu z + iS\theta - \frac{r^2}{2}\right),$$
$$\omega = \frac{1}{2}\left(5 + S - \lambda^2\right),\ \mu = \frac{1}{2}\left(\lambda^2 + 1 + S\right), \tag{35}$$

with arbitrary amplitude U_0 (the 2D solution of the GS type corresponds to $S = 0$ in Equation (35)). As well as in 1D solutions (32) and (33), μ takes only positive values in the 2D solution, hence it also represents states of the BIC/embedded type.

3.2. *The Nonlinear Half-Trapped System*

3.2.1. The Thomas–Fermi Approximation (TFA)

In the presence of the self-defocusing nonlinearity, $\sigma = -1$ in Equations (12) and (13), it is relevant to apply TFA to finding the GS of the half-trapped system, omitting the second derivatives in both equations [2], and keeping condition $\mu < 0$, which is necessary for the existence of a generic (non-BIC) localized state in the V component. Then, Equation (13) with $g = 0$ (the XPM coupling is omitted here) is solved as

$$U = (V/\lambda)\left(V^2 - \mu\right), \tag{36}$$

and Equation (12) amounts to an algebraic equation for the squared amplitude $W \equiv -V^2/\mu > 0$, *viz.*,

$$mW(W+1)^3 + \xi^2 W = \xi_0^2 - \xi^2, \tag{37}$$

where

$$m \equiv -\frac{2\mu}{\lambda^2} > 0,\ \xi \equiv -\frac{x}{\mu}, \tag{38}$$
$$\xi_0^2 \equiv -\frac{2}{\mu^3}\left[\lambda^2 - \mu(\omega + \mu)\right], \tag{39}$$

and the applicability condition for TFA is easily shown to be $m \ll 1$. The TFA solution exists under condition $\xi_0^2 > 0$ (see Equation (39)), i.e., in a finite interval (*bandgap*) of values of the propagation constant,

$$0 < -\mu < \sqrt{\lambda^2 + \omega^2/4} + \omega/2 \equiv -\mu_0. \tag{40}$$

Outside of the bandgap, i.e., at $\mu < \mu_0 < 0$, the TFA solution does not exist. In the bandgap, Equation (37) produces a usual GS profile, with a single value of W corresponding to each ξ^2 from region $\xi^2 < \xi_0^2$ (see Figure 12, which displays a typical example of the TFA-predicted GS and its comparison with the numerically found counterpart). The solution vanishes at the border points, $\xi = \pm\xi_0$, and is equal to zero at $\xi^2 > \xi_0^2$, so that the derivative of the TFA solution, $dW/d\xi$, is discontinuous at the border points, which is a usual peculiarity of the TFA [2,69].

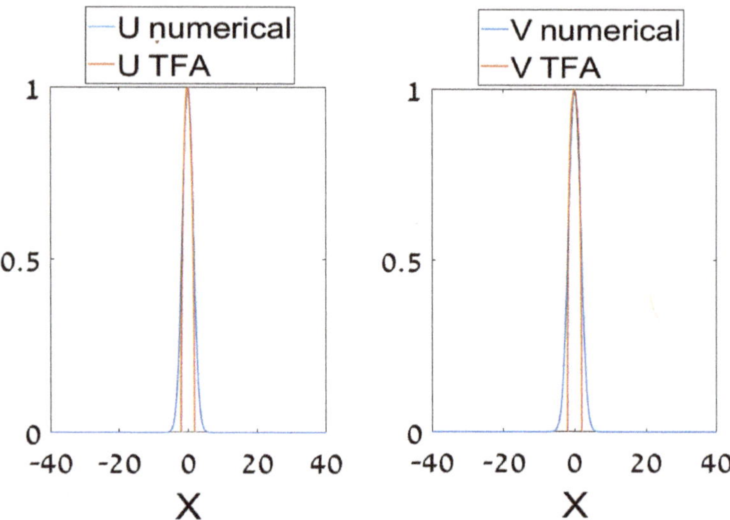

Figure 12. A typical example of the GS solution predicted by TFA (Thomas–Fermi approximation) for the half-trapped system, as per Equations (36)–(39), for $\sigma = -1$, $g = 0$, $\lambda = 8$, $\mu = -4$, and its comparison to the numerically found counterpart. The respective value of parameter m (see Equation (38)), which should be small for the applicability of TFA, is $m = 0.125$.

Note that in the limit of large λ, the bandgap's width, as given by Equation (40), is $-\mu_0 \approx \lambda + \omega/2$, which is close to the largest value of $-\mu$ predicted by Equation (31) for the GS ($n = 0$) in the linearized half-trapped system. Finally, the TFA solution may be cast in a simple explicit form close to the edge of the bandgap, i.e., at

$$0 < \delta\mu \equiv \mu - \mu_0 \ll -\mu_0, \tag{41}$$

$$\xi_0^2 \approx -\left(4/\mu_0^3\right)\sqrt{\lambda^2 + \omega^2/4}\delta\mu \tag{42}$$

In this case, Equations (37) and (36) simplify to

$$U^2 \approx \begin{cases} (\mu_0^2/2)(\xi_0^2 - \xi^2), \text{ at } \xi^2 < \xi_0^2, \\ 0 \text{ at } \xi^2 > \xi_0^2, \end{cases} \tag{43}$$

$$V^2 \approx \begin{cases} (\lambda^2/2)(\xi_0^2 - \xi^2), \text{ at } \xi^2 < \xi_0^2, \\ 0 \text{ at } \xi^2 > \xi_0^2. \end{cases} \tag{44}$$

Expressions (42)–(44) make it possible to calculate the total power of the GS (see Equation (2)),

$$P \approx \frac{16}{3}\left(\lambda^2 + \mu_0^2\right)\left(\lambda^2 + \frac{\omega^2}{4}\right)^{3/4}(-\mu_0)^{-7/2}(\delta\mu)^{3/2}. \tag{45}$$

Note that relation (45) satisfies the *anti-Vakhitov–Kolokolov criterion*, $dP/d(\delta\mu) > 0$, which is a necessary condition for stability of localized states supported by self-repulsive nonlinearity [70], as is the case in the present setting (the Vakhitov–Kolokolov criterion proper, $dP/d(\delta\mu) < 0$, is a well-known necessary condition for stability of states in models with self-attraction [71–73]).

3.2.2. Existence Boundaries for Nonlinear States

The shrinkage of the existence region for GS solutions, and its expansion for DM ones, in the linear version of the coupled half-trapped system, shown in Figure 8, suggests identifying existence boundaries of the same states in the full nonlinear system. Numerical data, necessary for the delineation of the existence region of the GSs and DMs in the nonlinear system, were collected by solving Equations (12) and (13) for spatially even and odd modes with the fixed total power, $P = 1$ (as per Equation (16)), while the linear-coupling coefficient, λ, and the nonlinearity coefficient, σ, were varied, the latter one taking both positive and negative values (for the self-attraction/repulsion). The results are summarized by the heatmap in Figure 13 (for the GS) and Figure 14 (for the DM), which show threshold values of the mismatch parameter, defined as per Equation (21).

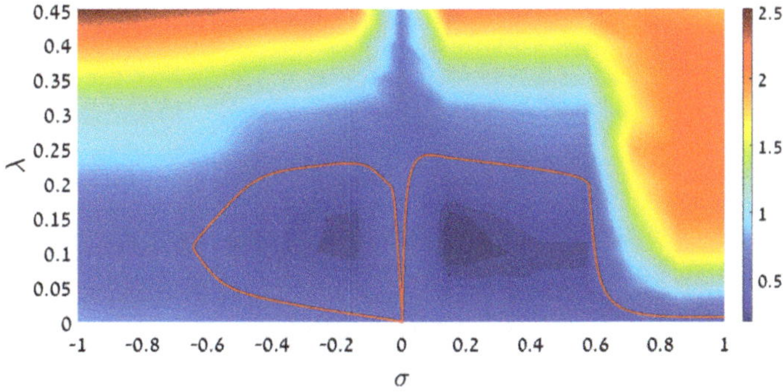

Figure 13. The heatmap of threshold values of the mismatch parameter, $(\omega_{\mathrm{GS}})_{\mathrm{thr}}$, in the half-trapped system, based on Equations (12) and (13). For given values of the nonlinearity and linear-coupling coefficients, σ and λ, the stable GS (ground state), subject to the normalization condition $P = 1$ (see Equation (16)), exist above the threshold, i.e., at $\omega \geq (\omega_{\mathrm{GS}})_{\mathrm{thr}}$. Positive and negative values of σ correspond to the attractive and repulsive sign of the self-interaction, respectively. The nontrivial region is one confined by red lines, in which the result is $(\omega_{\mathrm{GS}})_{\mathrm{thr}} < 1/2$, i.e., the nonlinearity and linear coupling help to maintain self-trapped GSs in the parameter area where the decoupled system, with $\lambda = 0$, cannot create such states.

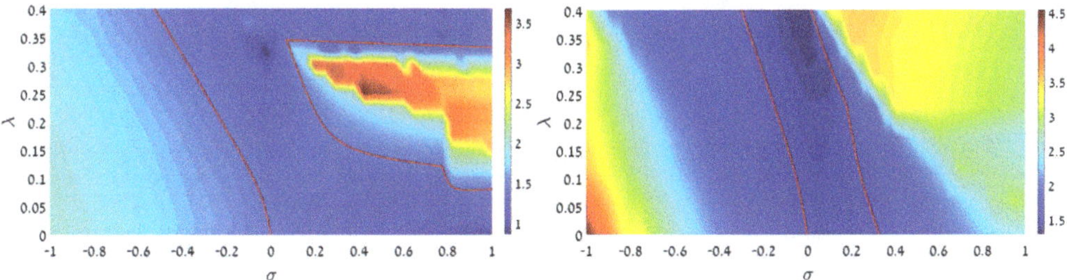

Figure 14. The same as in Figure 13, but for DMs (dipole modes), obtained as numerical solutions to Equations (12) and (13) with $g = 0$ and $g = 1$ (the left and right panels, respectively). In this case, the nontrivial region, located between the red boundaries, is one with $(\omega_{\mathrm{DM}})_{\mathrm{thr}} < 3/2$.

Nontrivial parametric areas for the GS and DM solutions are identified as domains with, respectively, $(\omega_{\mathrm{GS}})_{\mathrm{thr}} < 1/2$ and $(\omega_{\mathrm{DM}})_{\mathrm{thr}} < 3/2$. The former one is surrounded by red lines in Figure 13, and its counterpart for the DMs is located between red lines in Figure 14. In these areas, the coupled half-trapped system maintains stable localized GSs at $\omega < 1/2$, and DMs at $\omega < 3/2$, while in the absence of the coupling they may only exist at $\omega > 1/2$ and $\omega > 3/2$, respectively.

Note that the vertical cross-section of the heatmap in Figure 13, drawn through $\sigma = 0$ (along which the system remains linear), that starts from $\lambda = 0$, belongs to the area with $(\omega_{\mathrm{GS}})_{\mathrm{thr}} > 1/2$, in agreement with the result for the linear system, which shows that the coupling impedes the existence of the GS, see the left panel of Figure 8 and Equation (27). Furthermore, Figure 13 demonstrates that the SPM nonlinearity of either sign facilitates the creation of GS in parameter regions surrounded by red lines. Naturally, the self-attraction ($\sigma > 0$) helps to create such states starting from arbitrarily small values of λ, while the self-repulsion ($\sigma < 0$) can do it in the region separated by a gap from $\lambda = 0$. Nevertheless, at $\lambda > 0.24$ the detrimental effect of the linear coupling cannot be outweighed by the SPM nonlinearity.

In Figure 14, the vertical cross-section corresponding to the linear system ($\sigma = 0$) entirely belongs to the nontrivial area, with $(\omega_{\mathrm{DM}})_{\mathrm{thr}} < 3/2$, also in agreement with Equation (27) and the right panel of Figure 8. Figure 14 demonstrates that the nonlinearity, generally, impedes the maintenance of the localized DMs. This conclusion is supported, in particular, by the comparison of panels (a) and (b) in the figure, which shows that the addition of the attractive XPM nonlinearity with the SPM conspicuously reduces the remaining nontrivial area.

4. Conclusions

The objective of this work is to analyze new effects in the symmetric and asymmetric systems of linearly coupled fields, which are subject to the action of the HO (harmonic-oscillator) trapping potential and cubic self-attraction or repulsion. The system can be implemented in nonlinear optics and BEC. In the symmetric system, with identical HO potentials applied to both components, we focus on the consideration of JO (Josephson oscillations) in the system, by launching, in one component, an input in the form of the GS (ground state) or DM (dipole mode) of the HO potential. On the basis of systematically collected numerical data, we have identified two transitions in the system's dynamics, which occur with the increase of the input's power in the case of the self-attraction. The first is SSB (spontaneous symmetry breaking) between the linearly coupled components in the dynamical JO state. At a higher power, the nonlinearity causes a transition from regular JO, initiated by the GS input, to chaotic dynamics. This transition is identified through consideration of spectral characteristics of the dynamical regime. The input in the form of the DM undergoes the chaotization at essentially smaller powers than the dynamical regime initiated by the GS input, which is followed by the SSB at slightly higher powers. In the case of self-repulsion, SSB does not occur, while the chaotization takes place in a weak form, in a small part of the parameter space.

In the half-trapped system, with the HO potential acting on a single component, a nontrivial issue is identification of the system's linear spectrum, i.e., a parameter region in which the linearized system maintains trapped binary (two-component) modes. This problem is solved here analytically in the limit cases of weak and strong linear coupling, and in the numerical form in the general case. In particular, the linear coupling between the components leads to the shrinkage of the spectral band in which the GS exists, and expansion of the existence band for the DM. The existence region for trapped states in the full nonlinear system is identified numerically, and such states are constructed analytically by means of the TFA (Thomas–Fermi approximation). In addition, exceptional solutions of the linearized system of the BIC (bound-state-in-continuum), alias embedded, type were found in the exact analytical form, in both the 1D and 2D settings, the 2D solution being found with an arbitrary value of the vorticity.

The work may be extended by considering inputs in the form of higher-order HO eigenstates. Another relevant direction for the extension of the analysis is a systematic study of the 2D system.

Author Contributions: Formulation of the model and analytical considerations, B.A.M.; numerical computations, N.H. All authors have read and agreed to the published version of the manuscript.

Funding: This research was funded by Israel Science Foundation, grant number 1286/17.

Institutional Review Board Statement: Not applicable.

Informed Consent Statement: Not applicable.

Data Availability Statement: Not applicable.

Conflicts of Interest: The authors declare no conflict of interest.

References

1. Pethick, C.J.; Smith, H. *Bose-Einstein Condensation in Dilute Gases*; Cambridge University Press: Cambridge, UK, 2002.
2. Pitaevskii, L.P.; Stringari, S. *Bose-Einstein Condensation*; Oxford University Press: Oxford, UK, 2003.
3. Kevrekidis, P.G.; Frantzeskakis, D.J.; Carretero-Gonz ález, R. *Emergent Nonlinear Phenomena in Bose-Einstein Condensates: Theory and Experiment*; Springer: Heidelberg, Germany, 2008.
4. Schneider, B.I.; Feder, D.L. Numerical approach to the ground and excited states of a Bose-Einstein condensed gas confined in a completely anisotropic trap. *Phys. Rev. A* **1999**, *59*, 2232–2242. [CrossRef]
5. Adhikari, S.K. Numerical solution of the two-dimensional Gross-Pitaevskii equation for trapped interacting atoms. *Phys. Lett. A* **2000**, *265*, 91–96. [CrossRef]
6. Busch, T.; Anglin, J.R. Motion of dark solitons in trapped Bose-Einstein condensates. *Phys. Rev. Lett.* **2000**, *84*, 2298–2301. [CrossRef]
7. Kivshar, Y.S.; Alexander, T.J.; Turitsyn, S.K. Nonlinear modes of a macroscopic quantum oscillator. *Phys. Lett.* **2001**, *278*, 225–230. [CrossRef]
8. Alexander, T.J.; Bergé, L. Ground states and vortices of matter-wave condensates and optical guided waves. *Phys. Rev. E* **2002**, *65*, 026611. [CrossRef]
9. Huang, G.; Szeftel, J.; Zhu, S. Dynamics of dark solitons in quasi-one-dimensional Bose-Einstein condensates. *Phys. Rev. A* **2002**, *65*, 053605. [CrossRef]
10. Parker, N.G.; Proukakis, N.P.; Leadbeater, M.; Adams, C.S. Soliton-sound interactions in quasi-one-dimensional Bose-Einstein condensates. *Phys. Rev. Lett.* **2003**, *90*, 220401. [CrossRef]
11. Pelinovsky, D.E.; Frantzeskakis, D.J.; Kevrekidis, P.G. Oscillations of dark solitons in trapped Bose-Einstein condensates. *Phys. Rev. E* **2005**, *72*, 016615. [CrossRef]
12. Brazhnyi, V.A.; Konotop, V.V. Stable and unstable vector dark solitons of coupled nonlinear Schrödinger equations: Application to two-component Bose-Einstein condensates. *Phys. Rev. E* **2005**, *72*, 026616. [CrossRef]
13. Parker, N.G.; Proukakis, N.G.; Adams, C.S. Dark soliton decay due to trap anharmonicity in atomic Bose-Einstein condensates. *Phys. Rev. A* **2010**, *81*, 033606. [CrossRef]
14. Bland, T.; Parker, N.G.; Proukakis, N.P.; Malomed, B.A. Probing quasi-integrability of the Gross-Pitaevskii equation in a harmonic-oscillator potential. *J. Phys. B Atomic Mol. Opt. Phys.* **2018**, *51*, 205303. [CrossRef]
15. Raghavan, S.; Agrawal, G.P. Spatiotemporal solitons in inhomogeneous nonlinear media. *Opt. Commun.* **2000**, *180*, 377–382. [CrossRef]
16. Zezyulin, D.A.; Alfimov, G.L.; Konotop, V.V. Nonlinear modes in a complex parabolic potential. *Phys. Rev. A* **2010**, *81*, 013606. [CrossRef]
17. Charalampidis, E.G.; Kevrekidis, P.G.; Frantzeskakis, D.J.; Malomed, B.A. Dark-bright solitons in coupled NLS equations with unequal dispersion coefficients. *Phys. Rev. E* **2015**, *91*, 012924. [CrossRef]
18. Mayteevarunyoo, T.; Malomed, B.A.; Skryabin, D.V. One- and two-dimensional modes in the complex Ginzburg-Landau equation with a trapping potential. *Opt. Exp.* **2018**, *26*, 8849–8865. [CrossRef]
19. Leib, M.; Deppe, F.; Marx, A.; Gross, R.; Hartmann, M.J. Networks of nonlinear superconducting transmission line resonators. *New J. Phys.* **2012**, *14*, 075024. [CrossRef]
20. Kivshar, Y.S.; Agrawal, G.P. *Optical Solitons: From Fibers to Photonic Crystals*; Academic Press: San Diego, CA, USA, 2003.
21. Morsch, O.; Oberthaler, M. Dynamics of Bose-Einstein condensates in optical lattices. *Rev. Modern Phys.* **2006**, *78*, 179–215. [CrossRef]
22. Joannopoulos, J.D.; Johnson, S.G.; Winn, J.N.; Meade, R.D. *Photonic Crystals: Molding the Flow of Light*; Princeton University Press: Princeton, NJ, USA, 2008.
23. Skorobogatiy, M.; Yang, J. *Fundamentals of Photonic Crystal Guiding*; Cambridge University Press: Cambridge, UK, 2008.
24. Cerda-Mendez, E.A.; Sarkar, D.; Krizhanovskii, D.N.; Gavrilov, S.S.; Biermann, K.; Skolnick, M.S.; Santos, P.V. Exciton-polariton gap solitons in two-dimensional lattices. *Phys. Rev. Lett.* **2013**, *111*, 146401. [CrossRef]
25. Brazhnyi, V.A.; Konotop, V.V. Theory of nonlinear matter waves in optical lattices. *Modern Phys. Lett. B* **2004**, *18*, 627–651. [CrossRef]
26. Sakaguchi, H.; Malomed, B.A. Dynamics of positive- and negative-mass solitons in optical lattices and inverted traps. *J. Phys. B* **2004**, *37*, 1443–1459. [CrossRef]
27. Zakharov, V.E.; Manakov, S.V.; Novikov, S.P.; Pitaevskii, L.P. *Theory of Solitons: Inverse Scattering Method*; Nauka: Moscow, Russia, 1980; English translation: Consultants Bureau, New York, 1984.
28. Zakharov, V.; Dias, F.; Pushkarev, A. One-dimensional wave turbulence. *Phys. Rep.* **2004**, *398*, 1–65. [CrossRef]
29. Mazets, I.E.; Schmiedmayer, J. Thermalization in a quasi-one-dimensional ultracold bosonic gas. *New J. Phys.* **2010**, *12*, 055023. [CrossRef]
30. Cockburn, S.P.; Negretti, A.; Proukakis, N.P.; Henkel, C. Comparison between microscopic methods for finite-temperature Bose gases. *Phys. Rev. A* **2011**, *83*, 043619. [CrossRef]
31. Grisins, P.; Mazets, I.E. Thermalization in a one-dimensional integrable system. *Phys. Rev. A* **2011**, *84*, 053635. [CrossRef]

32. Thomas, K.F.; Davis, M.J.; Kheruntsyan, K.V. Thermalization of a quantum Newton's cradle in a one-dimensional quasicondensate. *Phys. Rev. A* **2021**, *103*, 023315. [CrossRef]
33. Merhasin, M.I.; Malomed, B.A.; Driben, R. Transition to miscibility in a binary Bose-Einstein condensate induced by linear coupling. *J. Phys. B* **2005**, *38*, 877–892. [CrossRef]
34. Nistazakis, H.E.; Frantzeskakis, D.J.; Kevrekidis, P.G.; Malomed, B.A.; Carretero-Gonzalez, R. Bright-dark soliton complexes in spinor Bose-Einstein condensates. *Phys. Rev. A* **2008**, *77*, 033612. [CrossRef]
35. Chen, Z.; Li, Y.; Malomed, B.A.; Salasnich, L. Spontaneous symmetry breaking of fundamental states, vortices, and dipoles in two and one-dimensional linearly coupled traps with cubic self-attraction. *Phys. Rev. A* **2016**, *96*, 033621. [CrossRef]
36. Decker, M.; Ruther, M.; Kriegler, C.E.; Zhou, J.; Soukoulis, C.M.; Linden, S.; Wegener, M. Strong optical activity from twisted-cross photonic metamaterials. *Opt. Lett.* **2009**, *34*, 2501–2503. [CrossRef] [PubMed]
37. Ballagh, R.J.; Burnett, K.; Scott, T.F. Theory of an output coupler for Bose-Einstein condensed atoms. *Phys. Rev. Lett.* **1997**, *78*, 1607–1611. [CrossRef]
38. Öhberg, P.; Stenholm, S. Internal Josephson effect in trapped double condensates. *Phys. Rev. A* **1999**, *59*, 3890–3895. [CrossRef]
39. Son, D.T.; Stephanov, M.A. Domain walls of relative phase in two-component Bose-Einstein condensates. *Phys. Rev. A* **2002**, *65*, 063621. [CrossRef]
40. Jenkins, S.D.; Kennedy, T.A.B. Dynamic stability of dressed condensate mixtures. *Phys. Rev. A* **2003**, *68*, 053607. [CrossRef]
41. Malomed, B.A. (Ed.) *Spontaneous Symmetry Breaking, Self-Trapping, and Josephson Oscillations*; Springer: Berlin/Heidelberg, Germany, 2013.
42. Hung, N.V.; Tai, L.X.T.; Bugar, I.; Longobucco, M.; Buzcy ński, R.; Malomed, B.A.; Trippenbach, M. Reversible ultrafast soliton switching in dual-core highly nonlinear optical fibers. *Opt. Lett.* **2020**, *45*, 5221–5224.
43. Mineev, V.P. The theory of the solution of two near-ideal Bose gases. *Zh. Eksp. Teor. Fiz.* **1974**, *67*, 263–272. [English translation: Sov. Phys. JETP **1974**, *40*, 132–136].
44. Malomed, B.A. Solitons and nonlinear dynamics in dual-core optical fibers. In *Handbook of Optical Fibers*; Peng, G.-D., Ed.; Springer: Berlin/Heidelberg, Germany, 2018; pp. 421–474.
45. Tratnik, M.V.; Sipe, J.E. Bound solitary waves in a birefringent optical fiber. *Phys. Rev. A* **1988**, *38*, 2011–2017. [CrossRef]
46. Paré, C.; Orjańczyk, M.F. Approximate model of soliton dynamics in all-optical couplers. *Phys. Rev. A* **1990**, *41*, 6287. [CrossRef]
47. Maimistov, A.I. Propagation of a light pulse in nonlinear tunnel-coupled optical waveguides. *Kvant. Elektron.* **1991**, *18*, 758. [English translation: Sov. J. Quantum Electron. **1991**, *21*, 687].
48. Uzunov, I.M.; Muschall, R.; Goelles, M.; Kivshar, Y.S.; Malomed, B.A.; Lederer, F. Pulse switching in nonlinear fiber directional couplers. *Phys. Rev. E* **1995**, *51*, 2527–2536. [CrossRef]
49. Abbarchi, M.; Amo, A.; Sala, V.G.; Solnyshkov, D.D.; Flayac, H.; Ferrier, L.; Sagnes, I.; Galopin, E.; Lemaître, A.; Malpuech, G.; et al. Macroscopic quantum self-trapping and Josephson oscillations of exciton polaritons. *Nat. Phys.* **2013**, *9*, 275. [CrossRef]
50. Milburn, G.J.; Corney, J.; Wright, E.M.; Walls, D.F. Quantum dynamics of an atomic Bose-Einstein condensate in a double-well potential. *Phys. Rev. A* **1997**, *55*, 4318. [CrossRef]
51. Smerzi, A.; Fantoni, S.; Giovanazzi, S.; Shenoy, S.R. Quantum coherent atomic tunneling between two trapped Bose-Einstein condensates. *Phys. Rev. Lett.* **1997**, *79*, 4950. [CrossRef]
52. Albiez, M.; Gati, R.; Fölling, J.; Hunsmann, S.; Cristiani, M.; Oberthaler, M.K. Direct observation of tunneling and nonlinear self-trapping in a single bosonic Josephson junction. *Phys. Rev. Lett.* **2005**, *95*, 010402. [CrossRef] [PubMed]
53. Shin, Y.; Jo, G.-B.; Saba, M.; Pasquini, T.A.; Ketterle, W.; Pritchard, D.E. Optical weak link between two spatially separated Bose-Einstein condensates. *Phys. Rev. Lett.* **2005**, *95*, 170402. [CrossRef] [PubMed]
54. Levy, S.; Lahoud, E.; Shomroni, I.; Steinhauer, J. The a.c. and d.c. Josephson effects in a Bose–Einstein condensate. *Nature* **2007**, *449*, 579–583. [CrossRef]
55. Chen, Z.; Li, Y.; Malomed, B.A. Josephson oscillations of chirality and identity in two-dimensional solitons in spin-orbit-coupled condensates. *Phys. Rev. Res.* **2020**, *2*, 033214. [CrossRef]
56. Sakaguchi, H.; Malomed, B.A. Symmetry breaking in a two-component system with repulsive interactions and linear coupling. *Comm. Nonlin. Sci. Num. Sim.* **2020**, *92*, 105496. [CrossRef]
57. Panoiu, N.-C.; Malomed, B.A.; Osgood, R.M., Jr. Semidiscrete solitons in arrayed waveguide structures with Kerr nonlinearity. *Phys. Rev. A* **2008**, *78*, 013801. [CrossRef]
58. Stillinger, F.H.; Herrick, D.R. Bound states in continuum. *Phys. Rev. A* **1975**, *11*, 446–454. [CrossRef]
59. Kodigala, A.; Lepetit, T.; Gu, Q.; Bahari, B.; Fainman, Y.; Kante, B. Lasing action from photonic bound states in continuum. *Nature* **2017**, *54*, 196–199. [CrossRef]
60. Midya, B.; Konotop, V.V. Coherent-perfect-absorber and laser for bound states in a continuum. *Opt. Lett.* **2018**, *43*, 607–610. [CrossRef]
61. Champneys, A.R.; Malomed, B.A.; Yang, J.; Kaup, D.J. "Embedded solitons": solitary waves in resonance with the linear spectrum. *Physica D* **2001**, *152–153*, 340–354. [CrossRef]
62. Ananikian, D.; Bergeman, T. Gross-Pitaevskii equation for Bose particles in a double-well potential: Two-mode models and beyond. *Phys. Rev. A* **2006**, *73*, 013604. [CrossRef]
63. Kartashov, Y.V.; Konotop, V.V.; Vysloukh, V.A. Dynamical suppression of tunneling and spin switching of a spin-orbit-coupled atom in a double-well trap. *Phys. Rev. A* **2018**, *97*, 063609. [CrossRef]

64. Kivshar, Y.S.; Malomed, B.A. Dynamics of solitons in nearly integrable systems. *Rev. Mod. Phys.* **1989**, *61*, 763–915. [CrossRef]
65. Iooss, G.; Joseph, D.D. *Elementary Stability Bifurcation Theory*; Springer: New York, NY, USA, 1980.
66. Landau, L.D.; Lifshitz, E.M. *Quantum Mechanics*; Nauka Publishers: Moscow, Russia, 1974.
67. Kartashov, Y.V.; Konotop, V.V.; Torner, L. Bound states in the continuum in spin-orbit-coupled atomic systems. *Phys. Rev. A* **2017**, *96*, 033619. [CrossRef]
68. dos Santos, M.C.P.; Malomed, B.A.; Cardoso, W.B. Double-layer Bose-Einstein condensates: A quantum phase transition in the transverse direction, and reduction to two dimensions. *Phys. Rev. E* **2020**, *102*, 042209. [CrossRef] [PubMed]
69. Fetter, A.L.; Feder, D.L. Beyond the Thomas-Fermi approximation for a trapped condensed Bose-Einstein gas. *Phys. Rev. A* **1998**, *58*, 3185–3194. [CrossRef]
70. Sakaguchi, H.; Malomed, B.A. Solitons in combined linear and nonlinear lattice potentials. *Phys. Rev. A* **2010**, *81*, 013624. [CrossRef]
71. Vakhitov, N.G.; Kolokolov, A.A. Stationary solutions of the wave equation in a medium with nonlinearity saturation. *Radiophys. Quant. Electron.* **1973**, *16*, 783–789. [CrossRef]
72. Bergé, L. Wave collapse in physics: Principles and applications to light and plasma waves. *Phys. Rep.* **1998**, *303*, 259–370. [CrossRef]
73. Fibich, G. *The Nonlinear Schrödinger Equation: Singular Solutions and Optical Collapse*; Springer: Heidelberg, Germany, 2015.

Article

Modulational Instability, Inter-Component Asymmetry, and Formation of Quantum Droplets in One-Dimensional Binary Bose Gases

Thudiyangal Mithun [1,2,*,†], Aleksandra Maluckov [3,*,†], Kenichi Kasamatsu [4,†], Boris A. Malomed [5,†] and Avinash Khare [6,†]

1 Center for Theoretical Physics of Complex Systems, Institute for Basic Science, Daejeon 34141, Korea
2 Department of Mathematics and Statistics, University of Massachusetts, Amherst MA 01003-4515, USA
3 Vinca Institute of Nuclear Sciences, University of Belgrade, P. O. B. 522,11001 Belgrade, Serbia
4 Department of Physics, Kindai University, Higashi-Osaka, Osaka 577-8502, Japan; kenichi@phys.kindai.ac.jp
5 Department of Physical Electronics, School of Electrical Engineering, Faculty of Engineering, and Center for Light-Matter Interaction, Tel Aviv University, Tel Aviv 69978, Israel; malomed@tauex.tau.ac.il
6 Department of Physics, Savitribai Phule Pune University, Pune 411007, India; avinashkhare45@gmail.com
* Correspondence: mthudiyangal@umass.edu (T.M.); sandram@vin.bg.ac.rs (A.M.)
† These authors contributed equally to this work.

Received: 9 December 2019; Accepted: 14 January 2020; Published: 18 January 2020

Abstract: Quantum droplets are ultradilute liquid states that emerge from the competitive interplay of two Hamiltonian terms, the mean-field energy and beyond-mean-field correction, in a weakly interacting binary Bose gas. We relate the formation of droplets in symmetric and asymmetric two-component one-dimensional boson systems to the modulational instability of a spatially uniform state driven by the beyond-mean-field term. Asymmetry between the components may be caused by their unequal populations or unequal intra-component interaction strengths. Stability of both symmetric and asymmetric droplets is investigated. Robustness of the symmetric solutions against symmetry-breaking perturbations is confirmed.

Keywords: quantum droplet; binary Bose–Einstein condensate; modulational instability

1. Introduction

The mean-field (MF) theory of weakly interacting dilute atomic gases rules out formation of a liquid state [1,2]. However, it has been recently shown that a liquid phase arises if one takes into account beyond-MF effects originating from quantum fluctuations around the MF ground state of weakly interacting binary (two-component) Bose gases [3]. A fundamental property that allows one to interpret this phase as a fluid is incompressibility: It maintains a limit density which cannot be made larger (see details below), hence adding more atoms leads to spatial expansion of the state. Another fundamental feature of this quantum-fluid phase is that it facilitates self-trapping of *quantum droplets* (QDs), which are stabilized by the interplay between the contact MF interaction and the beyond-MF Lee–Huang–Yang (LHY) correction [4]. Binary Bose–Einstein condensates (BECs) with competing intra- and inter-component MF interactions of opposite signs offer a remarkable possibility for the generation of QDs, as proposed by Petrov [3]. This possibility was further elaborated in various settings, including different effective dimensions [5–20]. In particular, the dynamics of QDs with the flat-top (FT) or Gaussian shape, which correspond to large or relatively small numbers of particles, respectively, was addressed in the framework of the one-dimensional (1D) reduction of the model [20]. The theoretical prediction was followed by experimental creation of QDs in mixtures of two different atomic states of ^{39}K, with quasi-2D [21,22] and fully 3D [23,24] shapes (see also recent reviews [25,26]). Very recently, the creation of especially long-lived QDs was reported in a heteronuclear ^{41}K-^{87}Rb system [27]. Another theoretically predicted and experimentally realized option for the creation of QDs makes use of the single-component condensate with dipole–dipole interactions [28–35]. It is relevant to mention that the formation of multiple droplets was also predicted and experimentally observed as an MF effect in strongly nonequilibrium (turbulent) states of BECs [36].

Collective modes of QDs are a subject of special interest, as they reveal internal dynamics of the droplets [20,24,32,37,38]. In particular, the stable existence of the QDs is secured if the particle-emission threshold lies below all excitation modes, hence a perturbation in the form of such modes will not cause decay of the droplet.

We aim to address issues that are related to the creation of QDs in the 1D setting and were not addressed in previous works. First, we consider modulational instability (MI) of spatially uniform plane-wave (PW) states, in the framework of the coupled system of Gross–Pitaevskii (GP) equations with the LHY corrections, for the two-component MF wave function of the binary condensate. This is the system which was originally derived in [5]. Recently, MI has been experimentally demonstrated in BECs with attractive interactions [39–41]. Other examples of the MI are provided by the binary BEC with the linear Rabi coupling or the spin-orbit coupling [41,42], and by a system combining the MF and LHY terms [43]. The linear-stability analysis, followed by direct simulations of the corresponding GP equations, shows that the lower branch of the PW states exhibits MI, the instability splitting the PW into a chain of localized droplet-like structures. Secondly, we address properties of the QDs in the binary condensate in the framework of the two-component GP system, without assuming effective inter-component symmetry, which reduces the system to a single-component GP equation. The asymmetry implies different MF self-repulsion coefficients in the two components, and/or unequal norms in them. Although properties of QDs have been studied by using the two-component GP system in some papers [6,11,14,15,17,18], the explicit asymmetry of the system parameters has not been addressed, except for [14] in which the situation for ^{39}K-^{39}K and ^{23}Na-^{87}Rb atomic mixtures have been considered. We conclude that the population difference between the components does not significantly affect density profiles of QDs in the system with equal MF self-repulsion strengths in the two components. On the other hand, we find that profiles of the QD solutions are essentially asymmetric when the self-repulsion coefficients are different in the components. Generally, the numerical findings corroborate stability of the known symmetric states against symmetry-breaking perturbations. We also address the MI of the two-component system, and demonstrate that chains of asymmetric QDs can be generated by the MI-induced nonlinear evolution.

The paper is organized as follows. In Section 2 we introduce the model and discuss conditions necessary for the formation of the droplets. Section 3.1 deals with the single-component version of the symmetric system. We consider various solutions admitted by it (PW, FT, periodic, etc.), and apply the linear-stability analysis of the PW solution to assess the MI, in a combination with direct simulations. In Section 3.2, we address the stability of asymmetric droplets, as well as the formation of droplets in the two-component asymmetric system via the MI. The paper is concluded by Section 4. Additional symmetric and asymmetric exact and approximate analytical solutions are presented in Appendices.

2. Model and Methods

We consider the 1D model of the two-component condensate with coefficients of the intra-component repulsion, $g_1 > 0$ and $g_2 > 0$, and inter-component attraction, $g_{12} < 0$. In the weak-interaction limit, the corresponding energy density, which includes the MF terms and LHY correction, was derived in [5]:

$$\mathcal{E}_{1\mathrm{D}} = \frac{\left(\sqrt{g_1}\rho_1 - \sqrt{g_2}\rho_2\right)^2}{2} + \frac{g\delta g\left(\sqrt{g_2}\rho_1 + \sqrt{g_1}\rho_2\right)^2}{(g_1+g_2)^2} - \frac{2\sqrt{m}\left(g_1\rho_1 + g_2\rho_2\right)^{3/2}}{3\pi\hbar}, \tag{1}$$

where m is the atomic mass (the same for both components), $\rho_j = |\Psi_j|^2$ $(j = 1,2)$ is the density of the j-th component, represented by the MF wave function Ψ_j, and

$$g \equiv \sqrt{g_1 g_2}, \qquad \delta g \equiv g_{12} + g. \tag{2}$$

The last term in Equation (1) represents the LHY correction. Derivation of Equation (1) assumes that the binary BEC is close to the point of the MF repulsion-attraction balance, with $|\delta g| \ll g$. In experiments, δg may be tuned to be both positive and negative [21–23].

Equation (1) is valid in the case of tight confinement applied in the transverse dimensions, which makes the setting effectively one-dimensional. In the 3D case, the LHY term $\sim -\rho^{3/2}$ (for $\rho_1 = \rho_2 \equiv \rho$) is replaced by one $\sim +\rho^{5/2}$. A detailed consideration of the crossover from 3D to 1D [12,44,45] in the two-component system is a problem which may be a subject of a separate work. Here, it is relevant to compare the symmetric version of Equation (1) for the energy density with that recently presented in [12]. It demonstrates that an accurately derived LHY contribution to the energy density of the 1D system contains, in addition to the $\rho^{3/2}$ term which was derived in [5], a term $\sim \rho^2$, which can be absorbed into the mean-field energy density, and a higher-order term $\sim \rho^3$, which was omitted in the analysis reported in [12]. A conclusion formulated in that work is that the energy density originally derived in [5] is literally valid if the ratio of the mean-field energy to that of the transverse confinement takes values ≤ 0.03. For typical experimental parameters, this implies that the difference between absolute values of scattering lengths of the mean-field intra-component repulsion and inter-component attraction should be ≤ 1 nm, which may be achieved in the experiment. The 1D QDs originate from the balance of the second term in Equation (1), corresponding to the weakly repulsive MF interaction, with $\delta g > 0$, and the LHY term, which introduces effective attraction in the 1D setting, on the contrary to the repulsion in the 3D setting [5,20].

The energy functional, $\int_{-\infty}^{+\infty} \mathcal{E}_{1D} dZ$, gives rise to the system of GP equations, which include the LHY correction,

$$\begin{aligned} i\hbar\frac{\partial \Psi_1}{\partial T} &= -\frac{\hbar^2}{2m}\frac{\partial^2 \Psi_1}{\partial Z^2} + (g_1 + G g_2)|\Psi_1|^2\Psi_1 - (1-G)g|\Psi_2|^2\Psi_1 - \frac{g_1\sqrt{m}}{\pi\hbar}\sqrt{g_1|\Psi_1|^2 + g_2|\Psi_2|^2}\Psi_1, \\ i\hbar\frac{\partial \Psi_2}{\partial T} &= -\frac{\hbar^2}{2m}\frac{\partial^2 \Psi_2}{\partial Z^2} + (g_2 + G g_1)|\Psi_2|^2\Psi_2 - (1-G)g|\Psi_1|^2\Psi_2 - \frac{g_2\sqrt{m}}{\pi\hbar}\sqrt{g_1|\Psi_1|^2 + g_2|\Psi_2|^2}\Psi_2, \end{aligned} \tag{3}$$

where T and Z are the time and coordinate measured in physical units, and parameter

$$G = \frac{2g\delta g}{(g_1 + g_2)^2}, \tag{4}$$

measures the deviation from the MF repulsion–attraction balance point, see Equation (2). The normalization of the components of the wave function is determined by numbers of bosons in each component:

$$N_j = \int_{-\infty}^{+\infty} |\Psi_j|^2 dZ. \tag{5}$$

Further, rescaling

$$\left(\frac{mg^2}{\hbar^3}\right) T \equiv t, \quad \left(\frac{mg}{\hbar^2}\right) Z \equiv z, \quad \left(\frac{\hbar}{\sqrt{mg}}\right) \Psi_{1,2} \equiv \psi_{1,2} \tag{6}$$

casts Equation (3) in the normalized form,

$$\begin{aligned} i\frac{\partial \psi_1}{\partial t} &= -\frac{1}{2}\frac{\partial^2 \psi_1}{\partial z^2} + (P + G P^{-1})|\psi_1|^2\psi_1 - (1-G)|\psi_2|^2\psi_1 - \frac{P}{\pi}\sqrt{P|\psi_1|^2 + P^{-1}|\psi_2|^2}\psi_1, \\ i\frac{\partial \psi_2}{\partial t} &= -\frac{1}{2}\frac{\partial^2 \psi_2}{\partial z^2} + (P^{-1} + G P)|\psi_2|^2\psi_2 - (1-G)|\psi_1|^2\psi_2 - \frac{1}{\pi P}\sqrt{P^{-1}|\psi_2|^2 + P|\psi_1|^2}\psi_2, \end{aligned} \tag{7}$$

where parameter

$$P \equiv \sqrt{\frac{g_1}{g_2}} = \frac{g_1}{g} \tag{8}$$

determines the asymmetry of the system, in the case of $P \neq 1$. Note that, as concerns stationary solutions with chemical potentials $\mu_{1,2}$, sought for as

$$\psi_{1,2}(z,t) = \exp(-i\mu_{1,2}t)\phi_{1,2}(z), \tag{9}$$

states with mutually proportional components, $\phi_1(z) = K\phi_2(z)$, are only possible in the fully symmetric case with $P = 1$, $\mu_1 = \mu_2$, and $K = 1$. In previous works [5,20], 1D solutions for QDs were considered only in the framework of the single GP equation which corresponds to symmetric system Equation (7) with $P = 1$ and $\psi_1 = \psi_2$.

3. Modulation Instability Versus QDs

In this section, we address MI of PWs in both symmetric and asymmetric GP systems, and relate it to formation of the QDs in the binary bosonic gas. To the best of our knowledge, this is the first work aiming to associate the MI with the formation of the 1D droplets in the system with unequal components. We first consider MI in the framework of the single-component reduction of the symmetric version of Equation (7), after briefly reviewing stationary solutions of the GP equation. Next, we extend the analysis for the two-component GP system, which makes it possible to produce asymmetric QDs, starting from the MI of asymmetric PW states.

3.1. The Single-Component GP Model

Under the single-component reduction of the binary system, with $g_1 = g_2 \equiv g$ and $\psi_1 = \psi_2 \equiv \psi$, Equation (1) simplifies to [5]

$$\varepsilon_{1D} \equiv \frac{\hbar^4}{m^2 g^3}\mathcal{E}_{1D} = \frac{\delta g}{g} n^2 - \frac{2^{5/2}}{3\pi} n^{3/2}, \tag{10}$$

with the single dimensionless density, $n = |\psi|^2 \equiv \left(\hbar^2/mg\right)\rho$. Assuming a spatially uniform state, the equilibrium density and the corresponding chemical potential are given by

$$n_0 = \frac{8}{9\pi^2}\left(\frac{g}{\delta g}\right)^2, \quad \mu_0 = -\frac{4}{9\pi^2}\frac{g}{\delta g}. \tag{11}$$

Density n_0 corresponds to the minimum of the energy per particle, $\partial_n\left[n^{-1}\varepsilon_{1D}(n)\right] = 0$, and μ_0 is negative for $\delta g/g > 0$. The corresponding single GP equation is

$$i\frac{\partial\psi}{\partial t} = -\frac{1}{2}\frac{\partial^2\psi}{\partial z^2} + \frac{\delta g}{g}|\psi|^2\psi - \frac{\sqrt{2}}{\pi}|\psi|\psi, \tag{12}$$

with normalization condition $\int_{-\infty}^{+\infty}|\psi(z)|^2 dz = N$, where $N \equiv N_1 = N_2$ is the number of atoms in each component.

Although coefficient $\delta g/g$ can be scaled out in Equation (12), as done in [20], we keep it here as a free parameter. This option is convenient for the subsequent consideration of the MI, treating $\delta g/g$ and density n as independent constants, which may be matched to experimentally relevant parameters.

Below, we address two stationary solutions of Equation (12). One is the QD bound state of a finite size, which was studied in detail in [5] and [20]. The other solution is the PW with uniform density. Here, we briefly recapitulated basic properties of these solutions for the completeness of the presentation. In Section 3.1.3 we address the MI of the PWs and associate it with the spontaneous generation of chains of localized modes. Additional families of exact analytical solutions of Equation (12) are given in Appendix A.

3.1.1. The Droplet Solution

As shown in [5,20,46], at $\delta g/g > 0$ Equation (12) gives rise to an exact soliton-like solution representing a QD, maintained by the balance between the effective cubic self-repulsion and quadratic attraction:

$$\psi(z,t) = \frac{Ae^{-i\mu t}}{1 + B\cosh(\sqrt{-2\mu}z)}, \quad A = \sqrt{n_0}\frac{\mu}{\mu_0}, \quad B = \sqrt{1 - \frac{\mu}{\mu_0}}. \tag{13}$$

This solution exists in a finite range of negative values of the chemical potential $\mu_0 < \mu < 0$, featuring the FT shape at $0 < \mu - \mu_0 \ll |\mu_0|$, with size $L \approx (-2\mu_0)^{-1/2} \ln\left[(1 - \mu/\mu_0)^{-1}\right]$ [5,20]. A typical density profile of the FT solution is displayed in the inset of Figure 1. At $\mu = \mu_0$, the size of the droplet diverges, and the solution carries over into the delocalized PW with uniform density, $n = n_0$. The fact that the density of the condensate filling the FT state cannot exceed the largest value, n_0, implies its incompressibility. For this reason, the condensate may be considered as a fluid, as mentioned above. With the increase of μ from μ_0 towards $\mu = 0$, the maximum density of the localized mode,

$$n_{\max} \equiv n(z=0) = n_0 \left(\frac{\mu}{\mu_0}\right)^2 \left(1 + \sqrt{1 - \frac{\mu}{\mu_0}}\right)^{-2}, \tag{14}$$

monotonously decreases from n_0 to 0. The QD's FWHM size, defined by condition $n\,(z = L_{\mathrm{FWHM}}/2) = n\,(z=0)\,/2$, also shrinks at first with increasing μ, attaining a minimum value $(L_{\mathrm{FWHM}})_{\min} \approx 2.36/\sqrt{-\mu_0}$ at $\mu/\mu_0 \approx 0.776$. Further increase of μ towards $\mu = 0$ makes the QD broader, its width diverging as $L_{\mathrm{FWHM}} \approx 1.71/\sqrt{-\mu}$ at $\mu \to -0$.

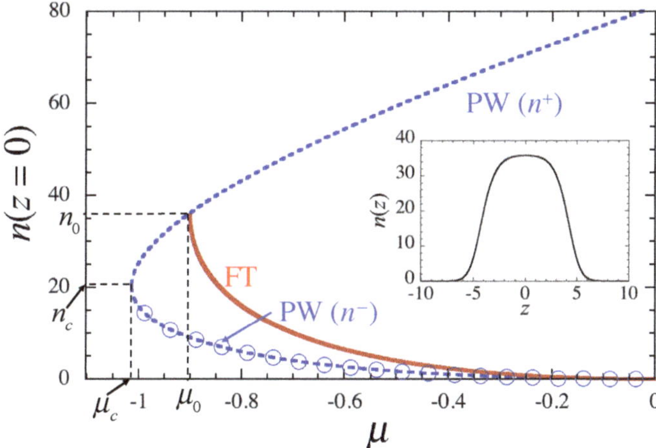

Figure 1. The maximum density $n_{\max} \equiv n(z=0)$ in the FT (flat-top) state, as per Equation (14), and the PW (plane-wave)/density are displayed as functions of μ by the red-solid and blue-dashed curves, respectively, for $\delta g/g = 0.05$. In this case, Equation (11) yields $n_0 = 36.025$ and $\mu_0 = -0.900633$. The PW solution includes upper and lower branches corresponding to $n^{\pm}$, as given by Equation (20), the lower one (marked by circles) being subject to the MI (modulational instability). The spinodal point is one with coordinates (μ_c, n_c). For other values of $\delta g/g$, the plot can be generated from the present one by rescaling. The inset shows the density profile of the FT solution for $\delta g/g = 0.05$ and $\mu = \mu_0 + 0.00001$, very close to the delocalization limit (the transition to PW).

The norm of the exact QD solutions given by Equation (13) is

$$N(\mu) = n_0\sqrt{-\frac{2}{\mu_0}}\left[\ln\left(\frac{1+\sqrt{\mu/\mu_0}}{\sqrt{1-\mu/\mu_0}}\right) - \sqrt{\frac{\mu}{\mu_0}}\right]. \tag{15}$$

It satisfies the well-known Vakhitov–Kolokolov (VK) necessary stability criterion [47],

$$\frac{dN(\mu)}{d\mu} = -\frac{n_0}{\mu_0^2}\sqrt{-\frac{\mu}{2}}\frac{1}{1-\mu/\mu_0} < 0, \tag{16}$$

due to $\mu_0 < 0$ and

$$0 < \mu/\mu_0 < 1. \tag{17}$$

Full stability of the QD family has been verified by direct simulations of the evolution of perturbed QDs in the framework of Equation (12).

It is relevant to mention that the exact solution of Equation (13) is valid too at $\delta g/g < 0$, when the cubic term in Equation (12) is self-attractive, like the quadratic one. In that case, μ_0 is positive, as per Equation (11), while the chemical potential of the self-trapped state remains negative, as the solution of Equation (13) may exist only at $\mu < 0$. Then, it follows from Equation (13) that the soliton-like mode exists for all values of $\mu < 0$ (unlike the finite interval Equation (17), in which the solution exists for $\delta g/g > 0$), and it does not feature the FT shape. Rather, with the increase of $-\mu$, it demonstrates a crossover between the KdV-soliton shape $\sim \text{sech}^2\left(\sqrt{-\mu/2}z\right)$ and the nonlinear-Schrödinger one, $\sim \text{sech}\left(\sqrt{-2\mu}z\right)$. For $\delta g/g < 0$, the $N(\mu)$ dependence for the soliton family carries over into the following form,

$$N(\mu)\Big|_{\delta g<0} = n_0\sqrt{\frac{2}{\mu_0}}\left[\sqrt{-\frac{\mu}{\mu_0}} - \arctan\left(\sqrt{-\frac{\mu}{\mu_0}}\right)\right], \tag{18}$$

which is an analytical continuation of Equation (15). This dependence also satisfies the VK criterion.

3.1.2. The Plane-Wave Solution

The PW solution of Equation (12)can be presented in a form $\psi(z,t) = \sqrt{n}\exp\left(iK_{\text{PW}}z - i\mu t\right)$ with wavenumber K_{PW} and constant density n, which determine the corresponding chemical potential:

$$\mu_{\text{PW}} = \frac{\delta g}{g}n - \frac{\sqrt{2}}{\pi}\sqrt{n} + \frac{1}{2}K_{\text{PW}}^2. \tag{19}$$

The Galilean invariance of Equation (12) implies that any quiescent solution $\psi_0(z,t)$ generates a family of moving ones, with arbitrary velocity c. Therefore, K_{PW} may be canceled by means of transformation $\psi_c(z,t) = \exp\left(icz - ic^2t/2\right)\psi_0(z-ct,t)$ with $c = -K_{\text{PW}}$.

For given μ, Equation (19) produces two different branches of the density as a function of μ (here, $K_{\text{PW}} = 0$ is set):

$$\sqrt{n^{\pm}(\mu)} = \frac{1}{\sqrt{2}\pi}\frac{g}{\delta g} \pm \sqrt{\frac{1}{2\pi^2}\left(\frac{g}{\delta g}\right)^2 + \frac{g}{\delta g}\mu}. \tag{20}$$

For $\delta g/g = 0.05$, these branches are shown in Figure 1. As follows from Equation (20), they exist (for $\delta g/g > 0$) above a minimum value of μ: $\mu_c = -(2\pi^2\delta g/g)^{-1} = (9/8)\mu_0$, the respective density being

$$n_c = n^{\pm}(\mu_c) = \frac{1}{2\pi^2}\left(\frac{g}{\delta g}\right)^2 = \frac{9}{16}n_0. \tag{21}$$

Values $\mu = \mu_c$ and $n = n_c$ correspond to the spinodal point [5], and $n^+(\mu_0) = n_0$ (see Equation (13)). Note that the above-mentioned existence region of the soliton solution in terms of the chemical potential, $\mu_0 < \mu < 0$, lies completely inside that of the PW state, which is $\mu_c \leq \mu$. Thus, the soliton always coexists with the PW (this fact is also obvious in Figure 1).

3.1.3. Modulational Instability of the Plane Waves

Here, we aim to analyze the MI of PW solutions in the framework of the single-component GP Equation (12) and demonstrate how the development of the MI can help to generate QDs. We perform the analysis for the PWs with zero wavenumber $K_{\rm PW} = 0$, which is sufficient due to the aforementioned Galilean invariance of the underlying equation.

A small perturbation is added to the stationary PW state as

$$\psi(z,t) = \left[\sqrt{n} + \delta\psi(z,t)\right] \exp\left(-i\mu t\right). \tag{22}$$

The substitution of this expression in Equation (12) and linearization with respect to perturbation $\delta\psi$ leads to the corresponding Bogoliubov–de Gennes equation,

$$i\frac{\partial}{\partial t}\delta\psi = -\frac{1}{2}\frac{\partial^2}{\partial z^2}\delta\psi + \frac{\delta g}{g}n(\delta\psi + \delta\psi^*) - \frac{\sqrt{n}}{\sqrt{2}\pi}(\delta\psi + \delta\psi^*). \tag{23}$$

By looking for perturbation eigenmodes with wavenumber k and frequency Ω,

$$\delta\psi = \zeta\cos(kz - \Omega t) + i\eta\sin(kz - \Omega t), \tag{24}$$

and real infinitesimal amplitudes ζ and η, Equation (23) yields a dispersion relation for the eigenfrequencies:

$$\Omega^2 = \frac{k^4}{4} + \left(\frac{\delta g}{g}n - \frac{\sqrt{n}}{\sqrt{2}\pi}\right)k^2. \tag{25}$$

The MI takes place when Ω acquires an imaginary part. As follows from Equation (25), this occurs when the density satisfies condition $n < [2\pi^2(\delta g/g)^2]^{-1} = n_c$ see (Equation (21)), which corresponds to branch n^- of the PW state. The instability region in terms of k is given by

$$k^2 < 4\left(\frac{\sqrt{n}}{\sqrt{2}\pi} - \frac{\delta g}{g}n\right) \equiv k_0^2. \tag{26}$$

The MI gain $\sigma \equiv |\mathrm{Im}\Omega|$ is plotted in Figure 2 versus $|k|$ and $\delta g/g$, for given density $n = 40$ in panel (a), and versus $|k|$ and n, for given $\delta g/g = 0.05$ in (b). It is easy to find from Equation (25) that the largest gain is attained at wavenumber

$$k_{\max} = \frac{k_0}{\sqrt{2}}, \tag{27}$$

with k_0 defined as per Equation (26). Note that Figure 2a includes the case of the self-attractive cubic nonlinearity, with $\delta g/g < 0$, which naturally displays much stronger MI, as in this case it is driven by both the quadratic and cubic nonlinear terms. In fact, the extension of the MI chart to $\delta g/g < 0$ makes it possible to compare the MI in the present system and its well-known counterpart in the setting with the fully attractive nonlinearity.

Comparing parameter values at which the QD solutions are predicted to appear, and the MI condition for the PW with the corresponding density, the MI is expected to provide a mechanism for the creation of the QDs. This is confirmed by direct simulations of the GP Equation (12), as shown in Figure 3. The PW with $n = 10$ is taken as the input, so that it is subject to the MI for $\delta g/g = 0.05$, as seen in Figure 2b. As shown in Figure 3, small initial perturbations trigger the emergence of multiple-QD patterns (chains) at $t \geq 100$. For these parameters, we get $k_{\max} = 0.6508$ and $\sigma(k_{\max}) = 0.2118$, which determines the wavelength of the fastest growing modulation, $\lambda = 2\pi/k_{\max} \approx 9.66$, and the growth-time scale, $\tau = 2\pi/\sigma(k_{\max}) \approx 30$. The number of the generated droplets in Figure 3 is consistent with estimate $L/\lambda \simeq 10$, where $L = 100$ is the size of the simulation domain. We have checked that the number of generated droplets is approximately given by L/λ for other values of parameters as well. This dynamical scenario is similar to those observed in other models in the course of the formation of soliton chains by MI of PWs [39,40]. The long-time evolution in Figure 3a shows that the number of the droplets becomes smaller due to merger of colliding droplets into a single one, which agrees with dynamical properties of 1D QDs reported in [20].

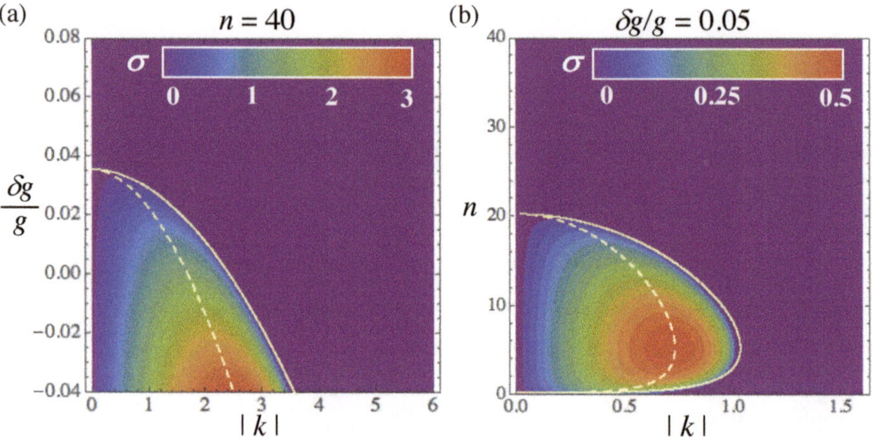

Figure 2. Color-coded values of the MI gain, $\sigma = \mathrm{Im}(\Omega)$, are displayed for fixed $n = 40$ in (**a**), and for fixed $\delta g/g = 0.05$ in (**b**). Note that panel (**a**) covers both signs of the cubic nonlinearity, $\delta g > 0$ and $\delta g < 0$. Solid and dashed white curves represent the MI boundary (Equation (26)) and the peak value of the MI gain (Equation (27)), respectively.

To implement this mechanism of the generation of a chain of solitons in the experiment, i.e., make the density smaller than the critical value n_c, one may either apply interaction quench (by means of the Feshbach resonance), suddenly decreasing $\delta g/g$, as was done in recent experimental works for different purposes [21–23,48]. Another option, which is specific to the 1D setting, is sudden decrease of density n by relaxing the transverse trapping.

3.2. The Two-Component Gross–Pitaevskii Model

In this section, we revert to the full two-component GP system Equation (7), aiming to explore the formation of QD states in it. The two-component setting may include parameter imbalance between the two components, as indicated theoretically [3] and observed experimentally [21–23,27]. Here, we present the analysis of asymmetric QDs in two cases: (i) the two-component GP system with different populations, $N_1/N_2 \neq 1$, and equal intra-component coupling strength, $g_1 = g_2$ (i.e., $P = 1$, see Equation (8)), and (ii) the system with different intra-component coupling strengths, $g_1 \neq g_2$ (i.e., $P \neq 1$). These options suggest a possibility to check the stability of the solutions of the symmetric system, reduced to the single-component form, against symmetry-breaking perturbations. That objective is relevant because, in the real experiment, scattering lengths of the self-interaction in the two components are never exactly equal [21–24]. We address, first, an asymmetric single-droplet solution, and, subsequently, MI of the PW states in the two-component system.

Because, as said above, solutions with mutually proportional components (written as $\phi_1 = K\phi_2$) are possible solely in the strictly symmetric setting, asymmetric QDs cannot be found in an exact analytical form. As shown in Appendix B (see Equations (A6)–(A12)), asymptotic analytical solutions can be obtained for strongly asymmetric states, with one equation replaced by its linearized version. In this section, we chiefly rely on numerical solution of Equation (7).

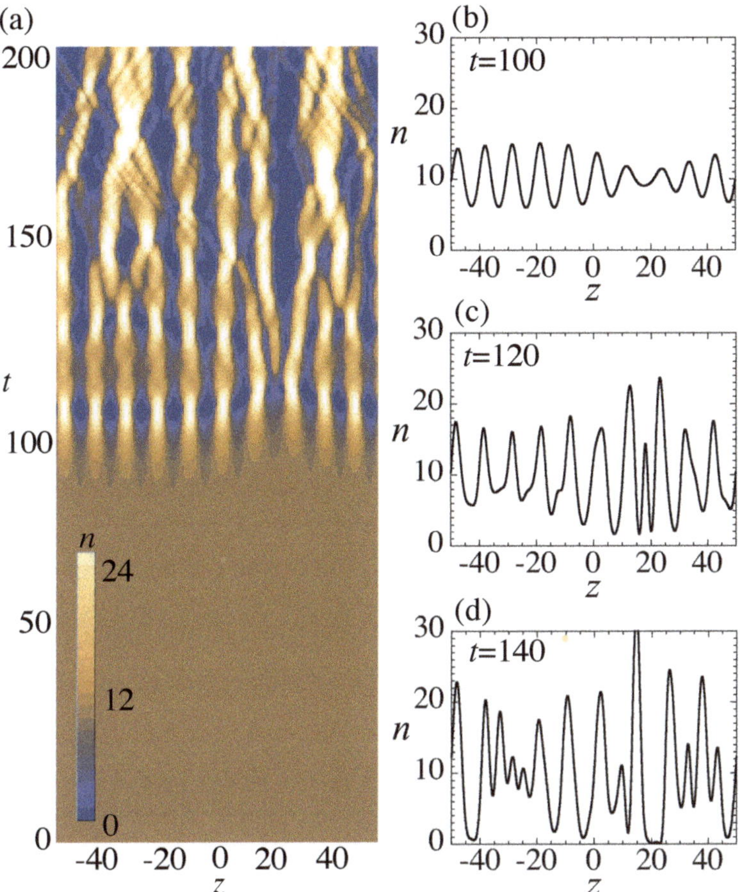

Figure 3. A typical example of the MI development, starting from an unstable PW state, with density $n = 10$ and $\delta g/g = 0.05$, which is subject to the MI, pursuant to Figure 2. In (**a**), the spatiotemporal pattern of the evolution of the condensate density is shown. In the right-hand panels, cross sections of the density profiles are displayed at $t = 100$ (**b**), $t = 120$ (**c**), and $t = 140$ (**d**). The simulations were performed in domain $-50 < z < +50$ with 2500 grid points and periodic boundary conditions.

3.2.1. Asymmetric QDs with Unequal Populations ($N_1 \neq N_2$) for $g_1 = g_2$ ($P = 1$)

In the system with $P = 1$ (see Equation (8)), we calculated the droplet states as stationary solutions of Equation (7) by means of the imaginary-time-evolution method with the Neumann's boundary conditions, under the constraint that the norm is fixed in the first component, $\int_{-\infty}^{+\infty} dz|\psi_1(z)|^2 = N_1$, while chemical potential μ_2 is fixed in the other one, allowing its norm N_2 to vary.

Figure 4 displays essential features of weakly asymmetric droplets for $\delta g/g = 0.05$ and fixed $N_1 = 100$. The symmetric (completely overlapping) solution with $N_1 = N_2$ is found at $\mu_1 = \mu_2 = -0.88878$. When μ_2 deviates from this value, profiles of the two components become slightly different, as shown in Figure 4a. The profiles of the droplet solution hardly change for different values of μ_2, but Figure 4b demonstrates that, at $\mu_2 \to -0$, ψ_2 develops small-amplitude extended tails, which are absent in ψ_1. Due to the contribution of the tails, the approach of $\mu_2 < 0$ towards zero leads to the increase of norm N_2, as seen in Figure 4c. Note that the growth of $N_2(\mu_2)$ at $\mu_2 \to -0$ is opposite to the decay of the QD's norm in the single-component model at $\mu \to -0$, cf. Equation (15). At $\mu_2 \geq 0$, the ψ_2

component undergoes delocalization, with its tails developing a nonzero background at $|z| \to \infty$, as seen in the density profile displayed in Figure 4b for $\mu_2 = 0$, and norm $N_2(\mu_2)$ diverging at $\mu_2 \to -0$ in Figure 4c.

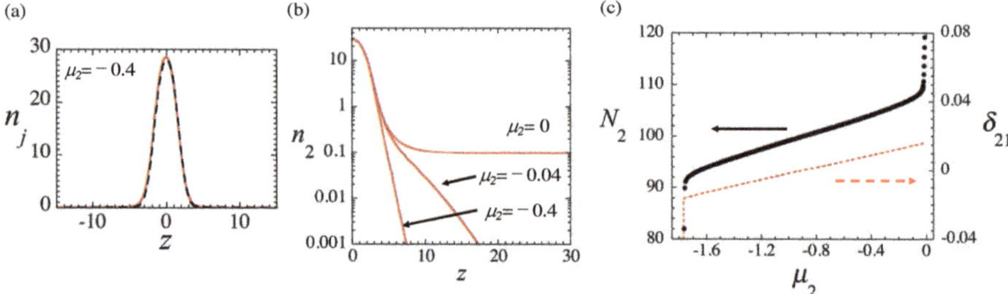

Figure 4. (**a**) Stationary weakly asymmetric (with respect to the two components) solutions of Equation (7), obtained for $\mu_2 = -0.4$ with fixed $N_1 = 100$. Dashed and solid curves display density profiles of the first (n_1) and second (n_2) components, respectively. (**b**) The semi-log plot of the density profiles of n_2 for $\mu_2 = -0.4, -0.04$, and 0 at $z > 0$. (**c**) Dependences of N_2 (black dots: the left vertical axis) and asymmetry parameter δ_{21}, defined as per Equation (28) (the red dashed line pertaining to the right vertical axis), on μ_2 for fixed $N_1 = 100$. The parameters are $P = 1$ ($g_1 = g_2$) and $\delta g/g = 0.05$. The symmetric point with $N_1 = N_2 = 100$ and $\delta_{21} = 0$ corresponds to $\mu_1 = \mu_2 = -0.88878$.

In Figure 4c, we also plot the parameter of the asymmetry between the two components, defined as

$$\delta_{21} = \frac{n_2(z=0) - n_1(z=0)}{n_2(z=0) + n_1(z=0)}. \tag{28}$$

It increases almost linearly with μ_2, although its absolute value does not exceed 0.02. Thus, the droplet tends to keep a nearly symmetric profile, with respect to the two components, in the symmetric system, even if the population imbalance is admitted. In fact, this circumstance makes the analysis self-consistent, as the use of the GP system with the LHY correction implies that the MF intra- and inter-component interactions nearly cancel each other, which is possible only if shapes of the two components are nearly identical.

3.2.2. Asymmetric QDs in the System with $P \neq 1$ ($g_1 \neq g_2$)

Next, we consider the QDs for $P \neq 1$, setting $P > 1$ without loss of generality. Then, the MF energy is minimized for $n_2 > n_1$; the situation with $n_1 > n_2$ can be considered too, replacing P by P^{-1}.

Following the procedure similar to that employed in Section 3.2.1, we produce QD solutions for $\delta g/g = 0.05$, $N_1 = 100$, and several different values of P, varying μ_2. In Figure 5a, we plot density profiles for three different values of P. Naturally, the difference of the two components increases with the increase of P. In Figure 5b we display N_2 and parameter δ_{21} (see Equation (28)) of the asymmetric QDs for $P = 1.25$ and 1.67. All these states have been checked to be stable in time-dependent simulations.

The density difference at the center of the droplet can be determined by the condition of the existence of the liquid phase in the free space. This condition is obtained by minimizing the grand-potential density $\mathcal{E}_{1D} - \mu_1\rho_1 - \mu_2\rho_2$ [5,14], which leads to the zero-pressure condition,

$$\begin{aligned} p(\rho_1, \rho_2) &= -\mathcal{E}_{1D} + \sum_{j=1,2} \left(\frac{\partial \mathcal{E}_{1D}}{\partial \rho_j} \right) \rho_j \\ &\equiv -\mathcal{E}_{1D} + \mu_1\rho_1 + \mu_2\rho_2 = 0. \end{aligned} \tag{29}$$

From this, we obtain relation

$$\frac{(\sqrt{g_1}\rho_1 - \sqrt{g_2}\rho_2)^2}{2} + \frac{g\delta g(\sqrt{g_2}\rho_1 + \sqrt{g_1}\rho_2)^2}{(g_1 + g_2)^2} - \frac{\sqrt{m}}{3\pi\hbar}(g_1\rho_1 + g_2\rho_2)^{3/2} = 0, \tag{30}$$

which can be rewritten in the scaled form as

$$\frac{P + GP^{-1}}{2}n_1^2 + \frac{P^{-1} + GP}{2}n_2^2 + (G-1)n_1n_2 = \frac{1}{3\pi}\left(Pn_1 + \frac{n_2}{P}\right)^{3/2}. \tag{31}$$

For given n_1, we solved Equation (31) to find the respective value of n_2, which is shown in Figure 6 for $\delta g/g = 0.05$ and several values of P. There are two branches of the solutions, that enclose the negative-pressure region, in which QDs may exist. The maximum value of n_j at the tip of the negative-pressure region corresponds to the density in the droplet's FT segment. The ascending negative-pressure region for each P nearly follows relation $n_2 = Pn_1$, which is derived by the minimization condition for the dominant first term in Equation (30) It is seen that a larger difference in the profiles of the two components occurs for larger P, as expected. Also, for given n_1, the negative-pressure region becomes wider with respect to n_2 for larger P (note that the figure displays a log–log plot).

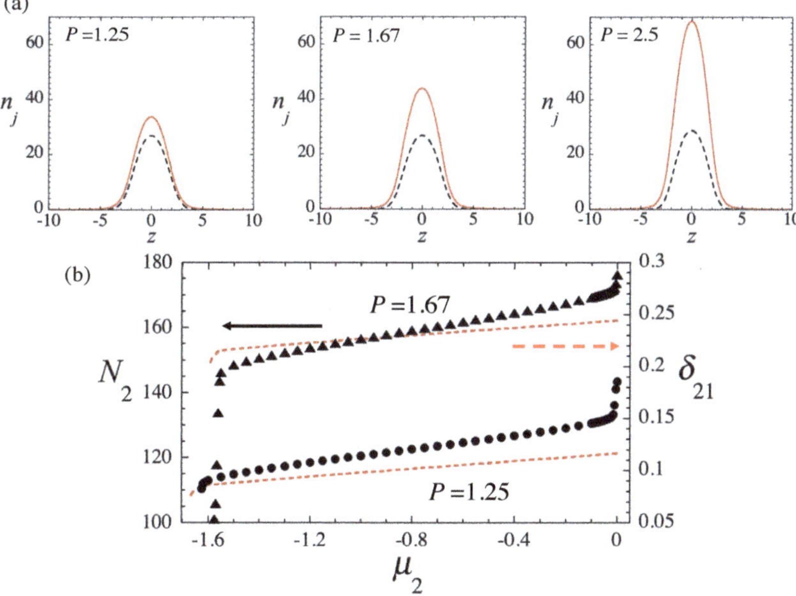

Figure 5. (**a**) Stationary solutions of Equation (7), obtained for $\delta g/g = 0.05$ and $N_1 = 100$. From the left panel to the right one, the parameter in Equation (8) is $P = 1.25$, 1.67, and 2.5, and the chemical potential for the second component is $\mu_2 = -0.018$, -0.011, and -0.006, respectively, just below the threshold above which the tails of ψ_2 extend to infinity. Dashed and solid curves represent the density of the first (n_1) and second (n_2) components. (**b**) Dependences of N_2 (black dots: the left vertical axis) and asymmetry parameter δ_{21}, defined as per Equation (28) (the red dashed line pertaining to the right vertical axis), on μ_2 for fixed $N_1 = 100$ and $P = 1.25$ or $P = 1.67$.

As the QDs have a finite norm, it is relevant to characterize the asymmetry in terms of the norm, rather than density. Here, we aim to find a largest value of the norm difference,

$$\Delta_{21} = (N_2 - N_1)/N_T, \tag{32}$$

where $N_T = N_1 + N_2$ is the total norm, which admits the existence of the QDs. For given N_1, we obtain the upper bound for N_2 above which the solution becomes delocalized, and calculate the corresponding critical value of Δ_{21}. The results are shown in Figure 7. For the system with $P = 1$, the curve demonstrates an empirical dependence $\Delta_{21} \propto N_T^{-\alpha}$ with exponent $\alpha \approx 0.58$. Accordingly, the asymmetry tends to vanish asymptotically for very "heavy" droplets, at $N_T \rightarrow \infty$. As the system becomes slightly asymmetric, with $P = 1.25$, exponent α is significantly reduced for small N_T, and converges to a certain finite value at $N_T \rightarrow \infty$. Thus, it is again confirmed that values $P > 1$ maintain conspicuous asymmetry between the QD's components. Finally, strongly asymmetric non-FT (Gaussian-shaped [20]) solutions can be obtained in an approximate analytical form for any value of P, as shown in Appendix C.

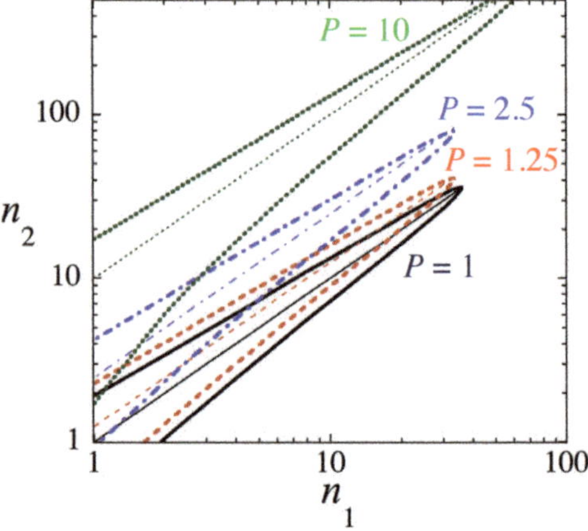

Figure 6. The negative-pressure region in the (n_1, n_2) plane for $\delta g/g = 0.05$ and values of asymmetry parameter in Equation (8) $P = 1$ (the solid curve), 1.25 (dashed), 2.5 (dashed-dotted), and 10 (dotted). Boundaries are determined by the zero-pressure condition, as given by Equation (31). The negative pressure, at which localized states may exist, occurs inside the boundaries. Thin lines represent relation $n_2 = Pn_1$.

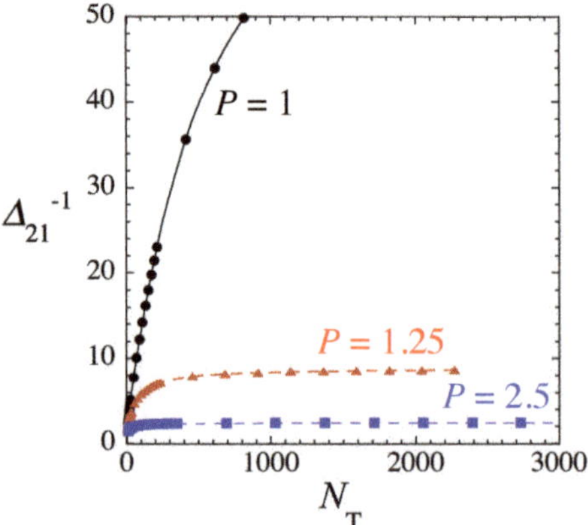

Figure 7. The inverse of the largest relative norm difference Δ_{21}, up to which the asymmetric droplets exist (see Equation (32)), shown as a function of the total number, N_T, at different values of asymmetry parameter of Equation (8). Here we set $\delta g/g = 0.05$.

3.2.3. The MI of the Asymmetric PW States

The MI of two-component asymmetric PWs is a relevant subject too. Such solutions are written as $\psi_j(z,t) = \sqrt{n_j} e^{-i\mu_j t}$, $(j = 1, 2)$. The substitution of this in Equation (7) yields

$$\begin{aligned} \mu_1 &= (P + GP^{-1})n_1 + (-1+G)n_2 - \frac{P}{\pi}\sqrt{Pn_1 + \frac{n_2}{P}}, \\ \mu_2 &= (P^{-1} + GP)n_2 + (-1+G)n_1 - \frac{1}{\pi P}\sqrt{Pn_1 + \frac{n_2}{P}}. \end{aligned} \tag{33}$$

Accordingly, in the symmetric system with $P = 1$, densities of the asymmetric PW state are expressed in terms of the chemical potentials as

$$n_j = \frac{1}{4}\left[\frac{1}{\pi^2 G^2} + \frac{\mu_1 + \mu_2}{G} + (-1)^{j+1}(\mu_1 - \mu_2)\right] \pm \frac{\sqrt{1 + 2\pi^2 G(\mu_1 + \mu_2)}}{4\pi^2 G^2}.$$

We introduce the perturbation around the PW states as

$$\psi_j(z,t) = \left[\sqrt{n_j} + \delta\psi_j(z,t)\right] e^{-i\mu_j t}, \tag{34}$$

$$\delta\psi_j = \zeta_j \cos(kz - \Omega t) + i\eta_j \sin(kz - \Omega t), \tag{35}$$

with infinitesimal amplitudes ζ_j and η_j, cf. Equation (24). The substitution of this in Equation (7) and the linearization with respect to ζ_j and η_j yields the dispersion equation for the perturbation:

$$\Omega_\pm^2 = \frac{k^2}{4}\left[k^2 + 2(P_1 + P_2 - Q_1 - Q_2)\right] \pm \frac{k^2}{2}\sqrt{(P_1 - P_2 - Q_1 + Q_2)^2 + 4(R - S)^2}, \tag{36}$$

where

$$\begin{aligned} P_1 &= (P + GP^{-1})n_1, \quad P_2 = (P^{-1} + GP)n_2, \\ Q_1 &= \frac{P^2 n_1}{2\pi\sqrt{Pn_1 + P^{-1}n_2}}, \quad Q_2 = \frac{P^{-2} n_2}{2\pi\sqrt{Pn_1 + P^{-1}n_2}}. \\ R &= (-1+G)\sqrt{n_1 n_2}, \quad S = \frac{\sqrt{n_1 n_2}}{2\pi\sqrt{Pn_1 + P^{-1}n_2}}, \end{aligned} \tag{37}$$

For $P = 1$ and $n_1 = n_2$, these results reproduce Equation (25) for the Ω_- branch. A parameter region in which at least one squared eigenfrequency $\Omega_\pm^2$ is negative gives rise to the MI of the two-component state.

3.2.4. The MI for $P = 1$

In Figure 8, we plot the gain spectrum $\sigma = \mathrm{Im}(\Omega)$ for the asymmetric PWs in the symmetric system with $P = 1$ and $\delta g/g = 0.05$, in the plane of wavenumber k and density ratio $n_{12} = n_2/n_2$. For the consistency with the single-component situation displayed in Figure 3, we here fix the total density as $(n_1 + n_2)/2 = 10$. For given n_{12}, the MI occurs at $|k| < k_0$, and the gain attains its maximum at $k = k_{\max} = k_0/\sqrt{2}$. The largest gain is obtained at equal densities, $n_{12} = 1$. Both the k-band of the instability and magnitude of the gain slowly decrease as the deviation of n_{12} from unity increases. This means that the MI occurs in the PW states with a large density difference, thus giving rise to the formation of solitons with large asymmetry even for equal intra-component MF interaction strengths, $P = 1$ (see Equation (8)).

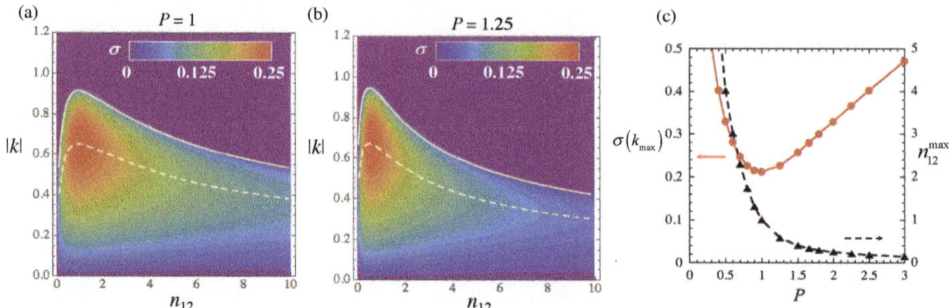

Figure 8. Color-coded values of the MI gain, $\sigma = \mathrm{Im}(\Omega)$, for asymmetric PWs, as calculated from Equation (36) in the plane of wave number $|k|$ and density ratio $n_{12} = n_1/n_2$, are displayed for (**a**) $P = 1$ and (**b**) $P = 1.25$ with fixed $\delta g/g = 0.05$ and $(n_1 + n_2)/2 = 10$. The solid and dashed white curves represent the MI boundary $k = k_0$ and the peak value of the MI gain at $k = k_{\max} = k_0/\sqrt{2}$, respectively. In (**c**), we plot $\sigma(k_{\max})$ (circles) and $n_{12}^{\max}$ (triangles) versus P.

In Figure 9 we display typical examples of the numerically simulated development of the MI in the symmetric two-component system with $P = 1$ and population imbalance. Figure 9a shows the evolution of central-point values of the density of the first component, $n_1(z = 0)$, for different values of the density ratio, $n_{12} = n_1/n_2$. Time required for the actual onset of the instability increases with the increase in n_{12}, as is clearly shown by the density-plot evolution in Figure 9d,e for $n_{12} = 1$ and Figure 9f,g for $n_{12} = 9$. This observation can be understood in terms of the MI gain σ, as shown in Figure 8c, where σ at $k = k_{\max}$ becomes smaller with increasing n_{12}.

Spatial profiles at fixed time, which are plotted in Figure 9b,c for these two cases, show fragmentation of the profiles into sets of localized structures. The decrease in the number of fragments with the increase of n_{12} is explained by the decrease of $k_{\max}$, see Figure 8a. For $n_{12} = 1$, the results are the same as in the single-component case, as coinciding profiles in the two components of the symmetric system are stable against spontaneous symmetry breaking. On the other hand, when $n_{12} \neq 1$ an in-phase two-component localized structure appears, keeping the initial density imbalance. Since one can select an arbitrary ratio of densities of the two components for the initial PW state, a highly asymmetric structure, like the one displayed in Figure 9c, may emerge even for $P = 1$, as a result of the MI-induced nonlinear evolution.

3.2.5. The MI for $P \neq 1$

Figure 8b represents the MI gain for $P = 1.25$ and a fixed total density, $(n_1 + n_2)/2 = 10$, in the case of slightly different strengths of the intra-component repulsion. The peak value of the MI gain is attained at $n_{12} = n_{12}^{\max} = 0.577$, below the equal-densities point $n_{12} = 1$. This is consistent with the fact that, at $P > 1$, unequal values $n_1 < n_2$ are suitable to the formation of an asymmetric soliton structure, as seen in Figure 5a. In Figure 8c, we plot the peak MI gain, $\sigma(k_{\max})$, along with the respective value of the density ratio, $n_{12}^{\max}$, as a function of P. Value $n_{12}^{\max}$ monotonously decreases as a function of P, while the peak gain attains a minimum at $P = 1$.

In Figure 10, we present the development of the MI in the two-component system for $P = 1.25$ and a fixed total density, $(n_1 + n_2)/2 = 10$. Figure 10a displays the evolution of the central-point density of the first component, $n_1(z = 0)$, for different values of the density ratio, $n_{12} = n_1/n_2$. It shows that time required for the development of the MI increases with the increase in the asymmetry of the density. This is also made evident by the density plots of the temporal evolution of the first component, shown in Figure 10e–g. This result is consistent with Equation (36), which shows a decrease of the MI gain with the increase of the asymmetry even for $P \neq 1$. Spatial profiles at fixed time, displayed in Figure 10b–d, show fragmentation of the profiles. Figure 10c clearly indicates that, even for $n_{12} = 1$, the MI generates asymmetric droplet-like structures similar to Figure 5a, where the complete overlapping of the two densities does not occur.

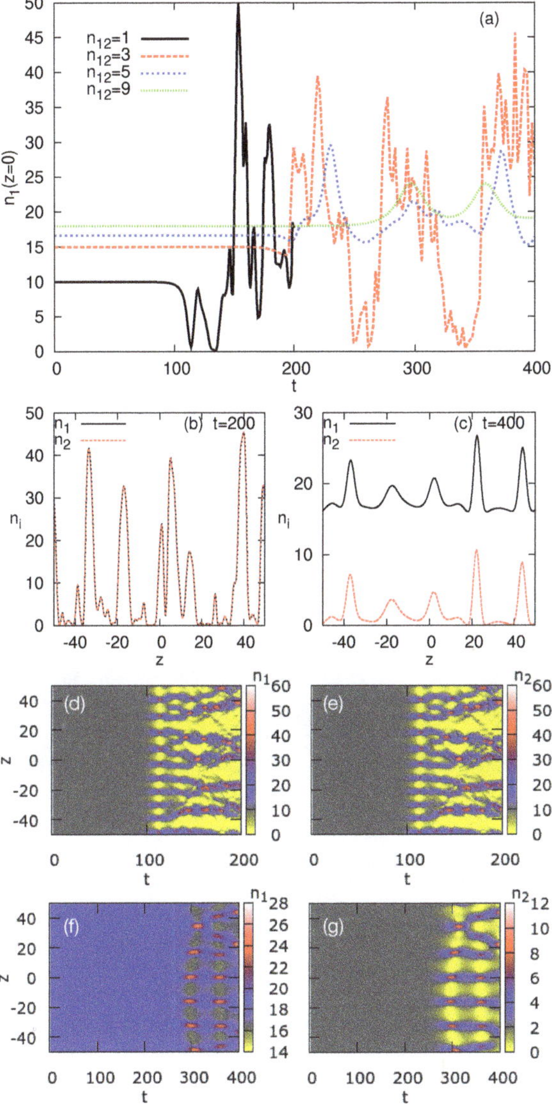

Figure 9. Numerically simulated development of the MI of asymmetric PW states in the two-component system, with $P = 1$ and $\delta g/g = 0.05$. The initial PW states are taken with fixed total density, $(n_1 + n_2)/2 = 10$. (**a**) The evolution of the central density of the first component, $n_1(z = 0)$, for different density ratios in the two components, $n_{12} = n_1/n_2$. (**b**,**c**) Snapshots of density profiles for the cases of (**b**) $n_{12} \equiv n_1/n_2 = 1$ at $t = 200$ and (**c**) $n_{12} = 9$ at $t = 400$. Panels (**d**,**e**) and (**f**,**g**) are top views of the spatiotemporal evolution of the densities, $n_1(z,t)$ and $n_2(z,t)$, for $n_{12} = 1$ and $n_{12} = 9$, respectively. Simulations were performed in the domain $-50 \leq z \leq +50$ with 2048 grid points, subject to periodic boundary conditions. In this figure and in Figure 10, the scaled time unit corresponds to ~ 1 µs in physical units.

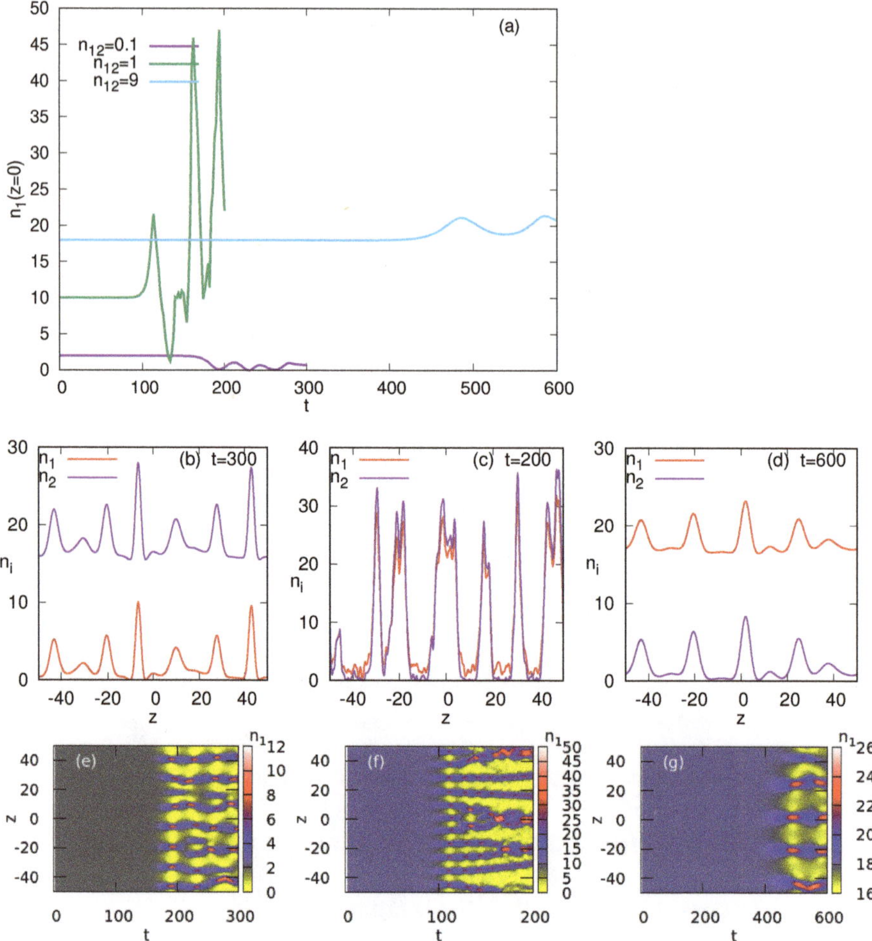

Figure 10. Numerically simulated development of the modulational instability in the two-component system with $\delta g/g = 0.05$ and $P = 1.25$. The initial PW states are taken with a fixed total density, $(n_1 + n_2)/2 = 10$. (**a**) The evolution of the central density of the first component, $n_1(z = 0)$, for different density ratios in the two components, $n_{12} = n_1/n_2$. (**b–d**) Snapshots of density profiles for the cases of (**b**) $n_{12} \equiv n_1/n_2 \sim 0.1$ at $t = 300$, (**c**) $n_{12} = 1$ at $t = 200$ and (**d**) $n_{12} = 9$ at $t = 600$. Panels (**e–g**) represent the top view of the spatiotemporal evolution of the densities, $n_1(z,t)$, corresponding to (**b–d**), respectively (the evolution of $n_2(z,t)$ shows similar patterns). Simulations were performed in the domain $-50 \leq z \leq +50$ with 2048 grid points, subject to periodic boundary conditions.

4. Conclusions

The main purpose of this work is to associate the MI (modulation instability) of plane waves (PWs) to the mechanism of the creation of QDs (quantum droplets) in the system described by the coupled GP (Gross–Pitaevskii) equations including the LHY (Lee–Huang–Yang) terms in the 1D setting. This system is the model of weakly interacting binary Bose gases with approximately balanced interactions between the intra-component self-repulsion and the inter-component attraction. We have investigated, analytically and numerically, the MI of the lower branch of PW states in both symmetric (effectively single-component) and asymmetric (two-component) GP systems, and ensuing formation of a chain of droplet-like states. In particular, numerical solution for QDs which are asymmetric with respect to the two components are obtained, both in the system with equal repulsion strengths but unequal populations in the two components, and in the one with different self-repulsion strengths. The results corroborate that the previously known symmetric states are robust against symmetry-breaking disturbances.

These predictions can be tested experimentally by preparing uniform binary Bose gases with equal or different densities of two components, and suddenly reducing the strength of the effective MF (mean-field) interaction by means of the Feshbach-resonance quench, in order to enhance the relative strength of the LHY terms. In particular, for typical values of physical parameters, an estimate of the characteristic time of the modulation instability growth for typical values of the physical parameters is ~ 1 μs. This time is much smaller than a typical lifetime of the droplet, which is $\gtrsim 100$ μs [21–23,27], thus making the observation of the MI feasible. The present analysis being restricted to the 1D setting, effects of the tight transverse confinement and crossover to the 3D configuration [12,44,45] deserves further consideration.

Author Contributions: Conceptualization and investigation, T.M., A.M., K.K., B.A.M., and A.K.; analytics, A.K. and B.A.M.; numerics, T.M., K.K., and A.M.; writing—original draft preparation, T.M.; writing—review and editing, A.M., K.K., B.A.M., and A.K. All authors have read and agreed to the published version of the manuscript.

Funding: This research received no external funding.

Acknowledgments: We appreciate valuable comments received from M. Modugno. T.M. acknowledges support from IBS (Project Code IBS-R024-D1). A.M. acknowledges support from the Ministry of Education, Science, and Technological Development of Republic of Serbia (project III45010) and the COST Action CA 16221. The work of K.K. is partly supported by the Japan Society for the Promotion of Science (JSPS) Grant-in-Aid for Scientific Research (KAKENHI Grant No. 18K03472). B.A.M. appreciates support from the Israel Science Foundation through grant No. 1287/17. A.K. thanks the Indian National Science Academy for the grant of INSA Scientist Position at Physics Department, Savitribai Phule Pune University).

Conflicts of Interest: The authors declare no conflict of interest.

Appendix A. Other Exact Solutions for the Single-Component GP Equation

Here we briefly list other types of exact solutions of the single-component Equation (12), in addition to the FT and PW solutions in Equations (13) and (22) which were considered in detail above (solutions to Equation (12) in the form of dark and anti-dark solitons were reported in [46]).The stability of a majority of these solutions is not addressed here, as it should be a subject for a separate work.

Appendix A.1. $\delta g/g > 0$

In the case of comparable quadratic self-attraction and cubic repulsion in Equation (12) with $\delta g > 0$, exact spatially-periodic solutions with odd parity can be expressed in terms of the Jacobi's elliptic sine, whose modulus q is an intrinsic parameter of the family:

$$\psi(z,t) = \exp\left(-i\mu_{\mathrm{sn}}t\right)\left[A\,\mathrm{sn}(\beta z,q) + B\right], \tag{A1}$$

where

$$B = \frac{\sqrt{2}}{3\pi}\frac{g}{\delta g} > 0, \quad A = \sqrt{\frac{2}{1+q^2}}B > 0, \quad \mu_{\mathrm{sn}} = -2\frac{\delta g}{g}B^2 < 0, \quad \beta^2 = \frac{2}{(1+q^2)}\frac{\delta g}{g}B^2.$$

In the limit of $q \to 1$, Equation (A1) goes over into the kink (the same as found in [46]),

$$\psi(z,t) = \exp\left(-i\mu_{\mathrm{kink}}t\right)\left[A\tanh(\beta z) + B\right], \tag{A2}$$

with parameters

$$A = B = \frac{\sqrt{2}}{3\pi}\frac{g}{\delta g} > 0, \quad \mu_{\mathrm{kink}} = -2\frac{\delta g}{g}B^2, \quad \beta^2 = \frac{\delta g}{g}B^2.$$

Appendix A.2. $\delta g/g < 0$

In the case when the inter-species MF attraction is stronger than the intra-species repulsion, resulting in $\delta g < 0$, spatially-periodic solutions are expressed in terms of even Jacobi's elliptic functions, $\mathrm{dn}(x,q)$ and $\mathrm{cn}(x,q)$. First, it is

$$\psi(z,t) = \exp\left(-i\mu_{\mathrm{dn}}t\right)\left[A\,\mathrm{dn}(\beta z,q) + B\right], \tag{A3}$$

with the elliptic modulus taking all values $0 < q < 1$, other parameters being

$$B = \frac{\sqrt{2}}{3\pi}\frac{g}{\delta g} < 0, \quad A = -\sqrt{\frac{2}{2-q^2}}B > 0, \quad \mu_{\mathrm{dn}} = -2B^2\frac{\delta g}{g} > 0, \quad \beta^2 = -\frac{2}{(2-q^2)}\frac{\delta g}{g}B^2.$$

The second solution is expressed in terms of the elliptic cosine, with $q^2 > 1/2$:

$$\psi(z,t) = \exp\left(-i\mu_{\mathrm{cn}}t\right)\left[A\,\mathrm{cn}(\beta z,q) + B\right], \tag{A4}$$

$$B = \frac{\sqrt{2}}{3\pi}\frac{g}{\delta g} < 0, \quad A = -\sqrt{\frac{2}{2q^2-1}}B > 0, \quad \mu_{cn} = -2\frac{\delta g}{g}B^2 > 0, \quad \beta^2 = -\frac{2}{(2q^2-1)}\frac{\delta g}{g}B^2.$$

In the limit of $q \to 1$, both solutions in Equations (A3) and (A4) carry over into a state of the "bubble" type [49], which changes the sign at two points (the same solution was reported as an "W-shaped soliton" in [46]):

$$\psi(z,t) = \exp\left(-i\mu_{\mathrm{bubble}}t\right)\left[A\mathrm{sech}(\beta z) + B\right], \tag{A5}$$

with parameters

$$B = \frac{\sqrt{2}}{3\pi}\frac{g}{\delta g} < 0, \quad A = -\sqrt{2}B > 0, \quad \mu_{\text{ bubble}} = \beta^2 = -2\frac{\delta g}{g}B^2 > 0.$$

Appendix B. Analytical Solutions for Strongly Asymmetric Fundamental and Dipole States

Here we consider analytical solutions of Equation (7) with strong asymmetry, $N_1 \ll N_2$, which can be found under small-amplitude conditions, $n_1(z=0) \ll n_2(z=0) \ll n_0$. Then, cubic terms may be neglected in Equation (7), and approximation $\sqrt{P|\psi_1|^2 + P^{-1}|\psi_2|^2} \approx P^{-1/2}\,|\psi_2|$ is used to simplify Equation (7) to the following equations for stationary states in Equation (9):

$$\mu_1\phi_1 = -\frac{1}{2}\frac{d^2\phi_1}{dz^2} - \frac{\sqrt{P}}{\pi}\phi_2\phi_1, \tag{A6}$$

$$\mu_2\phi_2 = -\frac{1}{2}\frac{d^2\phi_2}{dz^2} - \frac{1}{\pi P^{3/2}}\phi_2^2. \tag{A7}$$

Although this case is somewhat formal, in terms of the underlying concept of the quantum droplets, which is essentially based on the competition of residual MF and LHY terms, it is interesting to consider it too.

The soliton solution of Equation (A7) is obvious,

$$\phi_2(z) = \frac{3\pi}{2}(-\mu_2)\frac{P^{3/2}}{\cosh^2\left(\sqrt{-\mu_2/2}z\right)} \tag{A8}$$

where the solution in Equation (13) takes essentially the same form in the limit of $|\mu| \ll \mu_0$. Then, the substitution of Equation (A8) in Equation (A6) makes it tantamount to the linear Schrödinger equation with the Pöschl-Teller potential [50]. The ground-state (GS) solution of Equation (A6) for ϕ_1, with arbitrary amplitude $\phi_1^{(0)}$,

$$(\phi_1(z))_{GS} = \frac{\phi_1^{(0)}}{\left[\cosh\left(\sqrt{-\mu_2/2}z\right)\right]^{\gamma}}, \tag{A9}$$

exists with

$$\gamma = \frac{1}{2}\left(\sqrt{24P^2+1}-1\right), \tag{A10}$$

and eigenvalue

$$(\mu_1)_{GS} = \left(\sqrt{24P^2+1}-1\right)^2 \frac{\mu_2}{16}. \tag{A11}$$

In this case, the QD solutions are quasi-Gaussian objects [20]. Note that, in the symmetric system with $P = 1$, Equations (A10) and (A11) yield $\gamma = 2$ and $(\mu_1)_{GS} = \mu_2$, i.e., the eigenmode and eigenvalue coincide with their counterparts in the soliton solution in Equation (A8), while they are different in the asymmetric system, the GS level lying below or above the chemical potential of soliton in Equation (A8) at $g_1 > g_2$ and $g_1 < g_2$, respectively.

In Figure A1 we compare a typical asymptotic solution given by Equations (A8) and (A9) with a numerically obtained GS solution for the same values of the parameters. It is seen that the analytical and numerical results match well.

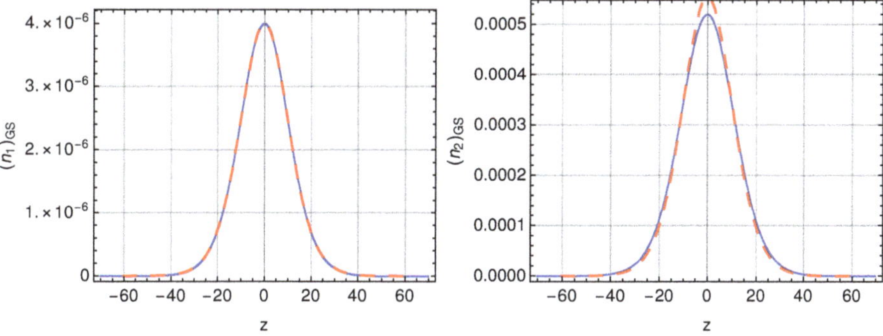

Figure A1. Comparison of the asymptotic analytical solutions, given by Equations (A8) and (A9), with their numerically obtained counterparts. The density of the first (n_1) and second (n_2) components are displayed in top and bottom panels, respectively. Solid blue lines represent the numerical results, while dashed red lines represent the analytical solution. Here, parameters are $\delta g/g = 0.05$, $N_1 = 0.0001067$, $N_2 = 0.0148044$ and $(\mu_2)_{GS} = \mu_2 = -0.005$.

Further, it is also possible to produce the first excited state of Equation (A6) in the form of the dipole (antisymmetric) mode with an arbitrary amplitude:

$$(\phi_1(z))_{dip} = \frac{\phi_1^{(0)} \sinh\left(\sqrt{-\mu_2/2}z\right)}{\left[\cosh\left(\sqrt{-\mu_2/2}z\right)\right]^{\gamma}}, \tag{A12}$$

where γ is the same as in Equation (A10), the respective eigenvalue being

$$(\mu_1)_{dip} = \left(\sqrt{24P^2+1}-3\right)^2 \frac{\mu_2}{16}, \tag{A13}$$

which is obviously higher than its GS counterpart in Equation (A11) (at $P = 1$, Equation (A13) yields $(\mu_1)_{dip} = \mu_2/4$, and $(\mu_1)_{dip}$ falls below μ_2 for $P > \sqrt{2}$). Unlike the GS, the dipole mode exists not at all values of P, but only for $P > \sqrt{1/3}$. Exactly at $P = \sqrt{1/3}$, one has $(\mu_1)_{\text{dip}} = 0$, and the dipole mode in Equation (A12), with $\gamma = 1$, is a delocalized one, $\sim \tanh\left(\sqrt{-\mu_2/2}z\right)$.

Linear Schrödinger Equation (A6) with the Pöschl-Teller potential may give rise to higher bound states of integer order ν as well, with eigenvalues

$$(\mu_1)_\nu = \left(\sqrt{24P^2+1} - (1+2\nu)\right)^2 \frac{\mu_2}{16}, \tag{A14}$$

where $\nu = 0$ and 1 correspond to Equations (A11) and (A13), respectively, the ν-th spate existing at $P^2 > \nu(\nu+1)/6$. The number of such solutions is always finite.

Unlike solutions considered in Appendices A and C, the stability of the solutions given by Equations (A8)–(A14) is obvious.

Appendix C. Other Exact Solutions in the Case of $N_1 \ll N_2$

Here we provide periodic solutions to the semi-linear system of Equations (A6) and (A7) in terms of Jacobi elliptic functions. In the limit of $q \to 1$, they go over into solutions given in the main text, in the form of Equations (A8), (A9), and (A12).

Appendix C.1. Solution of Equation (A7)

An exact periodic solution of Equation (A7) with the quadratic nonlinearity is

$$\phi_2 = A[\text{dn}^2(\beta z, q) + p], \tag{A15}$$

with

$$\beta^2 = -\frac{\mu_2}{2\sqrt{1-q+q^2}}, \quad A = -\frac{3\pi\mu_2 P^{3/2}}{2\sqrt{1-q+q^2}}, \quad p = \frac{-(2-q)+\sqrt{1-q+q^2}}{3}. \tag{A16}$$

In the limit of $q \to 1$, the solution in Equation (A15) goes over into the solution in Equation (A8). Note that p is vanishing in this limit, according to Equation (A16).

Appendix C.2. Solutions of Equation (A6)

We now show that, with ϕ_2 given by Equation (A15), linear Equation (A6) ϕ_1 has several particular solutions depending on the value of P.

Solutions For $P^2 = 1/3$

Appendix C.2.1. Solution I

It is easy to check that

$$\phi_1 = \phi_1^{(0)} \text{dn}(\beta z, q) \tag{A17}$$

is an exact solution to Equation (A6), provided that

$$P^2 = \frac{1}{3}, \quad \mu_1 = \left(\frac{\mu_2}{12}\right) \frac{2-q+2\sqrt{1-q+q^2}}{\sqrt{1-q+q^2}}.$$

Appendix C.2.2. Solution II

$$\phi_1 = \phi_1^{(0)} \text{cn}(\beta z, q) \tag{A18}$$

is an exact solution to Equation (A6), provided that

$$P^2 = \frac{1}{3}, \quad \mu_1 = \left(\frac{\mu_2}{12}\right) \frac{2q-1+2\sqrt{1-q+q^2}}{\sqrt{1-q+q^2}}.$$

In the limit of $q \to 1$, solutions I and II go over into the solution Equation (A9) with $\gamma = 1$ and $\mu_1 = \mu_2/4$.

Appendix C.2.3. Solution III

$$\phi_1 = \phi_1^{(0)}\mathrm{sn}(\beta z, q) \tag{A19}$$

is an exact solution to Equation (A6), provided that

$$P^2 = \frac{1}{3}, \quad \mu_1 = \left(\frac{\mu_2}{12}\right)\frac{2\sqrt{1-q+q^2}-(1+q)}{\sqrt{1-q+q^2}}.$$

In the limit of $q \to 1$, solution III goes over into the solution Equation (A12) with $\gamma = 1$ and $\mu_1 = 0$.

Solutions for $P^2 = 1$.

Appendix C.2.4. Solution IV

It is easy to check that

$$\phi_1 = \phi_1^{(0)}[\mathrm{dn}^2(\beta z, q) + p] \tag{A20}$$

is an exact solution to Equation (A6), provided that

$$P^2 = 1, \quad \mu_1 = \mu_2 .$$

Appendix C.2.5. Solution V

$$\phi_1 = \phi_1^{(0)}\mathrm{cn}(\beta z, q)\mathrm{dn}(\beta z, q) \tag{A21}$$

is an exact solution to Equation (A6), provided that

$$P^2 = 1, \quad \mu_1 = \left(\frac{\mu_2}{2}\right)\frac{q+\sqrt{1-q+q^2}}{\sqrt{1-q+q^2}}.$$

In the limit $q = 1$, solutions IV and V go over into solution Equation (A9) with $\gamma = 2$ and $\mu_1 = \mu_2$.

Appendix C.2.6. Solution VI

$$\phi_1 = \phi_1^{(0)}\mathrm{sn}(\beta z, q)\mathrm{dn}(\beta z, q) \tag{A22}$$

is an exact solution to Equation (A6), provided that

$$P^2 = 1, \quad \mu_1 = \left(\frac{\mu_2}{4}\right)\frac{3(1-q)+\sqrt{1-q+q^2}}{\sqrt{1-q+q^2}}.$$

Appendix C.2.7. Solution VII

$$\phi_1 = \phi_1^{(0)}\mathrm{sn}(\beta z, q)\mathrm{cn}(\beta z, q) \tag{A23}$$

is an exact solution to Equation (A6), provided that

$$P^2 = 1, \quad \mu_1 = \left(\frac{\mu_2}{4}\right)\frac{2\sqrt{1-q+q^2}-(2-q)}{\sqrt{1-q+q^2}}.$$

In the limit of $q \to 1$, solutions VI and VII go over into Equation (A12), with $\gamma = 2$ and $\mu_1 = \mu_2/4$.

References

1. Pitaevskii, L.; Stringari, S. *Bose–Einstein Condensation and Superfluidity*; Oxford University Press: Oxford, UK, 2016.
2. Pethick, C.; Smith, H. *Bose–Einstein Condensation in dilute Gases*; Cambridge University Press: Cambridge, UK, 2002.
3. Petrov, D.S. Quantum Mechanical Stabilization of a Collapsing Bose-Bose Mixture. *Phys. Rev. Lett.* **2015**, *115*, 155302.
4. Lee, T.D.; Huang, K.; Yang, C.N. Eigenvalues and Eigenfunctions of a Bose System of Hard Spheres and Its Low-Temperature Properties. *Phys. Rev.* **1957**, *106*, 1135–1145.
5. Petrov, D.S.; Astrakharchik, G.E. Ultradilute low-dimensional liquids. *Phys. Rev. Lett.* **2016**, *117*, 100401.
6. Li, Y.; Luo, Z.; Lio, Y.; Chen, Z.; Huang, C.; Fu, S.; Tan, H.; Malomed, B.A. Two-dimensional solitons and quantum droplets supported by competing self-and cross-interactions in spin-orbit-coupled condensates. *New J. Phys.* **2017**, *19*, 113043.
7. Cappellaro, A.; Macrí, T.; Bertacco, G.F.; Salasnich, L. Equation of state and self-bound droplet in Rabi-coupled Bose mixtures. *Sci. Rep.* **2017**, *7*, 13358.
8. Jørgensen, N.B.; Bruun, G.M.; Arlt, J.J. Dilute fluid governed by quantum fluctuations. *Phys. Rev. Lett.* **2018**, *121*, 173403.
9. Cikojević, V.; Dželalija, K.; Stipanović, P.; Markić, L.V.; Boronat, J. Ultradilute quantum liquid drops. *Phys. Rev. B* **2018**, *97* , 140502(R).
10. Cappellaro, A.; Macrí, T.; Salasnich, L. Collective modes across the soliton-droplet crossover in binary Bose mixtures. *Phys. Rev. A* **2018**, *97*, 053623.
11. Kartashov, Y.V.; Malomed, B.A.; Tarruell, L.; Torner, L. Three-dimensional droplets of swirling superfluids. *Phys. Rev.* **2018**, *98*, 013612.
12. Zin, P.; Pylak, M.; Wasak, T.; Gajda, M.; Idziaszek, Z. Quantum Bose-Bose droplets at a dimensional crossover. *Phys. Rev. A* **2018**, *98*, 051603(R).
13. Li, Y.; Chen, Z.; Luo, Z.; Huang, C.; Tan, H.; Pang, W.; Malomed, B.A. Two-dimensional vortex quantum droplets. *Phys. Rev. A* **2018**, *98*, 063602.
14. Ancilotto, F.; Barranco, M.; Guilleumas, M.; Pi, M. Self-bound ultradilute Bose mixtures within local density approximation. *Phys. Rev. A* **2018**, *98*, 053623.
15. Liu, B.; Zhang, H.-F.; Zhong, R.-X.; Zhang, X.-L.; Qin, X.-Z.; Huang, C.; Li, Y.-Y.; Malomed, B.A. Symmetry breaking of quantum droplets in a dual-core trap. *Phys. Rev. A* **2019** *99*, 053602.
16. Chiquillo, E. Low-dimensional self-bound quantum Rabi-coupled bosonic droplets. *Phys. Rev. A* **2019**, *99*, 051601(R).
17. Tononi, A.; Wang, Y.; Salasnich, L. Quantum solitons in spin-orbit-coupled Bose-Bose mixtures. *Phys. Rev. A* **2019**, *99*, 063618.
18. Kartashov, Y.V.; Malomed, B.A.; Torner, L. Metastability of Quantum Droplet Clusters. *Phys. Rev. Lett.* **2019**, *122*, 193902.
19. Zhang, X.; Xu, X.; Zheng, Y.; Chen, Z.; Liu, B.; Huang, C.; Malomed, B.A.; Li, Y. Semidiscrete quantum droplets and vortices. *Phys. Rev. Lett.* **2019**, *123*, 133901.
20. Astrakharchik, G.E.; Malomed, B.A. Dynamics of one-dimensional quantum droplets. *Phys. Rev. A* **2018**, *98*, 013631.
21. Cabrera, C.; Tanzi, L.; Sanz, J.; Naylor, B.; Thomas, P.; Cheiney, P.; Tarruell, L. Quantum liquid droplets in a mixture of Bose–Einstein condensates. *Science* **2018**, *359*, 301–304.
22. Cheiney, P.; Cabrera, C.R.; Sanz, J.; Naylor, B.; Tanzi, L.; Tarruell, L. Bright Soliton to Quantum Droplet Transition in a Mixture of Bose–Einstein Condensates. *Phys. Rev. Lett.* **2018**, *120* 135301 .
23. Semeghini, G.; Ferioli, G.; Masi, L.; Mazzinghi, C.; Wolswijk, L.; Minardi, F.; Modugno, M.; Modugno, G.; Inguscio, M.; Fattori, M. Self-Bound Quantum Droplets of Atomic Mixtures in Free Space. *Phys. Rev. Lett.* **2018**, *120*, 235301.
24. Ferioli, G.; Semeghini, G.; Masi, L.; Giusti, G.; Modugno, G.; Inguscio, M.; Gallemi, A.; Recati, A.; Fattori, M. Collisions of self-bound quantum droplets. *Phys. Rev. Lett.* **2019**, *122*, 090401.
25. Kartashov, Y.; Astrakharchik, G.; Malomed, B.; Torner, L. Frontiers in multidimensional self-trapping of nonlinear fields and matter. *Nat. Rev. Phys.* **2019**, *1*, 185–197.

26. Ferrier-Barbut, I. Ultradilute Quantum Droplets. *Phys. Today* **2019**, *72*, 46–52.
27. D'Errico, C.; Burchianti, A.; Prevedelli, M.; Salasnich, L.; Ancilotto, F.; Modugno, M.; Minardi, F.; Fort, C. Observation of quantum droplets in a heteronuclear bosonic mixture. *Phys. Rev. Research* **2019**, *1*, 033155.
28. Kadau, H.; Schmitt, M.; Wentzel, M.; Wink, C.; Maier, T.; Ferrier-Barbut, I.; Pfau, T. Observing the Rosenzweig instability of a quantum ferrofluid. *Nature* **2016**, *530*, 194–197.
29. Schmitt, M.; Wenzel, M.; Böttcher, F.; Ferrier-Barbut, I.; Pfau, T. Self-bound droplets of a dilute magnetic quantum liquid. *Nature* **2016**, *539*, 259–262.
30. Wächtler, F.; Santos, L. Quantum filaments in dipolar Bose–Einstein condensates. *Phys. Rev. A* **2016**, *93*, 061603(R).
31. Ferrier-Barbut, I.; Kadau, H.; Schmitt, M.; Wenzel, M.; Pfau, T. Observation of Quantum Droplets in a Strongly Dipolar Bose Gas. *Phys. Rev. Lett.* **2016**, *116*, 215301.
32. Wächtler, F.; Santos, L. Ground-state properties and elementary excitations of quantum droplets in dipolar Bose–Einstein condensates. *Phys. Rev. A* **2016**, *94*, 043618.
33. Baillie, D.; Blakie, P.B. Droplet Crystal Ground States of a Dipolar Bose Gas. *Phys. Rev. Lett.* **2018**, *121*, 195301.
34. Ferrier-Barbut, I.; Wenzel, M.; Schmitt, M.; Böttcher, F.; Pfau, T. Onset of a modulational instability in trapped dipolar Bose–Einstein condensates. *Phys. Rev. A* **2018**, *97* , 011604(R).
35. Cidrim, A.; Santos, F.E.A.d.; Henn, E.A.L.; Macrí, T. Vortices in self-bound dipolar droplets. *Phys. Rev. A* **2018**, *98*, 023618.
36. Yukalov, V.I.; Novikov, A.N.; Bagnato, V.S. Formation of granular structures in trapped Bose–Einstein condensates under oscillatory excitations. *Laser Phys. Lett.* **2014**, *11*, 095501.
37. Bulgac, A. Dilute Quantum Droplets. *Phys. Rev. Lett.* **2002**, *89*, 050402.
38. Baillie, D.; Wilson, R.M.; Blakie, P.B. Collective Excitations of Self-Bound Droplets of a Dipolar Quantum Fluid. *Phys. Rev. Lett.* **2017**, *119*, 255302.
39. Nguyen, J.H.V.; Luo, D.; Hulet, R.G. Formation of matter-wave soliton trains by modulational instability. *Science* **2018**, *356*, 422–426.
40. Everitt, P.J.; Sooriyabandara, M.A.; Guasoni, M.; Wigley, P.B.; Wei, C.H.; McDonald, G.D.; Hardman, K.S.; Manju, P.; Close, J.D.; Kuhn, C.C.N.; et al. Observation of a modulational instability in Bose–Einstein condensates. *Phys. Rev. A* **2017**, *96*, 041601.
41. Sanz, J.; Frölian, A.; Chisholm, C.S.; Cabrera, C.R.; Tarruell, L. Interaction control and bright solitons in coherently-coupled Bose–Einstein condensates. *arXiv* **2019**, arXiv:1912.06041.
42. Bhat, I.A.; Mithun, T.; Malomed, B.A.; Porsezian, K. Modulational instability in binary spin-orbit-coupled Bose–Einstein condensates. *Phys. Rev. A* **2015**, *92*, 063606.
43. Singh, D.; Parit, M.K.; Raju, T.S.; Panigrahi, P.K. Modulational instability in one-dimensional quantum droplets. *Res. Gate Prepr.* **2019**, doi:10.13140/RG.2.2.34638.82246.
44. Ilg, T.; Kumlin, J.; Santos, L.; Petrov, D.S.; Büchler, H.P. Dimensional crossover for the beyond-mean-field correction in Bose gases. *Phys. Rev. A* **2018**, *98*, 051604.
45. Edler, D.; Mishra, C.; Wächtler, F.; Nath, R.; Sinha, S.; Santos, L. Quantum Fluctuations in Quasi-One-Dimensional Dipolar Bose–Einstein Condensates. *Phys. Rev. Lett.* **2017**, *119*, 050403.
46. Triki, H.; Biswas, A.; Moshokoa, S.P.; Belić, M. Optical solitons and conservation laws with quadratic-cubic nonlinearity. *Optik* **2017**, *128*, 63–70.
47. Vakhitov, N.G.; Kolokolov, A.A. Stationary solutions of the wave equation in a medium with nonlinearity saturation. *Radiophys. Quantum Electron.* **1973**, *16*, 783–789.
48. di Carli, A.; Colquhoun, C.D.; Henderson, G.; Flannigan, S.; Oppo, G.-L.; Daley, A.J.; Kuhr, S.; Haller, E. Excitation modes of bright matter-wave solitons. *Phys. Rev. Lett.* **2019**, *123*, 123602.
49. Barashenkov, I.V.; Panova, E.Y. Stability and evolution of the quiescent and travelling solitonic bubbles. *Phys. D Nonlinear Phenom.* **1993**, *69*, 114–134.
50. Landau, L.D.; Lifshitz, E.M. *Quantum Mechanics*; Nauka Publishers: Moscow, Russia, 1989.

Article

Kramers Degeneracy and Spin Inversion in a Lateral Quantum Dot

Konstantin Pichugin [1], Antonio Puente [2] and Rashid Nazmitdinov [3,4,*]

[1] Kirensky Institute of Physics, Federal Research Center KSC Siberian Branch, Russian Academy of Sciences, 660036 Krasnoyarsk, Russia; knp@tnp.krasn.ru

[2] Departament de Física, Universitat de les Illes Balears, E-07122 Palma de Mallorca, Spain; toni.puente@uib.es

[3] Bogoliubov Laboratory of Theoretical Physics, Joint Institute for Nuclear Research, 141980 Dubna, Russia

[4] Faculty of Natural and Engineering Science, Dubna State University, 141982 Dubna, Russia

* Correspondence: rashid@theor.jinr.ru

Received: 19 October 2020; Accepted: 8 December 2020; Published: 10 December 2020

Abstract: We show that the axial symmetry of the Bychkov–Rashba interaction can be exploited to produce electron spin-flip in a circular quantum dot, without lifting the time reversal symmetry. In order to elucidate this effect, we consider ballistic electron transmission through a two-dimensional circular billiard coupled to two one-dimensional electrodes. Using the tight-binding approximation, we derive the scattering matrix and the effective Hamiltonian for the considered system. Within this approach, we found the conditions for the optimal realization of this effect in the transport properties of the quantum dot. Numerical analysis of the system, extended to the case of two-dimensional electrodes, confirms our findings. The relatively strong quantization of the quantum dot can make this effect robust against the temperature effects.

Keywords: quantum dot; Kramers degeneracy; spin-orbit interaction; tight-binding approach

---

1. Introduction

Spin-polarized transport in semiconductor nanostructures attracts a continuous experimental and theoretical attention due to great interests for both basic research and device applications (see for a review [1–3]). Indeed, apart fundamental aspects related to the origin of spin current in nanosystems, the inversion of spin polarization is necessary, for example, for operation of spin-based logic elements. The inversion of spin polarization can be achieved in an external AC field with the aid of the electron spin resonance (see, for example, [4]), or by the lifting the spin degeneracy by means of a magnetic field that induces the Zeeman splitting (e.g., [5]). The spin currents can be inverted also by mechanical strain of a silicene [6]. Periodically rippled graphene can as well invert the polarized spin current, by changing the electron flow direction through the system [7].

One of the main requirements for device operability is the efficient manipulation of spin-polarized currents in a semiconductor structure. An additional condition for device applications is that a polarized current should be generated by means of all-electrical methods. In particular, a remarkable progress has been achieved in all-electrical injection from ferromagnetic contacts [8,9] and (Ga,Mn)As [10]. Alternatively to the injection, a spin-orbit interaction (SOI) present in semiconductors provides a natural mechanism to manipulate the spin (e.g., [11–14]). In particular, the electrical field, caused by the structure inversion asymmetry of the heterostructure, gives rise to the Bychkov–Rashba term [15,16]. The strength of this interaction can be controlled by means of an applied electric field [17–19]. It should be mentioned that the Rashba interaction is one of the basic ingredients in the physics of Majorana fermions [20–22]. This physics is based on two superconducting electrodes electrically connected to a semiconductor nanowire with strong Rashba coupling, and a uniform magnetic field. The explosive activity in this direction includes among many proposals as well the use of high-temperature cuprate superconductors [23] and exotic pairs of parafermions without magnetic field [24] to create Majorana zero modes for quantum computing applications (e.g., [25–27]).

Several proposals rely on the SOI as the basic mechanism to achieve a spin filtering effect in low-dimensional semiconductor structures [28–31]. In fact, it might allow for an all-electrical spin-polarized current generation. The relatively small energy scale, produced by the SOI, presents, however, a major obstacle for technological applications. Indeed, the spin splitting induced by the SOI in typical semiconductor alloys can reach a few meV [17,18]. This scale stipulates certain restrictions on the choice of nanosystems that would be enabled to overcome the impact of thermal effects. It is well known that wide gap semiconductors (e.g., GaAs) possess a relatively weak spin-orbit interaction. In contrast, narrow gap semiconductors own strong spin-orbit couplings as well as g factors. These two factors guide the choice of most favorable materials, especially, in view of the Majorana physics. The latter question is, however, beyond the scope of the present paper, and we leave this problem for future. The main result of the present paper is that the *symmetry* of a circular quantum dot can be used to obtain the robust mechanism to inverse z-component of spin-polarized current for non-magnetic metallic contacts.

2. Symmetry of Rashba SOI

It is well known that the invariance of the SOI with respect to the time-reverse operation leads to the Kramers degeneracy (e.g., [32]). Therefore, any system, where the SOI is the only spin-dependent term, will exhibit this two-fold degeneracy. Explicitly, the time reversal symmetry is obtained by inverting both spin and momentum operators. The resulting states, although degenerated in energy, are distinguished by an opposite orbital motion and an opposite spin orientation.

To employ this fundamental feature, we consider a two-dimensional (2D) semiconductor quantum dot (QD) with a circular confinement and the Rashba spin-orbit interaction. In the effective mass approximation for the conduction band, the Hamiltonian can be written as $\hat{H} = \sum_{i=1}^{N} \hat{h}_i$, with the single-particle Hamiltonian taken in the form

$$\hat{h} = \hat{\mathcal{H}}_0 + \hat{\mathcal{H}}_R = \frac{\hat{p}_x^2 + \hat{p}_y^2}{2m^*} + V_{ext}(r) + \hat{\mathcal{H}}_R, \tag{1}$$

where m^* is the electron effective mass and $V_{ext}(r)$ is an external rotationally symmetric potential. We consider the limit of the weak Coulomb interaction, when the external potential dominates in electron properties (e.g., [33]). The Bychkov–Rashba interaction has the form: $\hat{\mathcal{H}}_R = \tilde{\alpha}\left(\hat{p}_y\sigma_x - \hat{p}_x\sigma_y\right)/\hbar$. The strength parameter $\tilde{\alpha}$ depends strongly on the material, reaching its maximum value for narrow gap III-V semiconductor alloys. For instance, typical values of $\tilde{\alpha}$ = 10–40 meV $\times$ nm have been experimentally determined for different InAs-based structures [28,29,34]. An important feature is that the Bychkov–Rashba interaction preserves the axial symmetry, i.e., $\left[\hat{\mathcal{H}}_R, \hat{J}_z\right] = 0$, where $\hat{J}_z = \hat{L}_z + \hat{s}_z$. Therefore, the full Hamiltonian obeys the conservation law $\left[\hat{H}, \hat{J}_z\right] = 0$.

In the dimensionless cylindrical coordinates $r = \sqrt{x^2+y^2}/d$ and $\phi = \arctan(y/x)$ the Hamiltonian (1) takes the form

$$\begin{aligned}\hat{h} &= -\Delta + V(r) + \alpha\left[\begin{pmatrix} 0 & e^{-\mathrm{i}\phi} \\ -e^{\mathrm{i}\phi} & 0 \end{pmatrix}\frac{\partial}{\partial r}\right. \\ &\left. -\frac{\mathrm{i}}{r}\begin{pmatrix} 0 & e^{-\mathrm{i}\phi} \\ e^{\mathrm{i}\phi} & 0 \end{pmatrix}\frac{\partial}{\partial\phi}\right].\end{aligned} \tag{2}$$

Here $\alpha = 2m^*\tilde{\alpha}d/\hbar^2$, $\Delta = \frac{\partial^2}{\partial r^2} + \frac{1}{r}\frac{\partial}{\partial r} + \frac{1}{r^2}\frac{\partial^2}{\partial \phi^2}$ is the Laplace operator, and the parameter d is a characteristic length in our system. Since the eigenstates of the Hamiltonian (2) are eigenstates of the J_z operator, they can be expressed in the following form

$$\mathbf{\Phi}_{nj} = \begin{pmatrix} u_{nj}(r)e^{\mathrm{i}(j-1/2)\phi} \\ v_{nj}(r)e^{\mathrm{i}(j+1/2)\phi} \end{pmatrix}, \tag{3}$$

where $n = 1, 2, ...$ and $j = 1/2, 3/2, ...$ stand for radial and the angular momentum quantum numbers, respectively. To simplify the eigenvalue problem, we represent the wavefunction (3) as a formal series with respect to the strength of the spin-orbit interaction α

$$\mathbf{\Phi}_{nj} = e^{\mathrm{i}(j-1/2)\phi} \sum_m \alpha^m \begin{pmatrix} u_{nj}^m(r) \\ v_{nj}^m(r)e^{\mathrm{i}\phi} \end{pmatrix}. \tag{4}$$

At $m = 0$ the wavefunction $f_{nj}^0(r)$ is the radial part of the Shrödinger equation solution without the spin-orbit interaction

$$-\Delta f_{nj}^0(r) + (V_{ext}(r) + \frac{(j-1/2)^2}{r^2}) f_{nj}^0(r) = E_{nj} f_{nj}^0(r). \tag{5}$$

We recall that the energy scale, produced by the effective external potential $V_{ext}(r)$, is larger than that produced by the spin-orbit interaction (cf [35]). Therefore, it is enough to consider the expansion of the wavefunction (4) up to the first order with the respect to the strength parameter. As a result, we obtain two differential equations for the coefficients $u_{nj}^m(r), v_{nj}^m(r)$, at $m = 1$:

$$\left(2r(f_{nj}^0)' + f_{nj}^0\right)(u_{nj}^1)' + r f_{nj}^0 (u_{nj}^1)'' = 0, \tag{6}$$

$$r^2 (f_{nj}^0)' \left(4(v_{nj}^1)' + 2\right) + f_{nj}^0 \left(r\left(2r(v_{nj}^1)'' + 2(v_{nj}^1)' - 2j + 1\right) - 4j v_{nj}^1\right) = 0 \tag{7}$$

The trivial solution of Equation (6) is $u_{nj}^1 = 0$, while we obtain $v_{nj}^1 = r/2$ to hold true Equation (7). Thus, the approximate eigenfunctions of the Hamiltonian (2) to the first order in α can be written in the form

$$\mathbf{\Phi}_{nj} \approx f_{nj}^0(r)e^{\mathrm{i}(j-1/2)\phi} \begin{pmatrix} 1 \\ -[\alpha r/2]e^{\mathrm{i}\phi} \end{pmatrix}. \tag{8}$$

Below we will use this function to find the optimal conditions for the electron spin-flip phenomenon in the QD.

3. Effective Hamiltonian Model

To analyze transport properties of the circular QD we employ the effective Hamiltonian method [36,37]. According to this method, the scattering system is described by the Hamiltonian that contains the structure with discrete spectrum (our QD), the continuum with the external scattering states (external electrodes), and the interaction between continuum states with QD's eigenstates. Evidently, once the system is opened, the discrete states of QD's own the widths, i.e., they transform to resonance states. The main object in such an investigation is the scattering matrix that describes the relation between the amplitudes of incoming states from electrodes and the amplitudes of the reflected states from, or transmitted through the structure into electrodes.

One of the efficient approaches to reach reliable numerical results on ballistic transport through mesoscopic system is based on the tight-binding model. Following [38] we model the scattering system as a two-dimensional billiard with two attached 1D electrodes in the tight-binding representation (see details in Appendix A). In this case, the scattering matrix that describes scattering from a channel C' to a channel C takes the form

$$S_{CC'}(E) = \delta_{CC'} - 2\pi i \psi_C^1(E)\psi_{C'}^1(E) \sum_{mn} W_{nC}^* \mathcal{F}_{nm}(E) W_{mC'} . \tag{9}$$

Here, $\psi_C^1 = \sqrt{\frac{\sin k}{\pi}}$ – the wave function of a semi-infinite long electrodes (without the SOI) at the contact point; E is the electron energy in the electrode in the 1D tight-binding model. The matrix $\mathcal{F}$ is defined as $\mathcal{F} = \left(E - H^{eff}(E)\right)^{-1}$, while the effective Hamiltonian has form

$$H_{nm}^{eff}(E) = \delta_{nm}\varepsilon_n - \exp(ik) \sum_C W_{nC} W_{mC}^* . \tag{10}$$

Here, the coefficients

$$W_{nC} = \Psi_n^*(\mathbf{r} = \mathbf{r}_C) \tag{11}$$

are the normalized eigenfunctions $\Psi_n(\mathbf{r})$, corresponding to the eigenvalue ε_n of our structure; $\mathbf{r}_C$—the coordinate of C-th electrode junction (see Figure A1 and discussion around it in Appendix A). Note, that the index C denotes the electrode and as well the spin orientation.

Assuming a weak coupling of our structure with the external electrodes, let us consider a pair of degenerate levels with energy $\varepsilon_p = \varepsilon_{p+1} \approx E$, i.e., near the energy E of a scattering electron. We assume also a strong confinement potential $V_{ext}(r)$, that allows the neglecting of resonance overlapping for the open system. As discussed above, the pair of corresponding eigenfunctions are time conjugated: $\Psi_{p+1} = \hat{\mathcal{T}}\Psi_p$, where $\hat{\mathcal{T}} = -i\sigma_y\hat{K}$ is the time-reverse operator. As a result, the following relations take place

$$W_{p+1c\uparrow} = -W_{pc\downarrow}^*, \; W_{p+1c\downarrow} = W_{pc\uparrow}^*. \tag{12}$$

From this property it follows immediately that the effective Hamiltonian (10) is a diagonal matrix due to the orthogonality condition $\sum_C W_{pC}^* W_{p+1C} = 0$. Consequently, we obtain for the S-matrix elements between two electrodes the following definitions:

$$S_{c\uparrow c'\uparrow} = \delta_{cc'} - X(E)\left(W_{pc\uparrow}W_{pc'\uparrow}^* + W_{pc\downarrow}^* W_{pc'\downarrow}\right), \tag{13}$$

$$S_{c\uparrow c'\downarrow} = -X(E)\left(W_{nc\uparrow}W_{pc'\downarrow}^* - W_{pc\downarrow}^* W_{pc'\uparrow}\right), \tag{14}$$

$$S_{c\downarrow c'\uparrow} = -X(E)\left(W_{nc\downarrow}W_{pc'\uparrow}^* - W_{pc\uparrow}^* W_{pc'\downarrow}\right), \tag{15}$$

$$S_{c\downarrow c'\downarrow} = \delta_{cc'} - X(E)\left(W_{pc\downarrow}W_{pc'\downarrow}^* + W_{pc\uparrow}^* W_{pc'\uparrow}\right), \tag{16}$$

where

$$X(E) = \frac{2i\sin k}{E - \varepsilon_p + \exp(ik)w_p^2}. \tag{17}$$

Here, we introduce the parameter

$$w_p^2 = \sum_c |W_{pc\uparrow}|^2 + |W_{pc\downarrow}|^2. \tag{18}$$

The resonant condition arises at the electron energy

$$E_{res} = \varepsilon_p - w_p^2 \cos k, \tag{19}$$

when the factor (17) reaches its maximal value $2/w_p^2$. Hereafter, we assume that the ballistic transport occurs at the resonance energy (19). From Equations (13)–(16) it follows that

$$|S_{c\uparrow c\downarrow}|^2 = |S_{c\downarrow c\uparrow}|^2 = 0, \tag{20}$$

$$|S_{c\uparrow c'\uparrow}|^2 = |S_{c\downarrow c'\downarrow}|^2, \tag{21}$$

$$|S_{c\uparrow c'\downarrow}|^2 = |S_{c\downarrow c'\uparrow}|^2. \tag{22}$$

There are a few remarks in order. First, it should be noted that Equations (20)–(22) lead us to the fact that at the transport through our system from electrode c' to c the spin polarization is

$$P_{cc'} = \frac{|S_{c\uparrow c'\uparrow}|^2 + |S_{c\uparrow c'\downarrow}|^2 - |S_{c\downarrow c'\uparrow}|^2 - |S_{c\downarrow c'\downarrow}|^2}{|S_{c\uparrow c'\uparrow}|^2 + |S_{c\uparrow c'\downarrow}|^2 + |S_{c\downarrow c'\uparrow}|^2 + |S_{c\downarrow c'\downarrow}|^2} = 0 \,.$$

This result is in the agreement with the statement that a nonzero spin polarization in system with the SOI cannot occur if there is only one open channel in electrodes [39,40].

Second, from Equation (20) it follows that the reflection coefficient with the spin-flip is always zero. The reflection coefficient without the spin-flip can be rewritten as

$$S_{c\sigma c\sigma} \;=\; 1 - X(E)\left(|W_{pc\sigma}|^2 + |W_{pc\sigma}|^2\right), \quad \sigma = \uparrow, \downarrow \,. \tag{23}$$

Finally, without loss of generality, we assume the equal coupling of the QD's states to both electrodes (1 and 2). As a result, taking into account the definition (18), we have

$$\left(|W_{p1\uparrow}|^2 + |W_{p1\downarrow}|^2\right) = \left(|W_{p2\uparrow}|^2 + |W_{p2\downarrow}|^2\right) = w_p^2/2 \,. \tag{24}$$

Evidently, at the resonance energy (19) the reflection (23) becomes zero, and, consequently, we obtain for the S-matrix

$$S(E_{res}) = \begin{pmatrix} 0 & T^{\dagger} \\ T & 0 \end{pmatrix}, \tag{25}$$

with 2×2 matrix

$$T = \begin{pmatrix} S_{2\uparrow 1\uparrow} & S_{2\uparrow 1\downarrow} \\ S^*_{2\uparrow 1\downarrow} & S^*_{2\uparrow 1\uparrow} \end{pmatrix}. \tag{26}$$

From the unitarity of S matrix $S^{\dagger}S = 1$ it follows that $T^{\dagger}T = 1$. The spin-orbit interaction converts a two-component spinor into another two-component spinor. In particular, it could change the incoming spin up electron state $|\uparrow\rangle$ to the outgoing spin down electron state $|\downarrow\rangle$ and vice versa. Consequently, to reach the ideal electron spin-flip phenomenon we require that the direct scattering matrix elements (non-spin-flip components) should be equal zero: $T_{11} = T_{22} = 0 \Rightarrow S_{2\sigma 1\sigma} = 0$. From Equations (13) and (16) it follows that this requirement holds if the following relation takes place for the eigenstates of our QD:

$$W_{p2\uparrow}W^*_{p1\uparrow} = -W^*_{p2\downarrow}W_{p1\downarrow} \,. \tag{27}$$

To see the consequences of this relation for our system, we apply this condition to the eigenstates of the QD of the radius R: $W^*_{pc\uparrow} = u_{nj}(r_c)\exp(\mathrm{i}(j-1/2)\phi_c)$, $W^*_{pc\downarrow} = v_{nj}(r_c)\exp(\mathrm{i}(j+1/2)\phi_c)$ [see Equation (3)]. In the tight-binding approximation the point r_c is located just before the quantum point contact (QPC) between the QD and the electrode (see Figure A1). At the QPC $r = R$, and the radial wave function $\psi(r)$ $[v(r)$ or $u(r)]$ takes the form at $r = r_c$:

$$\psi(r_c) \approx \psi(R) - \frac{d\psi(R)}{dR}a_0 + \ldots \tag{28}$$

Here a small quantity a_0 is the distance between lattice sites. Taking into account that $\psi(R) = 0$ at the Dirichlet boundary condition (a closed QD), we obtain:

$$\begin{aligned} W^*_{nc\uparrow} &= -a_0 u'_{nj}(R)\exp[\mathrm{i}(j-1/2)\phi_c], \\ W^*_{nc\downarrow} &= -a_0 v'_{nj}(R)\exp[\mathrm{i}(j+1/2)\phi_c] \end{aligned} \tag{29}$$

(prime denotes derivative over R). As a result, the condition (27) takes the form

$$[u'_{nj}(R)]^2\exp[-\mathrm{i}(j-1/2)(\phi_2-\phi_1)] = -[v'_{nj}(R)]^2\exp[\mathrm{i}(j+1/2)(\phi_2-\phi_1))] \,, \tag{30}$$

that leads us to the following equations:

$$[u'_{nj}(R)]^2 = [v'_{nj}(R)]^2, \tag{31}$$

$$\exp[2ij(\phi_2 - \phi_1)] = -1. \tag{32}$$

For the approximate eigenfunction (8) we obtain the condition $R\alpha \approx 2$, resolving Equation (31). Among solutions of Equation (32) there is one $\phi_2 - \phi_1 = \pi$ which is common for all possible states with the quantum number j. In other words, an electron with the spin up (down), injected from one electrode to the QD, exits from the opposite electrode with the spin down (up). Thus, by altering either the spin-orbit strength α or the QD's radius R within the condition $R\alpha \approx 2$, we obtain the spin-flip transmission through our structure.

To illuminate this analytical solution, we consider the simplest quantum well potential of the form $V_{ext}(r < R) = 0, V_{ext}(R) = \infty$. The solution of the eigenvalue problem for this potential provides the radial wavefunctions $[v(r)$ and $u(r)]$ in terms of the Bessel functions (e.g., [41]). In this case, the Dirichlet boundary condition for the wavefunction (3) yields the equation

$$J_{j-1/2}(\mu_+)J_{j+1/2}(\mu_-) = J_{j-1/2}(\mu_-)J_{j+1/2}(\mu_+)\,, \tag{33}$$

that defines the energy spectrum ε_{nj}. Here, $J_{j\pm1/2}(r)$ is the Bessel functions, and the parameter $\mu_\pm$ has the following structure

$$\mu_\pm = \left(\alpha/2 \pm \sqrt{\varepsilon_{nj} + (\alpha/2)^2}\right) R\,. \tag{34}$$

The application of Equation (31) leads to the transcendental equation

$$J_{j-1/2}(\mu_+) \pm J_{j+1/2}(\mu_+) = 0. \tag{35}$$

Plus or minus sign here are opposite to signs of $u'v'$-derivatives. In the case of the ideal spin-flip process the numerical solution of the transcendental Equation (35) for several lowest eigenvalues with quantum numbers j and n gives $R\alpha = 1.5\ldots2.5$ (see Table 1). Please note that the transport properties could be affected at the interface between the 1D lead and the 2D QD. However, the use of the QPC restricts the number of open channels between the electrode and the QD [42]. Consequently, we belief that our results will be valid for a realistic situation as well. To confirm our findings, we consider a 2D case below.

Table 1. Value of $R\alpha$ for quantum numbers n and j.

n	j	$R\alpha$
0	1/2	1.68
0	3/2	1.85
1	1/2	1.57
0	5/2	1.98
1	3/2	1.61
2	1/2	1.59
0	7/2	2.1

4. The 2D Model

The results, obtained with the aid of the one-dimension electrodes, serve to illustrate the basic principles of the spin-flip at the transmission through the QD. To elucidate these principles we have considered the particular situation, when direct transmission matrix elements were equal zero. In this section, to demonstrated the vitality and the validity of our findings we consider the 2D structure depicted in Figure 1. The QD is modelled by a constant potential (gray) with the SOI included. The ballistic electrons, propagating from one electrode to another, tunnel to the QD through the thin potential shell (dark gray). On the thin lines the Dirichlet boundary conditions are imposed. All electrodes have equal width d, while the circular QD has the radius $R = 2d$.

Figure 1. Sketch of the 2D device that consists of a circular lateral QD. The effective QD's confinement includes a constant potential with the Rashba interaction (gray region). Additional thin ($0.1d$) constant potential shell (dark gray region) controls the coupling between the QD and two electrodes.

We perform numerical calculations in the framework of the tight-binding approach on the square lattice $n = n_x\hat{x} + n_y\hat{y}$ ($\hat{x}$ and $\hat{y}$ are 2D vectors of elementary translations with length a_0 in x and y direction, a_0—lattice constant, n_x and n_y—integers). In the tight-binding approximation the system Hamiltonian (1) has the following form

$$\begin{aligned}
\hat{H} &= \hat{\mathcal{H}}_0 + \hat{\mathcal{H}}_R \\
\hat{\mathcal{H}}_0 &= \sum_{n,\sigma} \epsilon_{n\sigma} c^{\dagger}_{n\sigma} c_{n\sigma} - \sum_{\langle nm\rangle,\sigma} t c^{\dagger}_{n\sigma} c_{m\sigma} \\
\hat{\mathcal{H}}_R &= -\frac{\alpha}{2a_0} \sum_{n} \left\{ \mathrm{i} \left(c^{\dagger}_{n\uparrow} c_{n+\hat{y}\downarrow} + c^{\dagger}_{n\downarrow} c_{n+\hat{y}\uparrow} \right) - \right. \\
& \left. \left(c^{\dagger}_{n\uparrow} c_{n+\hat{x}\downarrow} - c^{\dagger}_{n\downarrow} c_{n+\hat{x}\uparrow} \right) \right\} + \mathrm{H.c.}.
\end{aligned} \tag{36}$$

Here, we use the following notations: $\epsilon_{n,\sigma} = 4t - V(n_x a_0, n_y a_0)$, $t = \hbar^2/2m^* a_0^2$; the indices $\langle nm\rangle$ stand for nearest neighbor sites n and m. We solve the Schrödinger equation in a discretized space, according to the method developed by Ando [43]. Examples of numerical treatment of quantum billiards within this approach can be found, for example, in Refs. [44,45]. In our calculations the electrode width $d = 40a_0$; while dimensionless units are defined as $E = E_F/E_0 = E_F 2m^* d^2/\hbar^2$, $\alpha = \tilde{\alpha}/(dE_0) = \tilde{\alpha} 2m^* d/\hbar^2$, where the characteristic length of the device d was chosen as the unit length. In these units the energy range $2(d/a_0)^2(1 - \cos(\pi a_0/d)) < E < 2(d/a_0)^2(1 - \cos(2\pi a_0/d))$ corresponds to one transverse mode in all contacts.

Figure 2 displays the numerical results for the direct process $|S_{2\uparrow 1\uparrow}|^2$ as a function of the electron energy and the strength of the Rashba interaction. The results are in a good agreement with those that have been obtained from the condition (35) (see Table 1).

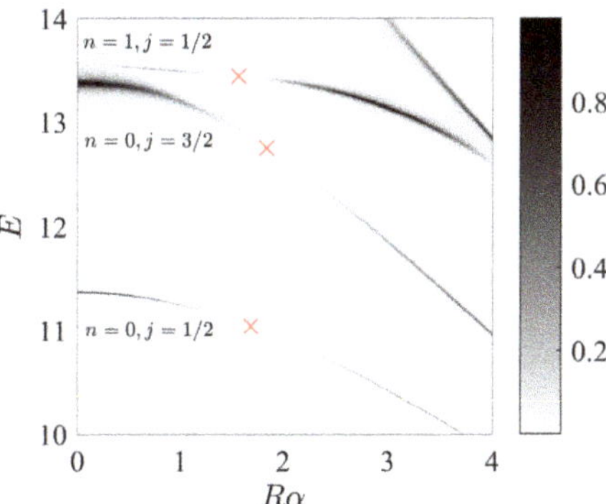

Figure 2. The probability for the direct transmission $|S_{2\uparrow 1\uparrow}|^2$ as function of the electron energy and the parameter $R\alpha$. Red crosses indicate $(E, R\alpha)$ that correspond to $|S_{2\uparrow 1\uparrow}|^2$=0 for: $n = 0, j = 1/2$, $n = 0, j = 3/2, n = 1, j = 1/2$.

It is noteworthy that the approximate condition $R\alpha \approx 2$ is unique in that there is no dependence neither on the electron energy, nor on the spin of the electron state. In other words, it can be fulfilled for a set of QD's electron states even at large opening of the QD. To model this case (a resonance overlapping regime), we remove the potential barrier between the QD and the electrodes. Indeed, the direct spin transmission $|S_{2\uparrow1\uparrow}|^2$ is suppressed strongly near $R\alpha \sim 1.6$ (see Figure 3a). In this case, the spin-flip process, averaged over energy (available at the one channel transport), becomes a dominant phenomenon, reaching about ~97% of the efficiency (see Figure 3b). In other words, the thermal effects affect only slightly the spin-flip process even in the regime of the large opening of the QD.

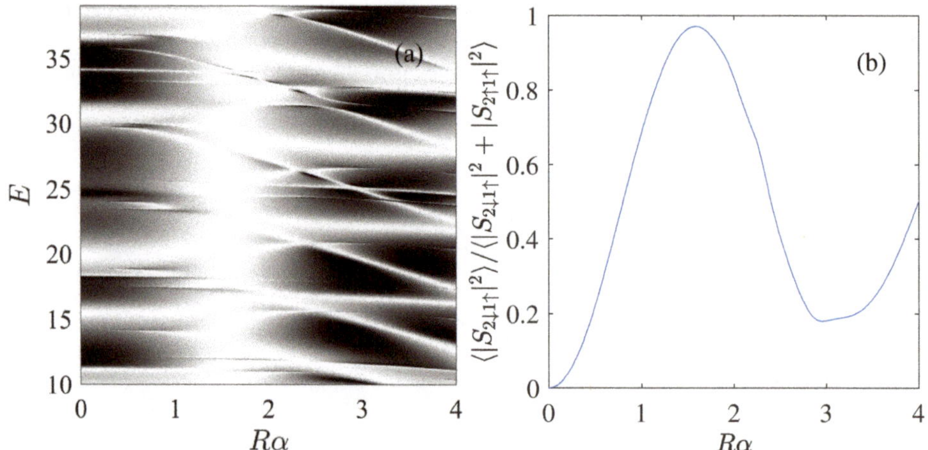

Figure 3. Quantum dot with radius $R = 2d$ without any additional potential. The probability for the direct transmission $|S_{2\uparrow1\uparrow}|^2$ as function of the electron energy and the parameter $R\alpha$ (**a**). Average effectivity of spin invertor $\int dE|S_{2\downarrow1\uparrow}|^2/\int dE(|S_{2\downarrow1\uparrow}|^2 + |S_{2\uparrow1\uparrow}|^2)$ over energy range $2(d/a_0)^2(1-\cos(\pi a_0/d)) < E < 2(d/a_0)^2(1-\cos(2\pi a_0/d))$ (**b**).

Larger the dot radius lesser the spin-orbit strength is required to hold the condition $R\alpha \approx 2$. However, with the increase of the dot radius the spacing between levels becomes smaller. Note, that in principle, QD levels will be affected by the coupling as well. In fact, they will be shifted with respect to those of the closed QD, especially, in the case of a strong electron-electron interaction [46]. Therefore, the discussed effect is most probable in a narrow gap semiconductor QD with a strong confinement potential and at the weak coupling regime. In this case, the resonance states will be well separated from each to other.

However, as discussed above, the overlapping of resonances and the thermal smearing decrease the efficiency of the spin-flip phenomenon as well. To evaluate the energy scaling we transform the equation $\alpha R = 2$ in dimensional units: $\tilde{R}\tilde{\alpha} = \hbar^2/m^* \approx 0.076/(m^*/m_e)$ nm^2 eV. In the case of InAs $\tilde{\alpha}$ = 40 meV nm with $m^* = 0.023\ m_e$ the desired QD radius should be $\tilde{R} \approx 80$ nm. For the QD' radius $\tilde{R}$ =100 nm the spacing between lowest levels is of the order $\Delta E \approx 5\hbar^2/2m^*\tilde{R}^2 \approx 0.8$ meV. The temperature smearing will be significant if $kT > \Delta E$. In other words, our device could operate with 100% efficiency at $T < 9K$, which is far from the typical temperature values ~100 mK for single-electron tunneling spectroscopy experiments (see, for example, the textbook [42]).

5. Summary

We suggest the mechanism of the z-component spin inversion with the aid of the circular lateral QD that symmetrically coupled to two electrodes. The effective confinement potential of the QD consists of the circular potential well. From our analysis of the ballistic electron transport through the QD with the Rashba SOI, it follows the Kramers degeneracy of the QD levels could lead to the destructive interference of the direct ($\sigma \to \sigma$) spin scattering process, while producing the spin-flip phenomenon. We found that the optimal conditions for the realization of the perfect spin-flip processes is subject to the condition $\tilde{\alpha}\tilde{R} \approx \hbar^2/m^*$. In fact, this condition depends quite weakly on the particular choice of the quantum level. We found that this effect is robust for the QD's states at the temperature less than 9 K.

Author Contributions: Software, investigation, writing–original draft preparation, K.P.; investigation, methodology, visualization, A.P.; conceptualization, validation, investigation, writing–review and editing, supervision, R.N. All authors have read and agreed to the published version of the manuscript.

Funding: This research received no external funding.

Acknowledgments: This research has been supported partially by the guest program of The University of Illes Balears.

Conflicts of Interest: The authors declare no conflict of interest.

Appendix A. Derivation of Effective Hamiltonian on 2D Square Lattice

The Hamiltonian of our scattering system can be presented as follows:

$$\hat{H} = \hat{H}_0 + \hat{V}\,, \tag{A1}$$

where the Hamiltonian $\hat{H}_0$ consists of three terms: two electrodes ($C = L, R$ or $1, 2$) with continuous spectra and a closed substructure with a discrete spectrum:

$$\begin{aligned} \hat{H}_0 &= \hat{H}_L + \hat{H}_B + \hat{H}_R \\ \hat{H}_B &= \sum_n E_n |n\rangle\langle n|, \\ \hat{H}_C &= \int dE E|E,C\rangle\langle E,C|. \end{aligned} \tag{A2}$$

The corresponding eigenstates are normalized:

$$\langle n|m\rangle = \delta_{nm}, \tag{A3}$$

$$\langle E,C|E',C\rangle = \delta(E-E'). \tag{A4}$$

The $\hat{V}$ operator connects the closed substructure ($\hat{H}_B$) with electrodes ($\hat{H}_{L,R}$). The stationary Shrödinger equation for the Hamiltonian $\hat{H}_0$ reads as

$$(E - \hat{H}_0)|\phi\rangle = 0\,, \tag{A5}$$

while we are interested in the solution for the total Hamiltonian (A1)

$$(E - \hat{H}_0)|\psi\rangle = \hat{V}|\psi\rangle\,. \tag{A6}$$

For the energy E different from the eigenvalue of the closed substructure E_n we can define the operator $(E + i\varepsilon - \hat{H})^{-1}$. Consequently, if the outgoing wave boundary condition is adopted, we can transform Equation (A6) to the Lippmann-Schwinger equation

$$|\psi\rangle = |\phi\rangle + \frac{1}{(E - \hat{H}_0)}\hat{V}|\psi\rangle\,. \tag{A7}$$

The formal solution of this equation reads as

$$\hat{F}(E)|\psi\rangle = \left(1 - \frac{1}{(E - \hat{H}_0)}\hat{V}\right)|\psi\rangle = |\phi\rangle\,. \tag{A8}$$

Following Refs. [36,37], with the aid of the basis states (A3) we construct projector operators for each term in the Hamiltonian (A2):

$$\hat{P}_B = \sum_n |n\rangle\langle n|, \tag{A9}$$

$$\hat{P}_C = \int dE|E,C\rangle\langle E,C|, \tag{A10}$$

with the properties

$$\hat{P}_s\hat{P}_{s'} = \delta_{ss'}\hat{P}_s, \tag{A11}$$

$$\hat{P}_s\hat{H}\hat{P}_{s'} = \delta_{ss'}\hat{H}_s, \tag{A12}$$

$$\sum_{s=L,B,R}\hat{P}_s = 1. \tag{A13}$$

By means of the projection operators the Lippmann-Schwinger Equation (A8) transforms to the following form

$$\left(\left(\sum_{s=L,B,R}\hat{P}_s\right)\hat{F}(E)\left(\sum_{s'=L,B,R}\hat{P}_{s'}\right)\right)\left(\sum_{s=L,B,R}\hat{P}_s\right)|\psi\rangle = \left(\sum_{s''=L,B,R}\hat{P}_{s''}\right)|\phi\rangle$$

$$\begin{pmatrix} \hat{P}_L\hat{F}\hat{P}_L & \hat{P}_L\hat{F}\hat{P}_B & \hat{P}_L\hat{F}\hat{P}_R \\ \hat{P}_B\hat{F}\hat{P}_L & \hat{P}_B\hat{F}\hat{P}_B & \hat{P}_B\hat{F}\hat{P}_R \\ \hat{P}_R\hat{F}\hat{P}_L & \hat{P}_R\hat{F}\hat{P}_B & \hat{P}_R\hat{F}\hat{P}_R \end{pmatrix}\begin{pmatrix} |\psi_L\rangle \\ |\psi_B\rangle \\ |\psi_R\rangle \end{pmatrix} = \begin{pmatrix} |\phi_L\rangle \\ |\phi_B\rangle \\ |\phi_R\rangle \end{pmatrix}. \tag{A14}$$

Here, each block of the matrix representation of the operator $F(E)$ (A14) has the following structure

$$\hat{P}_s\hat{F}\hat{P}_{s'} = \hat{P}_s\left(1-\frac{1}{E-\hat{H}_0}\hat{V}\right)\hat{P}_{s'} = \hat{P}_s\left(1-\frac{1}{E-\hat{H}_0}\left(\sum_{s''=L,B,R}\hat{P}_{s''}\right)\hat{V}\right)\hat{P}_{s'} \tag{A15}$$

$$= \left(\delta_{ss'} - \frac{1}{E-\hat{H}_s}\hat{P}_s\hat{V}\hat{P}_{s'}\right).$$

Our closed substructure is subjected to the Dirichlet boundary conditions on junction with electrodes, i.e., the following condition $\hat{P}_s\hat{V}\hat{P}_s = 0$ takes place. In other words, the operator $\hat{V}$ does not affect the structure of the isolated subsystem. As a result, we have

$$\hat{P}_s\hat{F}\hat{P}_s = 1. \tag{A16}$$

Another reasonable assumption is the absence of the direct connection between electrodes:

$$\hat{P}_C\hat{F}\hat{P}_{C'} = \delta_{CC'}. \tag{A17}$$

Taking into account the above arguments, we transform the Lippmann-Schwinger Equation (A14) to the form

$$\begin{pmatrix} 1 & -\frac{1}{E-\hat{H}_L}\hat{P}_L\hat{V}\hat{P}_B & 0 \\ -\frac{1}{E-\hat{H}_B}\hat{P}_B\hat{V}\hat{P}_L & 1 & -\frac{1}{E-\hat{H}_B}\hat{P}_B\hat{V}\hat{P}_R \\ 0 & -\frac{1}{E-\hat{H}_R}\hat{P}_R\hat{V}\hat{P}_B & 1 \end{pmatrix}\begin{pmatrix} |\psi_L\rangle \\ |\psi_B\rangle \\ |\psi_R\rangle \end{pmatrix} = \begin{pmatrix} |\phi_L\rangle \\ |\phi_B\rangle \\ |\phi_R\rangle \end{pmatrix}. \tag{A18}$$

Considering the initial state $|\phi\rangle$ in the form

$$\begin{pmatrix} |\phi_L\rangle \\ |\phi_B\rangle \\ |\phi_R\rangle \end{pmatrix} = \begin{pmatrix} \alpha_L|E,L\rangle \\ 0 \\ \alpha_R|E,R\rangle \end{pmatrix}, \tag{A19}$$

we present the scattering state $|\psi\rangle$ as

$$\begin{pmatrix} |\psi_L\rangle \\ |\psi_B\rangle \\ |\psi_R\rangle \end{pmatrix} = \begin{pmatrix} (\alpha_L r + \alpha_R t')|E,L\rangle \\ |\psi_B\rangle \\ (\alpha_L t + \alpha_R r')|E,R\rangle \end{pmatrix}, \tag{A20}$$

defining the scattering matrix elements

$$S = \begin{pmatrix} r & t' \\ t & r' \end{pmatrix} \tag{A21}$$

To get to the heart of the problem, let us consider, for example, the second equation from (A18), and corresponding wavefunctions $|\psi_L\rangle$, $|\psi_R\rangle$ from Equation (A20):

$$\left(1 - \frac{1}{E-\hat{H}_B}\sum_{C=L,R}\hat{P}_B\hat{V}\hat{P}_C\frac{1}{E-\hat{H}_C}\hat{P}_C\hat{V}\hat{P}_B\right)|\psi_B\rangle = \frac{1}{E-\hat{H}_B}\sum_{C=L,R}\hat{P}_B\hat{V}\hat{P}_C|\phi_C\rangle \tag{A22}$$

Multiplying the both sides of Equation (A22) by $E - \hat{H}_B$, we obtain

$$\left(E - \hat{H}^{Eff}\right)|\psi_B\rangle = \sum_{C=L,R}\hat{P}_B\hat{V}\hat{P}_C|\phi_C\rangle \tag{A23}$$

with the following definition of the effective Hamiltonian

$$\hat{H}^{Eff} = \hat{H}_B + \sum_{C=L,R}\hat{P}_B\hat{V}\hat{P}_C\frac{1}{E-\hat{H}_C}\hat{P}_C\hat{V}\hat{P}_B = \hat{H}_B + \sum_{C=L,R}\hat{X}^C. \tag{A24}$$

The eigenstates of H_B form the natural basis for the effective Hamiltonian. Taking into account this fact, we obtain by means of Equation (A23) the following system

$$\sum_m\left(E - \langle n|\hat{H}^{Eff}|m\rangle\right)\langle m|\psi_B\rangle = \sum_{C=L,R}\langle n|\hat{V}\int dE|E,C\rangle\langle E,C|\phi_C\rangle. \tag{A25}$$

Here, the matrix elements $\langle n|\hat{H}^{Eff}|m\rangle$ have the following structure

$$\langle n|\hat{H}^{Eff}|m\rangle = \delta_{nm}E_n + \sum_{C=L,R}\langle n|\hat{V}\hat{P}_C\frac{1}{E-\hat{H}_C}\hat{P}_C\hat{V}|m\rangle = \delta_{nm}E_n + \sum_{C=L,R}\hat{X}^C_{nm}, \tag{A26}$$

where the matrix elements of $\hat{X}^C$ operator are

$$\langle n|\hat{X}^C|m\rangle = \langle n|\hat{V}\left(\int dE'|E',C\rangle\langle E',C|\right)\frac{1}{E-\hat{H}_C}\left(\int dE''|E'',C\rangle\langle E'',C|\right)\hat{V}|m\rangle. \tag{A27}$$

Using the definition of $\hat{H}_C$ (A2), we can write

$$\hat{X}^C_{nm} = \int dE'\langle n|\hat{V}|E',C\rangle\frac{1}{E-E'}\langle E',C|\hat{V}|m\rangle. \tag{A28}$$

To obtain the matrix elements $\langle n|\hat{V}|E,C\rangle = \langle E,C|\hat{V}|n\rangle^*$ we make the transformation to the tight-binding representation by inserting the resolution of identity $1 = \sum_j |j\rangle\langle j|$ into the definition:

$$\langle n|\hat{V}|E,C\rangle = \langle n|\left(\sum_j|j\rangle\langle j|\right)\hat{V}\left(\sum_{j'}|j'\rangle\langle j'|\right)|E,C\rangle = \sum_{jj'}\langle n|j\rangle\langle j|\hat{V}|j'\rangle\langle j'|E,C\rangle. \tag{A29}$$

To visualize the idea of the tight-binding approach, Figure A1 displays the example of the 2D cavity connected to 1D electrodes. We can see that the matrix elements $\langle j|\hat{V}|j'\rangle = \langle j'|\hat{V}|j\rangle^*$ are nonzero only for $j = (N,1), j' = (N+1,1)$ at the connection to the right continuum and for $j = (0,1), j' = (1,1)$ at the connection to the left continuum. We introduce the coefficients

$$W_{nC} = \sum_{j\in B, j'\in C} \langle n|j\rangle\langle j|\hat{V}|j'\rangle, \tag{A30}$$

$$W_{nR} = \langle n|(N,1)\rangle\langle (N,1)|\hat{V}|(N+1,1)\rangle, \tag{A31}$$

$$W_{nL} = \langle n|(1,1)\rangle\langle (1,1)|\hat{V}|(0,1)\rangle, \tag{A32}$$

that are independent on energy. As a result, the matrix elements (A29) can be written in the form

$$\langle n|\hat{V}|E,C\rangle = W_{nC}\langle j_{in}^C|E,C\rangle, \tag{A33}$$

$$\langle n|\hat{V}|E,L\rangle = W_{nL}\langle (0,1)|E,L\rangle, \tag{A34}$$

$$\langle n|\hat{V}|E,R\rangle = W_{nR}\langle (N+1,1)|E,R\rangle, \tag{A35}$$

where j_{in}^C denotes the beginning of the semi-infinite electrode C. Taking into account the definition Equation (A33), the expression (A28) can be written in the form

$$\hat{X}_{nm}^C = W_{nC}W_{mC}^* \int dE' \frac{|\langle j_{in}^C|E',C\rangle|^2}{E-E'} \tag{A36}$$

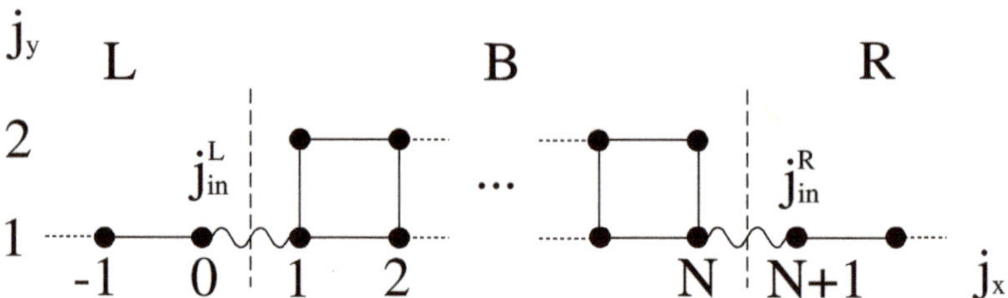

Figure A1. Connection between bounded system B and electrode C. The coupling operator $\langle j|\hat{V}|j'\rangle = \langle j'|\hat{V}|j\rangle^* = \langle (j_x,j_y)|\hat{V}|(j'_x,j'_y)\rangle$ is nonzero only for $j_x = 0$, $j'_y = 1$ and $j_y = j'_y = 1$ at the connection to the left electrode and for $j_x = N$, $j'_y = N+1$ and $j_y = j'_y = 1$ at the connection to the right electrode.

In the tight-binding representation the electron wave function $\langle j_{in}^C|E',C\rangle$ in the electrode C is (see Equation (A50))

$$\langle j_{in}^C|E,C\rangle = \frac{\sin(|k|)}{\sqrt{\pi|\sin(k)|}} = \frac{(1-(E/2)^2)^{1/4}}{\sqrt{\pi}}, \tag{A37}$$

where for the 1D tight-binding model the dispersion $E = -2\cos(k)$ is used. The integration in Equation (A36) over zone from $E = -2$ to $E = 2$ yields [see Equation (A52)] the expression for matrix elements of the operator $\hat{X}^C$

$$\langle n|\hat{X}^C|m\rangle = \frac{W_{nC}W_{mC}^*}{\pi} \int_{-2}^{2} dE' \frac{\sqrt{1-(E'/2)^2}}{E-E'} = -W_{nC}W_{mC}^* e^{ik}. \tag{A38}$$

The scattering matrix elements $S_{CC'}(E)$, describing the transition from a continuum C' to a continuum C at the incident energy E, are [37]

$$S_{CC'}(E) = \delta_{CC'} - 2\pi i\langle E,C|\hat{P}_C\hat{V}\hat{P}_B\left(E-\hat{H}^{eff}(E)\right)^{-1}\hat{P}_B\hat{V}\hat{P}'_C|E,C'\rangle. \tag{A39}$$

Inserting the resolution of identities $1 = \sum_j |j\rangle\langle j|$, $1 = \sum_n |n\rangle\langle n|$, by means of Equations (A19), (A23), (A30) and (A33), we obtain

$$S_{CC'}(E) = \delta_{CC'} - 2\pi i \psi_C^1(E)\psi_{C'}^1(E) \sum_{mn} W_{nC}^* \mathcal{F}_{nm}(E) W_{mC'}. \quad \text{(A40)}$$

Here, the function $\psi_C^1 = \langle j_{\rm in}^C | E, C\rangle$ (see Figure A1) are the wave functions of the semi-infinite electrodes at the QPC.

Appendix A.1. Normalization Constant of the Electrode Eigenfunctions

We consider the electron eigenstate in the electrode as

$$\langle j|E,C\rangle = a(k)\sin(kj) \quad \text{(A41)}$$

with the normalization constant $a(k)$ and $-\pi < k < \pi$. These functions must satisfy the equation

$$\sum_{j=1}^{\infty} \langle j|E,C\rangle\langle E',C|j\rangle = \delta(E-E') \quad \text{(A42)}$$

First, let us rewrite $\delta(E-E')$ using dispersion relation $E = -2\cos(k)$ at $-\pi < k \le \pi$:

$$\begin{aligned} \delta(E-E') &= \delta\left(-2(\cos(k)-\cos(k'))\right) = \delta\left(4\sin\left(\frac{k+k'}{2}\right)\sin\left(\frac{k-k'}{2}\right)\right) \\ &= \frac{\delta\left(\frac{k-k'}{2}\right)}{4|\sqrt{|\sin(k)|}\sqrt{|\sin(k')|}} + \frac{\delta\left(\frac{k+k'}{2}\right)}{4|\sqrt{|\sin(k)|}\sqrt{|\sin(k')|}}. \end{aligned} \quad \text{(A43)}$$

On the other hand, we have

$$\begin{aligned} \sum_{j=1}^{\infty} \langle j|E,C\rangle\langle E',C|j\rangle &= a(k)a^*(k')\sum_{j=1}^{\infty}\sin(kj)\sin(k'j) \\ &= \frac{a(k)a^*(k')}{2}\sum_{j=1}^{\infty}\left(\cos((k-k')j) - \cos((k+k')j)\right). \end{aligned} \quad \text{(A44)}$$

The Lagrange's trigonometric identity

$$\sum_{n=1}^{N}\cos(n\theta) = -\frac{1}{2} + \frac{\sin\left((2N+1)\frac{\theta}{2}\right)}{2\sin\left(\frac{\theta}{2}\right)} \quad \text{(A45)}$$

helps us to write down

$$\frac{1}{2} + \sum_{n=1}^{\infty}\cos(n\theta) = \lim_{N\to\infty}\frac{\sin\left((2N+1)\frac{\theta}{2}\right)}{2\sin\left(\frac{\theta}{2}\right)} = \frac{\theta/2}{2\sin\left(\frac{\theta}{2}\right)}\lim_{N\to\infty}\frac{\sin\left((2N+1)\frac{\theta}{2}\right)}{\theta/2} = \frac{\pi}{2}\delta\left(\frac{\theta}{2}\right). \quad \text{(A46)}$$

Taking into account the above results, we have for Equation (A44)

$$\sum_{j=1}^{\infty} \langle j|E,C\rangle\langle E',C|j\rangle = \frac{a(k)a^*(k')\pi}{4}\left(\delta\left(\frac{k-k'}{2}\right) - \delta\left(\frac{k+k'}{2}\right)\right). \quad \text{(A47)}$$

Finally, combining Equations (A43) and (A47), we have for Equation (A42)

$$\frac{1}{4|\sqrt{|\sin(k)|}\sqrt{|\sin(k')|}}\left(\delta\left(\tfrac{k-k'}{2}\right) + \delta\left(\tfrac{k+k'}{2}\right)\right) = \tfrac{a(k)a^*(k')\pi}{4}\left(\delta\left(\tfrac{k-k'}{2}\right) - \delta\left(\tfrac{k+k'}{2}\right)\right). \quad \text{(A48)}$$

Comparing the left and the right sides of this equation, we obtain

$$a(k) = \frac{\mathrm{sgn}(k)}{\sqrt{\pi|\sin(k)|}}. \tag{A49}$$

Consequently, Equation (A41) becomes

$$\langle j|E,C\rangle = \frac{\sin(|k|j)}{\sqrt{\pi|\sin(k)|}}. \tag{A50}$$

Appendix A.2. Integral Over Zone

We recall that $E = Re(E) + \mathrm{i}\epsilon$ and

$$\begin{aligned}
\lim_{\epsilon\to 0^+}\int_{-a}^{a} dx\frac{f(x)}{x+\mathrm{i}\epsilon} &= \lim_{\epsilon\to 0^+}\int_{-a}^{a} dx\frac{x-\mathrm{i}\epsilon}{x^2+\epsilon^2}f(x) = \lim_{\epsilon\to 0^+}\int_{-a}^{a} dx\left(\frac{xf(x)}{x^2+\epsilon^2} - \mathrm{i}\frac{\epsilon f(x)}{x^2+\epsilon^2}\right) \\
&= P\int_{-a}^{a} dx\frac{f(x)}{x} - \mathrm{i}\pi f(0),
\end{aligned} \tag{A51}$$

As a result, we have

$$\begin{aligned}
\int_{-2}^{2} dE'\frac{\sqrt{1-(E'/2)^2}}{E-E'} &= P\int_{-2}^{2} dE'\frac{\sqrt{1-(E'/2)^2}}{E-E'} - \mathrm{i}\pi\int_{-2}^{2} dE'\delta(E-E')\sqrt{1-(E'/2)^2} \\
&= \pi\left(\frac{E}{2} - \mathrm{i}\sqrt{1-(E/2)^2}\right) \\
&= -\pi e^{\mathrm{i}k}.
\end{aligned} \tag{A52}$$

Here we use the dispersion relation $E = -2\cos(k)$.

References

1. Wolf, S.A. Spintronics: A Spin-Based Electronics Vision for the Future. *Science* **2001**, *294*, 1488–1495. [CrossRef] [PubMed]
2. Žutić, I.; Fabian, J.; Sarma, D.S. Spintronics: Fundamentals and applications. *Rev. Mod. Phys.* **2004**, *76*, 323. [CrossRef]
3. Hanson, R.; Kouwenhoven, L.P.; Petta, J.R.; Tarucha, S.; Vandersypen, L.M.K. Spins in few-electron quantum dots. *Rev. Mod. Phys.* **2007**, *79*, 1217–1265. [CrossRef]
4. Engel, H.A.; Loss, D. Detection of Single Spin Decoherence in a Quantum Dot via Charge Currents. *Phys. Rev. Lett.* **2001**, *86*, 4648–4651. [CrossRef] [PubMed]
5. Dehghan, E.; Khoshnoud, D.S.; Naeimi, A. Logical spin-filtering in a triangular network of quantum nanorings with a Rashba spin-orbit interaction. *Phys. B Condens. Matter* **2018**, *529*, 21–26. [CrossRef]
6. Sattari, F.; Mirershadi, S. Spin-dependent transport properties in strained silicene with extrinsic Rashba spin-orbit interaction. *J. Magn. Magn. Mater.* **2018**, *445*, 6–10. [CrossRef]
7. Pudlak, M.; Nazmitdinov, R.G. Spin-dependent electron transmission across the corrugated graphene. *Physica E* **2020**, *118*, 113846. [CrossRef]
8. Lou, X.; Adelmann, C.; Crooker, S.A.; Garlid, E.S.; Zhang, J.; Reddy, K.S.M.; Flexner, S.D.; Palmstrm, C.J.; Crowell, P.A. Electrical detection of spin transport in lateral ferromagnet-semiconductor devices. *Nat. Phys.* **2007**, *3*, 197. [CrossRef]
9. Dash, S.P.; Sharma, S.; Patel, R.S.; de Jong, M.P.; Jansen, R. Electrical creation of spin polarization in silicon at room temperature. *Nature* **2009**, *462*, 491. [CrossRef]
10. Ciorga, M.; Einwanger, A.; Wurstbauer, U.; Schuh, D.; Wegscheider, W.; Weiss, D. Electrical spin injection and detection in lateral all-semiconductor devices. *Phys. Rev. B* **2009**, *79*, 165321. [CrossRef]
11. Valín-Rodríguez, M.; Nazmitdinov, R.G. Model for spin-orbit effects in two-dimensional semiconductors in magnetic fields. *Phys. Rev. B* **2006**, *73*, 235306. [CrossRef]

12. Fabian, J.; Matos-Abiague, A.; Ertler, C.; Stano, P.; Žutić, I. Semiconductor spintronics. *Acta Phys. Slovaca* **2007**, *57*, 565–907. [CrossRef]
13. Nazmitdinov, R.G.; Pichugin, K.N.; Valín-Rodríguez, M. Spin control in semiconductor quantum wires: Rashba and Dresselhaus interaction. *Phys. Rev. B* **2009**, *79*, 193303. [CrossRef]
14. Schliemann, J. Colloquium: Persistent spin textures in semiconductor nanostructures. *Rev. Mod. Phys.* **2017**, *89*, 011001. [CrossRef]
15. Bychkov, Y.A.; Rashba, E.I. Oscillatory effects and the magnetic susceptibility of carriers in inversion layers. *J. Phys. C Solid State Phys.* **1984**, *17*, 6039–6045. [CrossRef]
16. Rashba, E.I. Electron spin operation by electric fields: Spin dynamics and spin injection. *Phys. Low-Dimens. Syst. Nanostruct.* **2004**, *20*, 189–195. [CrossRef]
17. Nitta, J.; Akazaki, T.; Takayanagi, H.; Enoki, T. Gate Control of Spin-Orbit Interaction in an Inverted $In_{0.53}Ga_{0.47}As/In_{0.52}Al_{0.48}As$ Heterostructure. *Phys. Rev. Lett.* **1997**, *78*, 1335–1338. [CrossRef]
18. Grundler, D. Large Rashba splitting in $InAs$ quantum wells due to electron wave function penetration into the barrier layers. *Phys. Rev. Lett.* **2000**, *84*, 6074–6077. [CrossRef]
19. Hu, C.M.; Nitta, J.; Akazaki, T.; Takayanagi, H.; Osaka, J.; Pfeffer, P.; Zawadzki, W. Zero-field spin splitting in an inverted $In_{0.53}Ga_{0.47}As/In_{0.52}Al_{0.48}As$ heterostructure: Band nonparabolicity influence and the subband dependence. *Phys. Rev. B* **1999**, *60*, 7736–7739. [CrossRef]
20. Lutchyn, R.M.; Sau, J.D.; Das Sarma, S. Majorana Fermions and a Topological Phase Transition in Semiconductor-Superconductor Heterostructures. *Phys. Rev. Lett.* **2010**, *105*, 077001. [CrossRef]
21. Oreg, Y.; Refael, G.; von Oppen, F. Helical Liquids and Majorana Bound States in Quantum Wires. *Phys. Rev. Lett.* **2010**, *105*, 177002. [CrossRef] [PubMed]
22. Mourik, V.; Zuo, K.; Frolov, S.M.; Plissard, S.R.; Bakkers, E.P.A.M.; Kouwenhoven, L.P. Signatures of Majorana Fermions in Hybrid Superconductor-Semiconductor Nanowire Devices. *Science* **2012**, *336*, 1003–1007. [CrossRef] [PubMed]
23. Lucignano, P.; Mezzacapo, A.; Tafuri, F.; Tagliacozzo, A. Advantages of using high-temperature cuprate superconductor heterostructures in the search for Majorana fermions. *Phys. Rev. B* **2012**, *86*, 144513. [CrossRef]
24. Klinovaja, J.; Loss, D. Time-reversal invariant parafermions in interacting Rashba nanowires. *Phys. Rev. B* **2014**, *90*, 045118. [CrossRef]
25. Elliott, S.R.; Franz, M. Colloquium: Majorana fermions in nuclear, particle, and solid-state physics. *Rev. Mod. Phys.* **2015**, *87*, 137–163. [CrossRef]
26. Sarma, S.D.; Freedman, M.; Nayak, C. Majorana zero modes and topological quantum computation. *Quantum Inf.* **2015**, *1*, 15001. [CrossRef]
27. Aasen, D.; Hell, M.; Mishmash, R.V.; Higginbotham, A.; Danon, J.; Leijnse, M.; Jespersen, T.S.; Folk, J.A.; Marcus, C.M.; Flensberg, K.; et al. Milestones Toward Majorana-Based Quantum Computing. *Phys. Rev. X* **2016**, *6*, 031016. [CrossRef]
28. Koga, T.; Nitta, J.; Akazaki, T.; Takayanagi, H. Rashba spin-orbit coupling probed by the weak antilocalization analysis in $InAlAs/InGaAs/InAlAs$ quantum wells as a function of quantum well asymmetry. *Phys. Rev. Lett.* **2002**, *89*, 046801. [CrossRef]
29. Governale, M.; Boese, D.; Zülicke, U.; Schroll, C. Filtering spin with tunnel-coupled electron wave guides. *Phys. Rev. B* **2002**, *65*, 140403. [CrossRef]
30. Ohe, J.; Yamamoto, M.; Ohtsuki, T.; Nitta, J. Mesoscopic Stern-Gerlach spin filter by nonuniform spin-orbit interaction. *Phys. Rev. B* **2005**, *72*, 041308. [CrossRef]
31. Wang, X.F.; Vasilopoulos, P. Spin-dependent transmission in waveguides with periodically modulated strength of the spin-orbit interaction. *App. Phys. Lett.* **2003**, *83*, 940–942. [CrossRef]
32. Sakurai, J.J. *Modern Quantum Mechanics*; Addison-Wesley: Reading, MA, USA, 1994.
33. Heiss, W.D.; Nazmitdinov, R.G. Orbital magnetism in small quantum dots with closed shells. *J. Exp. Theor. Phys. Lett.* **1998**, *68*, 915. [CrossRef]
34. Könemann, J.; Haug, R.J.; Maude, D.K.; Fal'ko, V.I.; Altshuler, B.L. Spin-Orbit Coupling and Anisotropy of Spin Splitting in Quantum Dots. *Phys. Rev. Lett.* **2005**, *94*, 226404. [CrossRef] [PubMed]
35. Valín-Rodríguez, M.; Puente, A.; Serra, L. Role of spin-orbit coupling in the far-infrared absorption of lateral semiconductor dots. *Phys. Rev. B* **2002**, *66*, 045317. [CrossRef]

36. Mahaux, C.; Weidenmüller, H.A. *Shell-Model Approach to Nuclear Reactions;* North-Holland: Amsterdam, The Netherlands, 1969.
37. Dittes, F.M. The decay of quantum systems with a small number of open channels. *Phys. Rep.* **2000**, *339*, 215–316. [CrossRef]
38. Sadreev, A.F.; Rotter, I. S-matrix theory for transmission through billiards in tight-binding approach. *J. Phys. A Math. Gen.* **2003**, *36*, 11413–11433. [CrossRef]
39. Kiselev, A.A.; Kim, K.W. Prohibition of equilibrium spin currents in multiterminal ballistic devices. *Phys. Rev. B* **2005**, *71*, 153315. [CrossRef]
40. Zhai, F.; Xu, H.Q. Symmetry of Spin Transport in Two-Terminal Waveguides with a Spin-Orbital Interaction and Magnetic Field Modulations. *Phys. Rev. Lett.* **2005**, *94*, 246601. [CrossRef]
41. Bulgakov, E.N.; Sadreev, A.F. Spin polarization in quantum dots by radiation field with circular polarization. *J. Exp. Theor. Phys. Lett.* **2001**, *73*, 505. [CrossRef]
42. Ihn, T. *Semiconductor Nanostructures: Quantum States and Electronic Transport;* Oxford University Press: New York, NY, USA, 2010.
43. Ando, T. Quantum point contacts in magnetic fields. *Phys. Rev. B* **1991**, *44*, 8017–8027. [CrossRef]
44. Nazmitdinov, R.G.; Pichugin, K.N.; Rotter, I.; Šeba, P. Whispering gallery modes in open quantum billiards. *Phys. Rev. E* **2001**, *64*, 056214. [CrossRef] [PubMed]
45. Nazmitdinov, R.G.; Pichugin, K.N.; Rotter, I.; Šeba, P. Conductance of open quantum billiards and classical trajectories. *Phys. Rev. B* **2002**, *66*, 085322. [CrossRef]
46. Sandalov, I.; Nazmitdinov, R.G. Shell effects in nonlinear magnetotransport through small quantum dots. *Phys. Rev. B* **2007**, *75*, 075315. [CrossRef]

Article

On Symmetry Properties of The Corrugated Graphene System

Mihal Pudlak [1], Jan Smotlacha [2] and Rashid Nazmitdinov [2,3,*]

[1] Institute of Experimental Physics, Watsonova 47, 04001 Košice, Slovakia
[2] BLTP, Joint Institute for Nuclear Research, 141980 Dubna, Moscow Region, Russia
[3] BLTP, Dubna University, 141982 Dubna, Moscow Region, Russia
* Correspondence: rashid@theor.jinr.ru

Received: 20 February 2020; Accepted: 24 March 2020; Published: 3 April 2020

Abstract: The properties of the ballistic electron transport through a corrugated graphene system are analysed from the symmetry point of view. The corrugated system is modelled by a curved surface (an arc of a circle) connected from both sides to flat sheets. The spin–orbit couplings, induced by the curvature, give rise to equivalence between the transmission (reflection) probabilities of the transmitted (reflected) electrons with the opposite spin polarisation, incoming from opposite system sides. We find two integrals of motion that explain the chiral electron transport in the considered system.

Keywords: graphene; ripple; transport; symmetry

---

1. Introduction

It appears that graphene possesses a remarkable stretchability. For example, the DFT and molecular dynamics simulations predict that it can be stretched up to about 20–30%, without being damaged [1]. The experimental measurements demonstrate a good agreement with the theoretical estimations, while the robust engineering results indicate on sample-wide elastic strain ∼6% [2]. Evidently, transforming the flat surface to the curved one, one creates the strain that affects the graphene properties. This fact suggests that, by altering the stretchability, one might tune electronic and transport properties of the graphene sheet.

Recent experimental techniques enable demonstrating evidently a spatial variation of graphene and its direct consequences. For example, ripples can be formed by means of the electrostatic manipulation without any change of doping [3]. Periodically rippled graphene can be fabricated by the epitaxial technique (e.g., [4]). In this case, in contrast to free-standing graphene, a strong modification of the electronic structure of graphene is observed, which gives rise to localised phonon [5] and plasmon [6] modes. Periodic nanoripples can be created as well by means of the chemical vapour deposition [7]. It is found that ripples, acting as potential barriers, yield the localisation of charged carriers [8]. The potential surface variations could reach the figure of 20–30 meV. Similar independent prediction has been done in the study of Klein collimation by the rippled graphene superlattice [9]. In this model, the hybridisation between the π- and σ-orbitals creates the potential barrier between the flat and curved graphene pieces. The barrier value could reach $\Delta\varepsilon \approx 24$ meV at the ripple radius $R = 12$Å. This fact provides the confidence in the vitality and the validity of our model (outlined in Section 2) and following analyses its symmetry properties, presented in our paper.

Indeed, the lattice deformation changes the distance between ions, p_z orbital orientation, and is leading to shift of the on-site energies of p_z orbitals. This affects the effective Dirac equation that could simulate the low energy electron states as a result of a deformation-induced gauge field [10]. The surface curvature modulates also the hopping parameter in the tight-binding approach [10,11]. Moreover, it enhances as well the effect of the spin–orbit coupling [12], usually neglected in flat carbon-based systems.

Recently, it is predicted that rippled graphene could lead to the spin selectivity effect for the ballistic electrons [13,14] in virtue of the curvature-induced spin–orbit couplings (see details in [12,13,15–17]). As a result, it is shown that at the particular energy values the ballistic electrons with the one spin polarisation can travel through periodically repeated ripples without any reflection. At the same time, electrons with the opposite spin polarisation are fully reflected. Once we change the flow direction through the considered system, the situation becomes inverse. It is noteworthy that different experiments of a spin selective electron transmission through biomolecules has been discussed recently in Ref. [18]. In this review, the authors claimed that this phenomenon implies that chirality and spin may play an important role in biology.

In mesoscopic systems, symmetries are key points that allow to illuminate essential features of finite quantum systems (e.g., [19,20]). The basic goal of this paper is to elucidate the above discussed phenomenon from point of view of the symmetry properties of the considered system.

2. Basic Physics of The Corrugated Graphene

In our consideration, the corrugated graphene structure consists of a rippled graphene connected to two flat graphene sheets (see Figure 1).

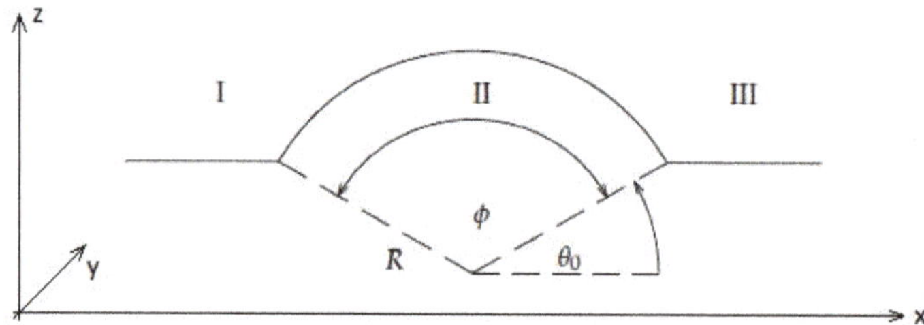

Figure 1. The corrugated graphene system. There are two flat surfaces: Region I, defined in the intervals $-\infty < x < -R\cos\theta_0$; and Region III, defined in the intervals $R\cos\theta_0 < x < \infty$. A ripple is modelled by an arc of a circle (Region II) of radius R, defined as $-R\cos\theta_0 < x < R\cos\theta_0$. At $\theta_0 = 0$, the ripple is a half of the nanotube, while at $\theta_0 = \pi/2$ the ripple does not exist. The angle $\phi = \pi - 2\theta_0$. Here, we have $-\infty < y < \infty$. We keep the translational invariance along the y-axis, which is chosen as the symmetry and the quantisation axis.

For analysis of the curved graphene surface, we recapitulate the major results [17] obtained for armchair CNTs. In this case, only the interaction between nearest neighbour atoms is considered. The analysis is done in an effective mass approximation for the point K, in the vicinity of the Fermi level $E = 0$. A similar approach can be applied for K' point.

The Hamiltonian of the nanotube has the following form in the effective mass approximation [12]

$$H_r = \gamma(\tau_x \hat{\pi}_x + \tau_y \hat{\pi}_y) \otimes I - \lambda_y \tau_y \otimes \sigma_x(\vec{r}) - \xi_x \tau_x \otimes \sigma_y \,. \tag{1}$$

Here, the operators $\hat{\pi}_x = -i\frac{\partial}{R\partial\theta}$, $\hat{\pi}_y = -i\frac{\partial}{\partial y}$, $\sigma_x(\vec{r}) = \sigma_x \cos\theta - \sigma_z \sin\theta$, and $\xi_x = 2\delta\gamma p/R$, $\lambda_y = \delta\gamma'/4R$. The Pauli matrices $\sigma_{x,y}$ act in the spin space. The matrices $\tau_{x,y}$ act on the sublattice degree of freedom. The Pauli matrix τ_i is called the "pseudospin", to distinguish it from the real electron spin.

The following notations are used: $\gamma = -\sqrt{3}V^{\pi}_{pp}a/2 = \sqrt{3}\gamma_0 a/2$, $\gamma' = \sqrt{3}(V^{\sigma}_{pp} - V^{\pi}_{pp})a/2 = \gamma_1 a$, $p = 1 - 3\gamma'/8\gamma$; where V^{σ}_{pp} and V^{π}_{pp} are the transfer integrals for σ- and π-orbitals, respectively, in a flat graphene. The distance d between atoms in the unit cell determines the length of the primitive translation vector $a = \sqrt{3}d \simeq 2.46$ Å. For numerical illustration, we assume that $\gamma_0 = -V^{\pi}_{pp} \approx 3$ eV and $\gamma_1 \approx 8$ eV (see, e.g., [12]). The intrinsic source of the spin–orbit coupling $\delta = \Delta/3\epsilon_{\pi\sigma}$ is defined by means of the quantity

$$\Delta = i\frac{3\hbar}{4m^2c^2}\left\langle x|\frac{\partial V}{\partial x}p_y - \frac{\partial V}{\partial y}p_x|y\right\rangle, \tag{2}$$

where V is the atomic potential and $\epsilon_{\pi\sigma} = \epsilon_{2p}^{\pi} - \epsilon_{2p}^{\sigma}$. Here, the energies ϵ_{2p}^{π} and ϵ_{2p}^{σ} denote the energies of π- and σ-orbitals orbitals, respectively. We recall that σ-orbitals are localised between carbon atoms, while π-orbitals are directed perpendicular to the nanotube surface.

With the aid of the method discussed in [13,17], one obtains the eigenvalues of the Hamiltonian in Equation (1)

$$E = \kappa E_{\pm}, \quad \kappa = \pm 1, \tag{3}$$

where $\kappa = +1(-1)$ is associated with the conductance (valence) band, and the energies $E_{\pm}$ are defined as

$$E_{\pm} = \sqrt{t_m^2 + t_y^2 + \lambda_y^2 + \lambda_x^2 \pm 2\sqrt{\lambda_x^2\left(t_m^2 + \lambda_y^2\right) + t_y^2\lambda_y^2}}. \tag{4}$$

Here, $t_m = m\gamma/R$, $t_y = \gamma k_y$, $\lambda_x = \gamma(1/2 + 2\delta p)/R$, and m is a magnetic quantum number (see details in [13,17]). Due to the curvature-induced spin–orbit coupling spin is no anymore a good quantum number. The eigenstates of Equation (1) are characterised by a quantum number $s = \pm 1$, and have the following form

$$\Phi_{m,k_y}^{s=\pm 1}(\theta, y) = e^{im\theta}e^{ik_y y}N_{\pm}\begin{bmatrix} \kappa(\cos\theta/2A_{\pm} - \sin\theta/2B_{\pm}) \\ \kappa(\sin\theta/2A_{\pm} + \cos\theta/2B_{\pm}) \\ \cos\theta/2C_{\pm} - \sin\theta/2D_{\pm} \\ \sin\theta/2C_{\pm} + \cos\theta/2D_{\pm} \end{bmatrix}, \tag{5}$$

where

$$D_{\pm} = \frac{\lambda_y\lambda_x \pm \sqrt{\lambda_x^2(t_m^2 + \lambda_y^2) + t_y^2\lambda_y^2}}{it_m\lambda_x - t_y\lambda_y}, \tag{6}$$

$$A_{\pm} = \frac{1}{E_{\pm}}\left(t_m - it_y + i(\lambda_y + \lambda_x)D_{\pm}\right), \tag{7}$$

$$B_{\pm} = \frac{1}{E_{\pm}}\left[(t_m - it_y)D_{\pm} + i\left(\lambda_y - \lambda_x\right)\right], \tag{8}$$

$$C_{\pm} = 1, \tag{9}$$

and $N_{\pm}$ is a normalisation constant

$$N_{\pm}^2 = \frac{t_y^2\lambda_y^2 + t_m^2\lambda_x^2}{2\left[\left(\lambda_y\lambda_x \pm \sqrt{\lambda_x^2\left(t_m^2 + \lambda_y^2\right) + t_y^2\lambda_y^2}\right)^2 + t_y^2\lambda_y^2 + t_m^2\lambda_x^2\right]}. \tag{10}$$

Generally, the relations $|A_{\pm}| = |D_{\pm}|$ and $|B_{\pm}| = |C_{\pm}|$ are fulfilled.

The solution for a flat graphene is well known (e.g., [21,22]). Near the center of each valley (the point K or K') electron dispersion is determined by the Dirac-type Hamiltonian

$$H_f = \gamma(\tau_x\hat{k}_x + \tau_y\hat{k}_y) \otimes I, \tag{11}$$

where again the Pauli matrices $\tau_{x,y}$ act on the sublattice degrees of freedom, I is 2×2 unity matrix acting in the spin space, with $\hat{\boldsymbol{k}} = -i(\partial/\partial_x, \partial/\partial_y)$. The eigenvalues and eigenstates of the flat graphene Hamiltonian are

$$E = \kappa\gamma\sqrt{k_x^2 + k_y^2}, \kappa = \pm 1, \tag{12}$$

$$\Psi_k^{\sigma}(x,y) = \frac{1}{2}\begin{pmatrix} \kappa e^{-i\varphi} \\ 1 \end{pmatrix} \otimes \begin{pmatrix} 1 \\ \sigma i \end{pmatrix} e^{i\mathbf{k}\cdot\mathbf{r}}, \qquad \sigma = \pm, \tag{13}$$

where $e^{-i\varphi} = (k_x - ik_y)/\sqrt{k_x^2 + k_y^2}$, $\mathbf{k} = (k_x, k_y)$, $\mathbf{r} = (x, y)$, and $k = \sqrt{k_x^2 + k_y^2}$. The spin degeneracy is taken into account. In our consideration, the states with the spin up and down are the eigenstates of the operator σ_y. The above describe solutions are used to calculate the electron transmission through the corrugated graphene. We recall that, in rippled graphene, the symmetries related to the spin degree of freedom and to the angular momentum are not conserved [13,17].

3. Symmetries

To illuminate specific symmetries of our system, we have to identify the corresponding operators that act within one valley only. Evidently, these operators should act on the A an B sublattices of the honeycomb lattice.

3.1. The Operator $\hat{S}_t = \tau_y \otimes i\sigma_y C$

The spin–orbit coupling implies that one of the symmetries should be related to the time-reverse symmetry operator $\mathcal{T} = i\sigma_y C$ with C the operator of complex conjugation (see, e.g., [23,24]). However, the operator $\mathcal{T}$ does not commute neither with the Hamiltonian of the flat graphene in Equation (11) or with the Hamiltonian of the ripple in Equation (1). Taking into account the "pseudospin" degree of freedom, we observe consequently that the operator $\hat{S}_t = \tau_y \otimes \mathcal{T}$ commutes with the both Hamiltonians

$$[\hat{S}_t, H_f] = 0, \quad [\hat{S}_t, H_r] = 0. \tag{14}$$

Let us investigate the properties of this operator with respect to the eigenfunctions of the Hamiltonian H_f(H_r) described in Section 2. For the sake of convenience, we introduce the following equivalent definitions: $\Psi_k^{\sigma}(x,y) \equiv \langle \mathbf{r}|\sigma,\mathbf{k}\rangle$, $\sigma = (+/-) \Leftrightarrow \sigma = (\uparrow / \downarrow)$. As a result, for the wave function, associated with the flat graphene sheet, we have

$$\left.\begin{aligned} \hat{S}_t|\uparrow \pm\mathbf{k}\rangle &= \pm e^{i\varphi}|\downarrow \mp\mathbf{k}\rangle \\ \hat{S}_t|\downarrow \pm\mathbf{k}\rangle &= \mp e^{i\varphi}|\uparrow \mp\mathbf{k}\rangle \end{aligned}\right\} \Rightarrow S_t^2 = 1, \tag{15}$$

i.e., the operator $\hat{S}_t$ has two eigenvalues $+1$ and -1. Since the phase $e^{i\varphi}$ does not affect our results, hereafter, we omit it in our calculations.

Any ket $|\psi\rangle$ can be expressed as

$$|\psi\rangle = \frac{1}{2}\left[(1+\hat{S}_t)\,|\psi\rangle + (1-\hat{S}_t)\,|\psi\rangle\right] = |\psi_+\rangle + |\psi_-\rangle, \tag{16}$$

where

$$|\psi_+\rangle = \frac{1+\hat{S}_t}{2}|\psi\rangle, \quad |\psi_-\rangle = \frac{1-\hat{S}_t}{2}|\psi\rangle, \tag{17}$$

with the property $\hat{S}_t|\psi_\pm\rangle = \pm|\psi_\pm\rangle$. In our particular case, we can form four types of the wave functions:

$$|\psi_\pm\rangle = \frac{1\pm\hat{S}_t}{2}|\uparrow +\mathbf{k}\rangle = \frac{1}{2}(|\uparrow +\mathbf{k}\rangle \pm |\downarrow -\mathbf{k}\rangle), \tag{18}$$

$$|\phi_\pm\rangle = \frac{1\pm\hat{S}_t}{2}|\downarrow +\mathbf{k}\rangle = \frac{1}{2}(|\downarrow +\mathbf{k}\rangle \mp |\uparrow -\mathbf{k}\rangle). \tag{19}$$

Thus, for the plane graphene sheet the full set of the operator $\hat{S}_t$ consists of the wave functions in Equations (18) and (19). These wave functions contain the equal mixture of the spin up and down states, associated with electrons that move in opposite directions of our structure.

Since for the curved graphene we use the eigenstates of the CNT, a complete set of quantum numbers consists of the magnetic quantum number m and the wave number k_y. In this case, we introduce the equivalent notations $\Phi^s_{m,k_y}(\theta) \equiv \langle\theta|s,m,k_y\rangle$. The action of the operator $\hat{S}_t$ on the wave function in Equation (5), associated with the nanotube, yields

$$\left.\begin{array}{rl} \hat{S}_t|s=+,\pm m,\pm k_y\rangle = & iB^*_+|s=+,\mp m,\mp k_y\rangle \\ \hat{S}_t|s=-,\pm m,\pm k_y\rangle = & iB^*_-|s=-,\mp m,\mp k_y\rangle \end{array}\right\} \Rightarrow S_t^2 = 1\,, \tag{20}$$

since $|B_\pm|^2 = 1$ [see Equation (8)]. Applying the same arguments [see Equations (16) and (17)], we obtain

$$|\chi_\pm\rangle = \frac{1\pm\hat{S}_t}{2}|s=+,m,k_y\rangle = \frac{1}{2}(|s=+,m,k_y\rangle \pm iB^*_+|s=+,-m,-k_y\rangle)\,, \tag{21}$$

$$|\varphi_\pm\rangle = \frac{1\pm\hat{S}_t}{2}|s=-,m,k_y\rangle = \frac{1}{2}(|s=-,m,k_y\rangle \pm iB^*_-|s=-,-m,-k_y\rangle)\,. \tag{22}$$

Having the symmetry properties of the wave functions of the different elements of our structure, we are ready to elucidate the symmetry properties of the transmission and reflection probabilities. Equating the wave functions Ψ, Φ at points $x=-R\cos\theta_0$ (the boundary between Regions I and II), and $x=R\cos\theta_0$ (the boundary between Regions II and III), we define the unknown reflection and transmission amplitudes r^β_α, t^β_α (α, $\beta = \uparrow, \downarrow$). In these amplitudes, the upper (bottom) index denotes the spin polarisation of the incoming (outgoing) (reflected and transmitted) electron. For the sake of discussion, in our model, electrons move from the left to the right direction. As a result, at the boundary between Regions I and II for the electron, which moves from the left side with the spin up polarisation, we have:

$$\Psi^+_{k_x,k_y}(x) + r_L(\phi)^\uparrow_\uparrow\Psi^+_{-k_x,k_y}(x) + r_L(\phi)^\uparrow_\downarrow\Psi^-_{-k_x,k_y}(x) = \tag{23}$$

$$a_+\Phi^+_{m_+,k_y}(\theta) + b_+\Phi^+_{-m_+,k_y}(\theta) + a_-\Phi^-_{m_-,k_y}(\theta) + b_-\Phi^-_{-m_-,k_y}(\theta)\,; \quad x=-R\cos\theta_0\,, \theta=-\phi/2\,.$$

The unknown coefficients $a_{(+/-)}$, $b_{(+/-)}$ characterise transport properties of the electron transfer across the rippled region. The wave numbers m_+ and m_- are determined by the equation

$$m_\pm = \frac{R}{\gamma}\sqrt{E^2 - t_y^2 - \lambda_y^2 + \lambda_x^2 \mp 2\sqrt{\lambda_x^2(E^2-t_y^2) + \lambda_y^2 t_y^2}}\,, \tag{24}$$

where $E=\gamma k$ ($E>0$) is the electron energy. At the boundary between Regions II and III, we have the following conditions

$$a_+\Phi^+_{m_+,k_y}(\theta) + b_+\Phi^+_{-m_+,k_y}(\theta) + a_-\Phi^-_{m_-,k_y}(\theta) + b_-\Phi^-_{-m_-,k_y}(\theta) = \tag{25}$$

$$t_L(\phi)^\uparrow_\uparrow\Psi^+_k(x) + t_L(\phi)^\uparrow_\downarrow\Psi^-_k(x);\ x=R\cos\theta_0,\ \theta=\phi/2.$$

The system of Equations (23) and (25) determine the coefficients $r_L(\phi)^\beta_\alpha$ and $t_L(\phi)^\beta_\alpha$. Acting by the operator $\hat{S}_t$ on these equations, we obtain:

- at the boundary between Region I and II

$$\begin{aligned} &\Psi^-_{-k_x,-k_y}(x) - r_L(\phi)^{\uparrow *}_\uparrow\Psi^-_{k_x,-k_y}(x) + r_L(\phi)^{\uparrow *}_\downarrow\Psi^+_{k_x,-k_y}(x) = \\ &ia^*_+B^*_+(m_+,k_y)\Phi^+_{-m_+,-k_y}(\theta) + ib^*_+B^*_+(-m_+,k_y)\Phi^+_{m_+,-k_y}(\theta) + \\ &+ia^*_-B^*_-(m_-,k_y)\Phi^-_{-m_-,-k_y}(\theta) + ib^*_-B^*_-(-m_-,k_y)\Phi^-_{m_-,-k_y}(\theta)\,; \quad x=R\cos\theta_0,\ \theta=-\phi/2; \end{aligned} \tag{26}$$

- at the boundary between Region II and III

$$
\begin{aligned}
& ia_{+}^{*}B_{+}^{*}(m_{+},k_y)\Phi^{+}_{-m_{+},-k_y}(\theta)+ib_{+}^{*}B_{+}^{*}(-m_{+},k_y)\Phi^{+}_{m_{+},-k_y}(\theta)+\\
& +ia_{-}^{*}B_{-}^{*}(m_{-},k_y)\Phi^{-}_{-m_{-},-k_y}(\theta)+ib_{-}^{*}B_{-}^{*}(-m_{-},k_y)\Phi^{-}_{m_{-},-k_y}(\theta)=\\
& t_L(\phi)^{\uparrow *}_{\uparrow}\Psi^{-}_{-k}(x)-t_L(\phi)^{\uparrow *}_{\downarrow}\Psi^{+}_{-k}(x)\,; \quad x=-R\cos\theta_0,\ \theta=\phi/2\,.
\end{aligned}
\tag{27}
$$

Next, we consider the equations that determine the coefficients $r_R(\phi)^{\beta}_{\alpha}$ and $t_R(\phi)^{\beta}_{\alpha}$ for the electron that moves from the right to the left side with the down spin polarisation. Namely, we have:

- at the boundary between Regions II and III

$$
\begin{aligned}
& \Psi^{-}_{-k_x,-k_y}(x)+r_R(\phi)^{\downarrow}_{\downarrow}\Psi^{-}_{k_x,-k_y}(x)+r_R(\phi)^{\downarrow}_{\uparrow}\Psi^{+}_{k_x,-k_y}(x)=\\
& \tilde{a}_{+}\Phi^{+}_{m_{+},-k_y}(\theta)+\tilde{b}_{+}\Phi^{+}_{-m_{+},-k_y}(\theta)+\tilde{a}_{-}\Phi^{-}_{m_{-},-k_y}(\theta)+\tilde{b}_{-}\Phi^{-}_{-m_{-},-k_y}(\theta)\,; \quad x=R\cos\theta_0\,,\theta=\phi/2;
\end{aligned}
\tag{28}
$$

- at the boundary between Regions I and II

$$
\begin{aligned}
& \tilde{a}_{+}\Phi^{+}_{m_{+},-k_y}(\theta)+\tilde{b}_{+}\Phi^{+}_{-m_{+},-k_y}(\theta)+\tilde{a}_{-}\Phi^{-}_{m_{-},-k_y}(\theta)+\tilde{b}_{-}\Phi^{-}_{-m_{-},-k_y}(\theta)=\\
& t_R(\phi)^{\downarrow}_{\downarrow}\Psi^{-}_{-k_x,-k_y}(x)+t_R(\phi)^{\downarrow}_{\uparrow}\Psi^{+}_{-k_x,-k_y}(x)\,; \quad x=-R\cos\theta_0\,, \quad \theta=-\phi/2\,.
\end{aligned}
\tag{29}
$$

From the comparison of Equations (26) and (27) with Equations (28) and (29), it follows that the coefficients $r_R(\phi)^{\beta}_{\alpha}$ and $t_R(\phi)^{\beta}_{\alpha}$ of Equations (28) and (29) can be expressed in the following form

$$
r_R(\phi)^{\downarrow}_{\downarrow}=-r_L(-\phi)^{\uparrow *}_{\uparrow}\,; \quad r_R(\phi)^{\downarrow}_{\uparrow}=r_L(-\phi)^{\uparrow *}_{\downarrow}\,;
\tag{30}
$$

and

$$
t_R(\phi)^{\downarrow}_{\downarrow}=-t_L(-\phi)^{\uparrow *}_{\uparrow}\,; \quad t_R(\phi)^{\downarrow}_{\uparrow}=-t_L(-\phi)^{\uparrow *}_{\downarrow}\,.
\tag{31}
$$

As a result, we obtain for the reflection probabilities

$$
|r_R(\phi)^{\downarrow}_{\downarrow}|^2=|r_L(-\phi)^{\uparrow}_{\uparrow}|^2; \quad |r_R(\phi)^{\downarrow}_{\uparrow}|^2=|r_L(-\phi)^{\uparrow}_{\downarrow}|^2,
\tag{32}
$$

while for the transmission probabilities we have

$$
|t_R(\phi)^{\downarrow}_{\downarrow}|^2=|t_L(-\phi)^{\uparrow}_{\uparrow}|^2; \quad |t_R(\phi)^{\downarrow}_{\uparrow}|^2=|t_L(-\phi)^{\uparrow}_{\downarrow}|^2.
\tag{33}
$$

We have similar probabilities for the electron motion with the spin up polarisation from the right to the left side

$$
|r_R(\phi)^{\uparrow}_{\uparrow}|^2=|r_L(-\phi)^{\downarrow}_{\downarrow}|^2; \quad |r_R(\phi)^{\uparrow}_{\downarrow}|^2=|r_L(-\phi)^{\downarrow}_{\uparrow}|^2,
\tag{34}
$$

and, consequently, for the transmission probabilities

$$
|t_R(\phi)^{\uparrow}_{\uparrow}|^2=|t_L(-\phi)^{\downarrow}_{\downarrow}|^2; \quad |t_R(\phi)^{\uparrow}_{\downarrow}|^2=|t_L(-\phi)^{\downarrow}_{\uparrow}|^2.
\tag{35}
$$

From these results, it follows that the operator $\hat{S}_t$ does not involve the other valley, i.e., QED. However, it interchanges the sign of the vector $\boldsymbol{k}$ and the electron spin polarisation in the flat part of the considered graphene structure. In the rippled graphene region, it changes $m_{\pm}\to -m_{\pm}$ and the sign of the k_y component: $k_y\to -k_y$. We conclude that the operator S_t acts like a time-reversal operator in the single valley.

3.2. The Operator $\hat{S}_{ch}=\tau_x\otimes\sigma_y$

We recall that it was found in Refs. [13,14] that the electron scattering in the superlattice, created by periodically repeated elements, has a curious behaviour. Note that in this way we mimic periodically rippled graphene as a set curvatures between flat graphene areas (e.g., see Figure 1 and the corresponding discussion in Ref. [14]). One element of the superlattice gives rise to the dominance of the electron transmission with a certain spin polarisation. While this effect is small for a few ripples, it defines the perfect transmission for electrons with the one spin polarisation and the perfect reflection for electrons with the opposite spin polarisation in the case of hundreds of ripples. As a result, we find the optimal angle values for the ripple that ensures the perfect transmission at relatively large number of elements $N \gg 1$ in the superlattice. Note that the transmission depends on the ripple radius and the spin–orbit coupling strengths (see details in [14]). It is noteworthy that transmitted electrons with different spin orientation choose different channels characterised by the quantum number $s = \pm 1$ [13]. This fact implies the existence of the additional symmetry that is fundamental for this feature.

To give an insight into this symmetry, we consider the case $k_y = 0$, when the spectrum in Equation (4) and the eigenspinors are particular simple. In this case, Equations (3) and (4) are reduced to the form

$$E = \pm(\mathcal{D} \pm \lambda_x), \tag{36}$$

where $\mathcal{D} = \sqrt{\lambda_y^2 + t_m^2}$ and $t_m = \frac{\gamma}{R} m$. For the rippled (arc) piece (see Figure 1), there are four eigenenergies

$$E = \begin{cases} \mathcal{E}_1 = \lambda_x + \mathcal{D} \\ \mathcal{E}_2 = \lambda_x - \mathcal{D} \\ \mathcal{E}_3 = -\lambda_x + \mathcal{D} \\ \mathcal{E}_4 = -\lambda_x - \mathcal{D} \end{cases} \tag{37}$$

The connection between the energy and the quantum number m can be formulated in the form

$$m \Rightarrow m_s = \pm \frac{R}{\gamma} \sqrt{(sE - \lambda_x)^2 - \lambda_y^2}, \quad s = \pm 1, \tag{38}$$

that determines four possible values of the quantum number m. Here, we introduce the additional quantum number s that characterises our eigenstates. Note that in Equation (38) the sign of the quantum number s depends on the sign of the energy E. In particular, the following relations take place

$$E > 0 :\Longleftrightarrow \begin{cases} s = +1, & \mathcal{E}_1 > 0 \\ s = +1, & \text{if } \mathcal{E}_2 > 0 \\ s = -1, & \text{if } \mathcal{E}_3 > 0 \end{cases} \tag{39}$$

On the other hand, at $E < 0$, the branches $\mathcal{E}_3$ and $\mathcal{E}_4$ have the quantum number $s = +1$, while the branch $\mathcal{E}_2$ is characterised by the quantum number $s = -1$.

At a fixed electron energy $E(E > 0)$, Equations (37), (38), and (39), establish the connection between the energies $\mathcal{E}_j (j = 1, 2, 3)$ and the magnetic quantum number m with the quantum number s

$$\mathcal{E}_1, \mathcal{E}_2 \to \pm t_{m_{+1}} \Longleftrightarrow s = +1, \tag{40}$$

$$\mathcal{E}_3 \to \pm t_{m_{-1}} \Longleftrightarrow s = -1. \tag{41}$$

We recall that the angular momentum is not conserved. As a result, the eigenfunction is the mixture of the eigenfunctions in Equation (5) at a given energy (see Figure 2). The energy branches in Equation (37) with the same quantum number s repel each other, while there is a crossing of the branches with the different s (see Figures 2 and 3). The anticrossings yield the energy gaps $= 2\lambda_y$ indicated by the arrows (see the insets in Figures 2 and 3). As a result, the energy gaps give rise to evanescent modes at energies $\lambda_x - \lambda_y < |E| < \lambda_x + \lambda_y$ in our system.

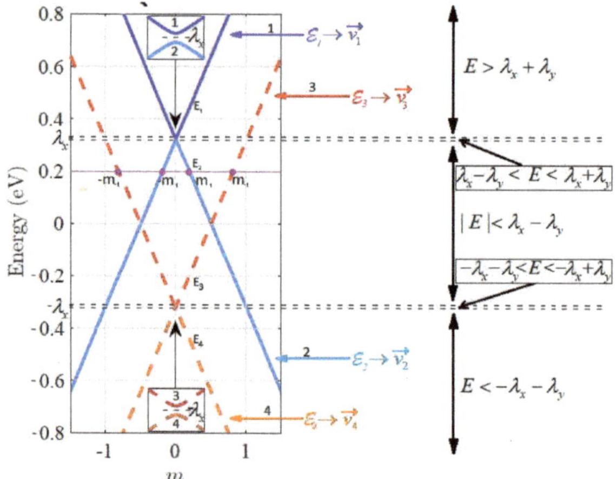

Figure 2. The spectrum in Equation (4) ($k_y = 0$) versus the magnetic quantum number m. The non-quantised values $\pm m_{s=\pm1}$ at the energy $E = 0.2$ eV (thin horizontal line that mimics the Fermi energy) are indicated at the crossing of the energy branches with different s. Symbols E_1, E_2, E_3, E_4 are used to guide the eyes on the formal solutions (straight lines) defined by Equation (36), irrespective of the sign of the quantum number s. In contrast, there is anticrossing of the energy branches with the same quantum number $s = +1$ for the pair $(\mathcal{E}_1, \mathcal{E}_2)$ at $E > 0$. Similar anticrossing occurs at $E < 0$, when the pair $(\mathcal{E}_3, \mathcal{E}_4)$ has the same quantum number $s = +1$. These anticrossings are caused by the term λ_y in the Hamiltonian in Equation (1), which creates the energy gaps $2\lambda_y$ near the energy $E = \pm\lambda_x$ (see [13,17]). The following parameters are used: $R = 10$ Å, $\delta = 0.01$, $p = 0.1$, $\gamma = (4.5 \cdot 1.42)$ eV$\cdot$ Å, $\gamma' = \frac{8}{3}\gamma$. These parameters define the values of the spin–orbit coupling strengths: $\lambda_y = \delta\gamma'/4R = 0.0043$ eV, $\lambda_x = \gamma(1/2 + 2\delta p)/R = 0.32$ eV (see Section 2).

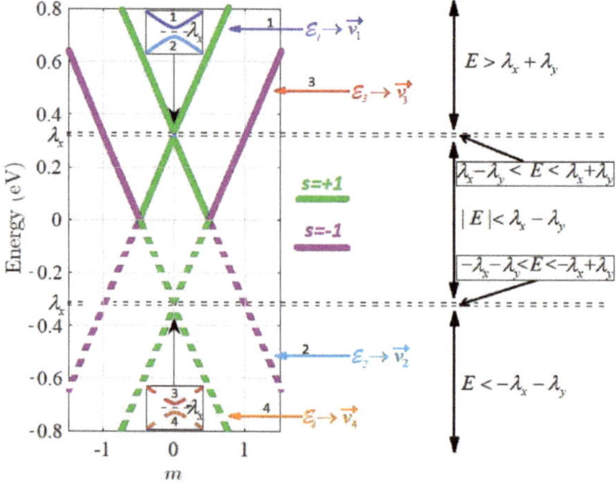

Figure 3. The same as in Figure 2. Solid lines are associated with the positive energy states ($E > 0$), while the negative energies are denoted by dashed lines (see text). Once the energy changes the sign, it affects the sign of the corresponding quantum number s. There are anticrossing of the energy branches with the same quantum number $s = +1$ for the pair $(\mathcal{E}_1, \mathcal{E}_2)$ at $E > 0$ and for the pair $(\mathcal{E}_3, \mathcal{E}_4)$ at $E < 0$.

At $k_y = 0$, the eigenvectors in Equation (5) of the Hamiltonian in Equation (1) transform to the following form (see details in [13,17]):

$$\Phi_{j,m}(\theta) = \left(I \otimes e^{-i\frac{\sigma_y}{2}\theta}\right) \vec{v}_j(m)\ e^{im\theta}\,, j = 1, ..4\,. \tag{42}$$

Here

$$\vec{v}_1 = \begin{pmatrix} -it_m \\ -A_- \\ -iA_- \\ -t_m \end{pmatrix}, \vec{v}_2 = \begin{pmatrix} -it_m \\ A_+ \\ iA_+ \\ -t_m \end{pmatrix}, \vec{v}_3 = \begin{pmatrix} -it_m \\ A_+ \\ -iA_+ \\ t_m \end{pmatrix}, \vec{v}_4 = \begin{pmatrix} -it_m \\ -A_- \\ iA_- \\ t_m \end{pmatrix}, \tag{43}$$

where

$$A_\pm = \pm\lambda_y + \sqrt{t_m^2 + \lambda_y^2}\,. \tag{44}$$

The choice of the components of the wave function in Equation (42) depends on the energy interval in Equation (37), available for electrons. For example (see Figure 2), for $E > \lambda_x \quad \Rightarrow \quad E = \mathcal{E}_1 \vee E = \mathcal{E}_3$, the base eigenfunctions are: $\sim \exp(\pm i m_{+1}\theta)\vec{v}_1(\pm t_{m_{+1}}); \sim \exp(\pm i m_{-1}\theta)\vec{v}_3(\pm t_{m_{-1}})$, respectively. As discussed above, the quantum number s, characterising the components of the wave function in Equation (42), is associated with the specific energy (see Equations (40) and (41)). The wave function in Equation (13) of the flat graphene at $k_y = 0$ is convenient to present in the form:

$$\Psi_k^\sigma(x) = \frac{1}{2}\begin{pmatrix} \tau\frac{k}{|k|} \\ 1 \end{pmatrix} \otimes \begin{pmatrix} 1 \\ \sigma i \end{pmatrix} e^{ikx}, \qquad \sigma = \pm\,. \tag{45}$$

As above, we use the following definitions: $\Psi_k^\sigma(x) \equiv \langle x|\sigma, k\rangle$, $\sigma = (+/-) \Leftrightarrow \sigma = (\uparrow / \downarrow)$. The symbol $\tau = -(+)$ is ascribed to the valence (conduction) band. We study the case $E \geq 0$; the opposite case can be analysed with the same method. As above, the positive value of the wave number $k \equiv k_x$ corresponds to the direction of the electron motion from the left to the right side of the considered system.

In the case $k_y = 0$, there is one more operator $\hat{S}_{ch} = \tau_x \otimes \sigma_y$ that commutes with both Hamiltonians in Equations (1) and (11), describing the flat and the rippled pieces of our system, respectively:

$$[\hat{S}_{ch}, H_f] = 0\,, \quad [\hat{S}_{ch}, H_r] = 0\,. \tag{46}$$

For the wave function, associated with the flat graphene sheet, we have

$$\left.\begin{array}{rl} \hat{S}_{ch}|\uparrow \pm k\rangle = & \pm|\uparrow \pm k\rangle \\ \hat{S}_{ch}|\downarrow \pm k\rangle = & \mp|\downarrow \pm k\rangle \end{array}\right\} \Rightarrow \alpha = \pm 1\,. \tag{47}$$

Thus, the operator $\hat{S}_{ch}$ has the eigenvalue $\alpha = +1(-1)$, acting on the wave function of the electron, traveling from the left side of our structure with spin up (down) polarisation. The eigenvalues are equal in value but opposite in sign if the operator $\hat{S}_{ch}$ acts on the wave function of the electron, traveling from the right side of our structure with spin up (down) polarisation. On the other hand, the action of the operator $\hat{S}_{ch}$ on the electron wave functions, associated with the rippled piece of our system, is as follows:

$$\hat{S}_{ch}\Phi_{1,m}(\theta) = -\Phi_{1,m}(\theta)\,, \tag{48}$$
$$\hat{S}_{ch}\Phi_{2,m}(\theta) = -\Phi_{2,m}(\theta)\,, \tag{49}$$
$$\hat{S}_{ch}\Phi_{3,m}(\theta) = +\Phi_{3,m}(\theta)\,, \tag{50}$$
$$\hat{S}_{ch}\Phi_{4,m}(\theta) = +\Phi_{4,m}(\theta)\,. \tag{51}$$

In this case, the operator $\hat{S}_{ch}$ has the eigenvalue $\alpha = +1(-1)$ as well, providing us the connection between the eigenstates of the flat and rippled pieces of the considered system.

Now, we are ready to discuss the list of eigenfunctions, responsible for the transport properties in different energy range.

- $E > \lambda_x$.
  1. *Quantum numbers*: $\alpha = +1, s = -1$. These quantum numbers determine the available set of the wave functions: $\Psi_k^+(x)$, $\Psi_{-k}^-(x)$,$\Phi_{3,m}(\theta)$. If the corresponding symmetries are responsible for the transport properties, there are only the following options.
     (a) The electron is moving from the left side of our structure (flat graphene sheet) with the spin up polarisation $[\Psi_k^+(x) \equiv |\uparrow +k\rangle]$. In this case, in the rippled graphene region (see Figure 2), there is one open channel, defined by the wave function $\Phi_{3,m}$. The wave function $\Psi_{-k}^-(x) \equiv |\downarrow -k\rangle$ describes the reflection with the electron spin–flip.
     (b) The electron is moving from the right side (flat graphene sheet) with the spin down polarisation $[\Psi_{-k}^-(x) \equiv |\downarrow -k\rangle]$. In this case, in the rippled graphene region, there is only the transmission channel, defined by the wave function $\Phi_{3,m}$. The wave function $\Psi_k^+(x) \equiv |\uparrow +k\rangle$ describes the reflection with the electron spin–flip.

     As a result, we expect the equivalence between the left/right transmission probabilities with the opposite spin polarisations. Indeed, this expectation is consistent with Equation (33), obtained from the different arguments at $k_y = 0$.
  2. *Quantum numbers*: $\alpha = -1, s = +1$. The available set of the wave functions: $\Psi_k^-(x)$, $\Psi_{-k}^+(x)$, $\Phi_{1,m}(\theta)$. In this case, the symmetries dictate the following options.
     (a) The electron is moving from the left side (flat graphene sheet) with the spin down polarisation $[\Psi_k^-(x) \equiv |\downarrow +k\rangle]$. In this case, in the rippled graphene region (see Figure 2), there is one open channel, defined by the wave function $\Phi_{1,m}(\theta)$. The wave function $\Psi_{-k}^+(x) \equiv |\uparrow -k\rangle$ describes the reflections with the electron spin–flip.
     (b) The electron is moving from the right side (flat graphene sheet) with the spin up polarisation $[\Psi_{-k}^+(x) \equiv |\uparrow -k\rangle]$. In this case, in the rippled graphene region for this electron there is only the transmission channel, defined by the wave function $\Phi_{1,m}(\theta)$. The wave function $\Psi_k^-(x) \equiv |\downarrow +k\rangle$ describes the reflections with the electron spin–flip.

     Again, we expect the equivalence between the left/right transmission probabilities with the opposite spin polarisations. Indeed, this expectation is consistent with Equation (35), obtained from different arguments at $k_y = 0$.
- $E < \lambda_x$.
  1. *Quantum numbers*: $\alpha = -1, s = +1$. In this case, the available set includes the following wave functions: $\Psi_{-k}^+(x)$, $\Psi_k^-(x)$, $\Phi_{2,m}(\theta)$. The symmetries dictate the following options.
     (a) The electron is moving from the left side (flat graphene sheet) with the spin down polarisation $[\Psi_k^-(x) \equiv |\downarrow +k\rangle]$. In the rippled graphene region, there is only the transmission channel, defined by the wave function $\Phi_{2,m}(\theta)$. The wave function $\Psi_{-k}^+(x) \equiv |\uparrow -k\rangle$ describes the reflections with the electron spin–flip.
     (b) The electron is moving from the right side (flat graphene sheet) with the spin up polarisation $[\Psi_{-k}^+(x) \equiv |\uparrow -k\rangle]$. In the rippled graphene region, there is only the transmission channel, defined by the wave function $\Phi_{2,m}(\theta)$. The wave function $\Psi_k^-(x) \equiv |\downarrow +k\rangle$ describes the reflections with the electron spin–flip.

Again, the expected equivalence between the left/right transmission probabilities with the opposite spin polarisations is consistent with Equation (35), obtained from different arguments at $k_y = 0$.

2. *Quantum numbers*: $\alpha = +1$, $s = -1$. In this case, the available set includes the following wave functions: $\Psi_k^+(x)$, $\Psi_{-k}^-(x)$, $\Phi_{3,m}(\theta)$. This situation is completely equivalent to the case discussed at $E > \lambda_x$, Point 1.

Thus, at $k_y = 0$, the symmetry, associated with the operator $\hat{S}_{ch}$, determines the following transport properties through the rippled graphene piece: (i) at the transmission, it preserves the electron spin polarisation, while forbids the spin–flip; and (ii) the reflection occurs only with the spin–flip.

For $k_y \neq 0$, the operator $\hat{S}_{ch}$ does not commute with the Hamiltonians. In this case, the discussed symmetry is broken. It results in constraint release on the reflections and transmissions mechanisms in our system (i.e., [25]).

3.3. The Relation Between the Operators $\hat{S}_t$ and $\hat{S}_{ch}$

Note that these two symmetry operator commute if $k_y = 0$. Evidently, they have the common basis, while having different eigenvalues. Let us analyse this situation in details. One can readily see that the eigenfunctions in Equations (18) and (19), associated with the flat graphene piece, are common eigenstates for the both operators:

$$\hat{S}_t|\psi_\pm\rangle = \pm|\psi_\pm\rangle\,, \quad \hat{S}_t|\phi_\pm\rangle = \pm|\phi_\pm\rangle\,, \tag{52}$$

$$\hat{S}_{ch}|\psi_\pm\rangle = +|\psi_\pm\rangle\,, \quad \hat{S}_{ch}|\phi_\pm\rangle = -|\phi_\pm\rangle\,. \tag{53}$$

For the rippled graphene piece, we have to consider only the case $k_y = 0$. In this case, it is convenient to construct the common basis from the set in Equation (42). Taking into account the properties in Equations (48)–(51), we introduce the following superpositions

$$\Phi_j = \frac{1}{2}[\Phi_{j,m}(\theta) + \Phi_{j,-m}(\theta)]\,, \quad j = 1,..,4\,. \tag{54}$$

As a result, we obtain

$$\hat{S}_t\Phi_{1,2} = \Phi_{1,2}\,, \quad \hat{S}_t\Phi_{3,4} = -\Phi_{3,4}\,, \tag{55}$$

$$\hat{S}_{ch}\Phi_{1,2} = -\Phi_{1,2}\,, \quad \hat{S}_{ch}\Phi_{3,4} = \Phi_{3,4}\,. \tag{56}$$

Thus, the eigenfuctions in Equation (54) form the complete set for the both symmetry operators in the case of the rippled graphene piece.

4. Summary

Evidently, symmetries play an essential role in our understanding different phenomena in mesoscopic physics. In graphene physics, they become especially apparent in the transport properties of the corrugated systems. We find two symmetry operators that explain the chiral behaviour of the ballistic electron transport through the rippled graphene. This unusual behaviour has emerged due to the curvature-induced spin–orbit coupling. In particular, the symmetry operator $\hat{S}_t$ (see Section 3.1) elucidates the equivalence between the transport characteristics of the ballistic electrons travelling from opposite sides of our system that have different type of polarisations. This operator acts as a time-reversal operator in the single valley system, considered in our paper. The other symmetry operator $\hat{S}_{ch}$ (see Section 3.2) enables us to explain the selection of open energy channels for the ballistic electrons travelling through the rippled graphene subsystem at the direct incident of the electron flow ($k_y = 0$). This symmetry explains the dominance of different electron spin polarisations that depend on the direction of the electron flow. This selection becomes increasingly important at multiple periodic repetition of the corrugated graphene structure, considered in our paper (see also [14]). From our preliminary analysis, it follows that similar symmetry preserves if we consider the down side of the CNT. This problem is, however, beyond the scope of the present studies and will be discussed in detail in forthcoming paper.

Author Contributions: All authors contributed equally to this work. All authors have read and agreed to the published version of the manuscript.

Funding: This work is partially supported by the Slovak Academy of Sciences in the framework of VEGA Grant No. 2/0009/19, and by Votruba-Blokhintsev program BLTP, JINR.

Conflicts of Interest: The authors declare no conflict of interest.

References

1. Kumar, S.; Parks, D.M. Strain Shielding from Mechanically Activated Covalent Bond Formation during Nanoindentation of Graphene Delays the Onset of Failure. *Nano Lett.* **2015**, *15*, 1503–1510. [CrossRef]
2. Cao, K.; Feng, S.; Han, Y.; Gao, L.; Hue, Ly, T.; Xu, Z.; Lu, Y. Elastic straining of free-standing monolayer graphene. *Nat. Commun.* **2020**, *11*, 284. [CrossRef]
3. Alyobi, M.M.; Barnett, C.J.; Rees, P.; Cobley, R.J. Modifying the electrical properties of graphene by reversible point-ripple formation. *Carbon* **2019**, *143*, 762–768. [CrossRef]
4. Vázquez de Parga, A.L.; Calleja, F.; Borca, B.; Passeggi, M.C.G.; Hinarejos, J.J.; Guinea, F.; Miranda, R. Periodically Rippled Graphene: Growth and Spatially Resolved Electronic Structure. *Phys. Rev. Lett.* **2008**, *100*, 056807. [CrossRef]
5. Maccariello, D.; Al Taleb, A.; Calleja, F.; Vázquez de Parga, A.L.; Perna, P.; Camarero, J.; Gnecco, E.; Farías, D.; Miranda, R. Observation of Localized Vibrational Modes of Graphene Nanodomes by Inelastic Atom Scattering. *Nano Lett.* **2016**, *16*, 2–7. [CrossRef]
6. Politano, A.; Chiarello, G. Plasmon modes in graphene: Status and prospect. *Nanoscale* **2014**, *6*, 10927–10940. [CrossRef]
7. Ni, G.X.; Zheng, Y.; Bae, S.; Kim, H.R.; Pachoud, A.; Kim, Y.S.; Tan, C.L.; Im, D.; Ahn, J.H.; Hong, B.H.; et al. Quasi-Periodic Nanoripples in Graphene Grown by Chemical Vapor Deposition and Its Impact on Charge Transport. *ACS Nano* **2012**, *6*, 1158–1164. [CrossRef]
8. Vasić, B.; Zurutuza, A.; Gajić, R. Spatial variation of wear and electrical properties across wrinkles in chemical vapour deposition graphene. *Carbon* **2016**, *102*, 304–310. [CrossRef]
9. Pudlak, M.; Nazmitdinov, R.G. Klein collimation by rippled graphene superlattice. *J. Phys. Cond. Matter* **2019**, *31*, 495301. [CrossRef]
10. Guinea, F.; Katsnelson, M.I.; Vozmediano, M.A.H. Midgap states and charge inhomogeneities in corrugated graphene. *Phys. Rev. B* **2008**, *77*, 075422. [CrossRef]
11. Katsnelson, M.; Geim, A. Electron scattering on microscopic corrugations in graphene. *Philos. Trans. R. Soc. A Math. Phys. Eng. Sci.* **2008**, *366*, 195–204. [CrossRef] [PubMed]
12. Ando, T. Spin-Orbit Interaction in Carbon Nanotubes. *J. Phys. Soc. Jpn.* **2000**, *69*, 1757–1763. [CrossRef]
13. Pudlak, M.; Pichugin, K.N.; Nazmitdinov, R.G. Cooperative phenomenon in a rippled graphene: Chiral spin guide. *Phys. Rev. B* **2015**, *92*, 205432. [CrossRef]
14. Smotlacha, J.; Pudlak, M.; Nazmitdinov, R.G. Spin transport in a rippled graphene periodic chain. *J. Phys. Conf. Ser.* **2019**, *1416*, 012035. [CrossRef]
15. Izumida, W.; Sato, K.; Saito, R. Spin–Orbit Interaction in Single Wall Carbon Nanotubes: Symmetry Adapted Tight-Binding Calculation and Effective Model Analysis. *J. Phys. Soc. Jpn.* **2009**, *78*, 074707. [CrossRef]
16. Del Valle, M.; Margańska, M.; Grifoni, M. Signatures of spin-orbit interaction in transport properties of finite carbon nanotubes in a parallel magnetic field. *Phys. Rev. B* **2011**, *84*, 165427. [CrossRef]
17. Pichugin, K.N.; Pudlak, M.; Nazmitdinov, R.G. Spin-orbit effects in carbon nanotubes—Analytical results. *Eur. Phys. J. B* **2014**, *87*, 124. [CrossRef]
18. Michaeli, K.; Kantor-Uriel, N.; Naaman, R.; Waldeck, D.H. The electron's spin and molecular chirality—How are they related and how do they affect life processes? *Chem. Soc. Rev.* **2016**, *45*, 6478–6487. [CrossRef]
19. Nazmitdinov, R.G. From Chaos to Order in Mesoscopic Systems. *Phys. Part. Nucl. Letts.* **2019**, *16*, 159–169. [CrossRef]

20. Birman, J.; Nazmitdinov, R.; Yukalov, V. Effects of symmetry breaking in finite quantum systems. *Phys. Rep.* **2013**, *526*, 1–91. [CrossRef]
21. Foa Torres, L.E.F.; Roche, S.; Charlier, J.C. *Introduction to Graphene-Based Nanomaterials: From Electronic Structure to Quantum Transport*; Cambridge University Press: Cambridge, UK, 2014. [CrossRef]
22. Katsnelson, M.I. *Graphene: Carbon in Two Dimensions*; Cambridge University Press: Cambridge, UK, 2012. [CrossRef]
23. Sakurai, J.J. *Modern Quantum Mechanics, Revised Edition*; Addison-Wesley: Reading, MA, USA, 1994.
24. Bohr, A.; Mottelson, B. *Nuclear Structure*; Number v. 1 in Nuclear Structure; World Scientific: Singapore, 1998.
25. Busa, J.; Pudlak, M.; Nazmitdinov, R.G. On Electron Scattering through a Single Corrugated Graphene Sructure. *Phys. Part. Nucl. Lett.* **2019**, *16*, 729–733. [CrossRef]

Article

Structural Properties of Vicsek-like Deterministic Multifractals

Eugen Mircea Anitas [1,2,*], Giorgia Marcelli [3], Zsolt Szakacs [4], Radu Todoran [4] and Daniela Todoran [4]

1 Bogoliubov Laboratory of Theoretical Physics, Joint Institute for Nuclear Research, 141980 Dubna, Russian
2 Horia Hulubei, National Institute of Physics and Nuclear Engineering, 77125 Magurele, Romania
3 Faculty of Mathematical, Physical and Natural Sciences, Sapienza University of Rome, 00185 Rome, Italy; marcelli.giorgia@gmail.com
4 Faculty of Science, Technical University of Cluj Napoca, North University Center of Baia Mare, Baia Mare, 430122 Maramures, Romania; szakacsz@yahoo.com (Z.S.); todoran_radu@yahoo.com (R.T.); todorandaniela05@yahoo.com (D.T.)
* Correspondence: anitas@theor.jinr.ru

Received: 13 May 2019; Accepted: 15 June 2019; Published: 18 June 2019

Abstract: Deterministic nano-fractal structures have recently emerged, displaying huge potential for the fabrication of complex materials with predefined physical properties and functionalities. Exploiting the structural properties of fractals, such as symmetry and self-similarity, could greatly extend the applicability of such materials. Analyses of small-angle scattering (SAS) curves from deterministic fractal models with a single scaling factor have allowed the obtaining of valuable fractal properties but they are insufficient to describe non-uniform structures with rich scaling properties such as fractals with multiple scaling factors. To extract additional information about this class of fractal structures we performed an analysis of multifractal spectra and SAS intensity of a representative fractal model with two scaling factors—termed Vicsek-like fractal. We observed that the box-counting fractal dimension in multifractal spectra coincide with the scattering exponent of SAS curves in mass-fractal regions. Our analyses further revealed transitions from heterogeneous to homogeneous structures accompanied by changes from short to long-range mass-fractal regions. These transitions are explained in terms of the relative values of the scaling factors.

Keywords: fractals; small-angle scattering; form factor; structural properties; dimension spectra; pair distance distribution function

1. Introduction

Recent progress in materials science and nanotechnology has opened new possibilities in the production of new types of nano- and micro-structured materials with improved functions and properties [1–7], thus providing links to both deterministic classical mechanics and chaotic quantum mechanics [8,9]. Since the physical characteristics greatly depend on their structure, one of the main challenges in the field of materials science is to understand the correlation between them on a broad spectrum of length scales, starting from atomic level. In particular, for many nanoscale structures, quantum-like properties are frequently observable, thus displaying many interesting nano-effects [10–12]. However, for some materials, such as electrospun nanofibers, it has been shown that these effects depend strongly upon their fractal structure [9].

Therefore, ongoing research is carried out to obtain structures with exact self-similar (ESS) properties [13–19], where an intrinsic pattern repeats itself *exactly* under scaling. This procedure usually leads to highly symmetric fractal structures, such as Cantor dust, Sierpinski carpet or Menger sponge [20]. In the field of chaotic deterministic systems they are known as strange attractors, since they may represent a set of infinite unconnected points (Cantor dust), a pathological curve (Weierstrass function) which is continuous everywhere but differentiable nowhere, or generally any geometric shape that cannot be easily described with a simple set operations of basic geometric objects.

ESS nano- and micro-materials attract a lot of attention due to their improved physical characteristics as compared with classical ones, and which arise mainly due to their symmetry and self-similar properties [21]. For example, the mechanical performance of 3D printed biomimetic Koch fractals interlocking can be effectively increased via fractal design [22], while the radiative heat flux can be kept at a short range (as compared to non-fractal structures) in ESS-based materials [23]. However, one of the main challenges in building such fractal materials is finding the suitable mixing composition between the embedding matrix and the fractal's material. To date, only a few materials have been successfully used to create such ESS structures, including dicarbonitrile [24] and bys-terpyridine molecules [13], single/poly crystalline silicone [14,15] or alkyletene dimers [25].

The structural properties of this new generation of nano- and micro-scale materials can effectively be determined using the small-angle scattering (SAS) of X-rays (SAXS) or neutrons (SANS) [26,27]. This widely used material-morphology investigation method has the advantage of sampling a statistically significant macroscopic volume. For ESS structures, the main advantage of SAS relies on its ability to distinguish between mass and surface fractals through the value of the scattering exponent τ in the fractal region [28–31]. More recently, it has been shown that SAS can also differentiate between ESS and statistically self-similar (SSS) structures [32] as well as between regular and fat fractals, that is, those with positive Lebesgue measure [33].

However, the behavior of the SAS intensity curves obtained from multifractals, that is, structures obtained by intermixing fractals with several scaling factors, are not yet completely understood. This is mainly due to the fact that the vast majority of physical materials are heterogeneous at nano- and micro-scales, thus requiring models with at least two scaling factors. Although a first step in this direction was done by obtaining the SAS spectrum from a multifractal structure generated using the chaos game algorithm [34], an expression for the scattering intensity was derived only recently [32]. Moreover, in Reference [34] it has been shown that, for the investigated structure, the oscillations in the fractal region are not very pronounced leading to difficulties in recovering the scaling factors from experimental data. This is an intrinsic consequence of the model's construction procedure, which involves a random variable in generating the positions of the fractal scattering units, as well as the presence of multiple scaling factors.

The purpose of this work is to provide a description of how to extract the scaling factor(s) from SAS data and how to relate them to the degree of fractal's heterogeneity. To this aim, in Section 2.1 we briefly describe the multifractals together with the moment method used to calculate the dimension spectra. Section 2.2 presents the main concepts of SAS with a focus on the form factor, pair distance distribution function (pddf) and their properties. In Section 3.1 we describe the construction process of the Vicsek-like multifractal model with two variable scaling factors [35,36] and show how one can obtain analytically the box-counting fractal dimension. In Section 3.2 we calculate numerically the corresponding dimension spectra and compare them with the theoretical results. In Section 3.3 we obtain the coefficients of the pddf. Finally, Section 3.4 presents an analytical expression of the form factor together with the influence of the polydispersity. Here, we also relate the behavior of the form factor with information from dimension spectra from Section 3.2.

2. Theoretical Background

2.1. Multifractals

Multifractals are non-uniform structures with rich scaling and self-similar properties that can change at every point [37,38]. A common procedure used to investigate their properties is to calculate their dimension spectrum. For this purpose several methods can be used, such as moment method [39], multifractal detrended fluctuation analysis [40] or wavelet transform modulus maxima [41]. Here, we make use of the moment method due to its simplicity of implementation in a computer code as well as for its general applicability. An advantage of this method is that it is particularly well suited for analysis of images acquired by various methods including atomic force microscopy, scanning electron microscopy, computed tomography etc.

For this purpose, let us consider first an object S covered by a grid of boxes $B_i(l)$ of size l and a measure $\mu(B)$ determined by the probability of hitting the object in the box B_i. The corresponding partition function Z_s is defined by [41]:

$$Z_s(l) = \sum_{i=1}^{N} p_i^s(l), \tag{1}$$

where $N \propto 1/l^2$ is the number of boxes, i indexes each individual box, and $p_i = \mu(B)$ represent probabilities with r_i fragmentation ratios, such that $\sum_{i=1}^{N} p_i = 1$.

In terms of the partition function, the generalized dimension spectrum D_s can be written as [41]:

$$D_s \equiv \frac{1}{1-s} \lim_{l \to 0} \frac{\ln Z_s(l)}{-\ln l}, \tag{2}$$

where we take into account that Z_s has a power-law behavior in the limit $l \to 0$ and $N \to \infty$, so that $Z_s \propto l^{D_s(s-1)}$. Therefore:

$$D_s = \lim_{l \to 0} \frac{1}{1-s} \frac{\ln \sum_{i=1}^{N} p_i^s(l)}{-\ln l}, \tag{3}$$

with $p_i \equiv N_i(l)/N$ being the relative weight of the i-th box. In practice, dimension spectra can be obtained from images acquired using various techniques, such as atomic force microscopy, scanning electron microscopy, computed tomography and so on. Thus the quantity N_i in defining the probabilities p_i, is given by the number of non-white pixels in the i-th box, while N is the total number of pixels in the image.

The function D_s is a monotonically decreasing one, with the horizontal asymptotes $\alpha_{max} = \lim_{s \to -\infty} D_s$ and $\alpha_{min} = \lim_{s \to \infty} D_s$. The quantities α_{max} and α_{min} describe the scaling properties of the most rarefied, and respectively of the most dense regions in the fractal. Thus, the object is homogeneous if $\alpha_{max} = \alpha_{min}$, so that it becomes a single scale fractal, and the corresponding D_s spectrum is a line. In particular, at $s = 0$ one recover the well-known box-counting dimension, since it gives:

$$D_0 = \lim_{l \to 0} \frac{\log N(l)}{-\log l}, \tag{4}$$

with $s \equiv D_0$ and $l = \delta$. Here, $N(l)$ is the number of boxes in the minimal cover. At $s = 1$, D_1 gives a description of how the morphology increases as $l \to 0$, and thus is called the information dimension. After applying L'Hopital's rule, D_1 can be written as:

$$D_1 = \lim_{l \to 0} \frac{\sum_{i=1}^{N} p_i \log p_i}{-\log l}, \tag{5}$$

which is related to Shannon's entropy and measures how the information scales with $1/l$. The higher the values of fractal dimension D_1, the more uniform the density. At $s = 2$, Equation (3) gives D_2, which is called the *two-point correlation dimension* and is a measure of the correlation between pairs of points in each box. It describes how the data are scattered, with higher values of D_2 corresponding to higher compactness.

2.2. Small-Angle Scattering

SAS technique is based on the interaction of an incident beam with the electrons in the sample (for SAXS), respectively with the atomic nuclei (for SANS). Since neutrons interact with the magnetic moments, they also provide important information about the sample material's magnetic properties. Therefore, SAS describes spatial density-density correlations in materials through a Fourier transform, leading to the determination of the differential elastic cross section $d\sigma/d\Omega$. When this quantity is normalized with regard to the irradiated volume V', it gives the scattering intensity [26] $I(q) = (1/V')d\sigma/d\Omega$ as a function of the momentum transfer $q = 4\pi\lambda^{-1}\sin\theta$, where λ is the incident beam's wavelength and 2θ the scattering angle. Although in this low-resolution technique the phase information is lost, a more detailed structural description can be obtained when it is complemented by data obtained from other methods, such as protein crystallography [42], nuclear magnetic resonance [43], or when numerical procedures are used to recover the phase, as in coherent SAXS [44]. Let us consider a volume V' irradiated by a beam of light, X-ray, or neutrons, which contains the matrix, with a scattering length density (SLD) ρ_p, together with a large number of randomly oriented, non-interacting multifractals, with uncorrelated positions. Denoting by ρ_m the SLD of the fractals, after subtracting the matrix density, we can consider a system of scattering units "frozen" in vacuum. It has a scattering density of $\Delta\rho = \rho_m - \rho_p$, called the scattering contrast. Thus, denoting by n the concentration of the fractals, the scattering intensity is given by [26]:

$$I(q) = n|\Delta\rho|^2 V^2 \left\langle |F(\boldsymbol{q})|^2 \right\rangle, \tag{6}$$

where V is each fractal's volume, $F(\boldsymbol{q})$ is the form factor:

$$F(\boldsymbol{q}) \equiv \frac{1}{V}\int_V e^{-i\boldsymbol{q}\cdot\boldsymbol{r}} d\boldsymbol{r}, \tag{7}$$

obeying the boundary condition $F(0) = 1$. The symbol $\langle \cdots \rangle$ stands for ensemble averaging over all orientations of the fractal, which, for an arbitrary two-dimensional function f, is defined as:

$$\langle f(q_x, q_y) \rangle = \frac{1}{2\pi}\int_0^{2\pi} f(q, \phi) d\phi. \tag{8}$$

For a mass fractal with fractal dimension D_m, total length L, composed of p basic units, each of size l, separated by the distance d, so that $l \lesssim d \lesssim L$, the normalized form factor in Equation (7) can be written as:

$$\left\langle |F^{(m)}(\boldsymbol{q})|^2 \right\rangle \simeq \begin{cases} 1, & q \lesssim 2\pi/L, \\ (qL/2\pi)^{-D_m}, & 2\pi/L \lesssim q \lesssim 2\pi/h, \\ 1/p, & 2\pi/h \lesssim q \lesssim 2\pi/l, \\ (1/p)(qL/2\pi)^{-4}, & 2\pi/l \lesssim q, \end{cases} \tag{9}$$

where p is of the order of $(L/h)^{D_m}$. The four intervals in the definition of this piecewise function delimit the main structural regions on a double logarithmic scale: the plateau at low q is the Guinier region, the simple power-law decay is the fractal region, a second plateau, and finally the Porod region represented by a power-law decay with scattering exponent -4, or respectively -3 for 2D structures. Because we are dealing with a model for which the distances between the objects is of the same order of magnitude as their size, so that $d/l \simeq 1$, the second plateau from Equation (9) will not be observed. This shall be of no concern to us, since we are extracting structural information from the fractal region.

In calculating the SAS curves one makes use of the following properties of the form factor defined in Equation (7):

- $F(\boldsymbol{q}) \to F(\beta\boldsymbol{q})$ if the particle's length is scaled as $L \to \beta L$,
- $F(\boldsymbol{q}) \to F(\boldsymbol{q})e^{-i\boldsymbol{q}\cdot\boldsymbol{a}}$ if the particle is translated by the vector $\boldsymbol{r} \to \boldsymbol{r} + \boldsymbol{a}$,
- $F(\boldsymbol{q}) = [V_I F_I(\boldsymbol{q}) + V_{II} F_{II}(\boldsymbol{q})] / (V_I + V_{II})$, if the particle can be decomposed as a union of two non-overlapping subsets I and II.

It is known that for a single-scale mass fractal consisting of k units (here disks will be used) at the m-th iteration, each with a form factor F_0, the scattering intensity can be written as [30]:

$$I(q) = I(0)S(q)\left\langle |F_0(\boldsymbol{q})|^2 \right\rangle / k^m, \tag{10}$$

where $S(q)$ is the structure factor, defined by:

$$S(q) \equiv \langle \rho_q \rho_{-q} \rangle / k^m, \tag{11}$$

and is related to the pddf $p(r)$ through:

$$S(q) = 1 + (k^m - 1) \int_0^{+\infty} \mathrm{d}r p(r) \frac{\sin qr}{qr}. \tag{12}$$

For single scale mass-fractals, the function $p(r)$ appearing in Equation (12) is the probability density of finding the distance r between the centers of two arbitrarily taken disks inside the fractal, and is defined by the following expression [30]:

$$p(r) = \frac{2}{k^m (k^m - 1)} \sum_{r_p} C_p \delta(r - r_p), \tag{13}$$

where the symbol δ is the Dirac's delta function, $r_{jk} \equiv |\boldsymbol{r}_j - \boldsymbol{r}_k|$ is the relative distance between the centers of disk j and k, and C_p are the number of distances separated by r_p.

In a physical system, the scatterers have almost always a certain degree of polydispersity. Thus, in order to take into account this effect we consider that their size obey a distribution function $D_{\mathrm{N}}(l)$, defined in such a way that $D_{\mathrm{N}}(l)\mathrm{d}l$ gives the probability of the size of the fractal to be found in the interval $(l, l + \mathrm{d}l)$. In particular, we consider a log-normal distribution of the type:

$$D_{\mathrm{N}}(l) = \frac{1}{\sigma l (2\pi)^{1/2}} e^{-\frac{\left(\log(l/\mu) + \sigma^2/2\right)^2}{2\sigma^2}}, \tag{14}$$

with relative variance $\sigma_{\mathrm{r}} = (\langle l^2 \rangle_D - \mu^2)^{1/2}/\mu$, mean value $\mu = \langle l \rangle_D$, and variance $\sigma = \left(\log(1 + \sigma_{\mathrm{r}}^2)\right)^{1/2}$. Therefore, by using Equations (6) and (14) one obtains the polydisperse form factor averaged over the distribution function [26]:

$$I(q)/I(0) = \int_0^{\infty} \langle |F(\boldsymbol{q})|^2 \rangle A_m^2(l) D_{\mathrm{N}}(l) \mathrm{d}l, \tag{15}$$

where A_m is the corresponding area at mth iteration.

Thus, the scattering intensity given by Equation (15) leads to a simple power-law decay $I(q) \propto q^{-\tau}$, where $\tau = D_{\mathrm{m}}$ for mass fractals, and $\tau = 2d - D_{\mathrm{s}}$ for surface fractals. Recall that for mass fractals embedded in a d-dimensional Euclidean space we have $D_{\mathrm{s}} = D_{\mathrm{m}} < d$ and $D_{\mathrm{p}} = d$, while for surface fractals $D_{\mathrm{m}} = D_{\mathrm{p}} = d$ and $d - 1 < D_{\mathrm{s}} < d$. Here, D_{s} is the fractal dimension of the set's boundary, D_{m} is the set's "mass" fractal dimension, and D_{p} is the "pore" dimension of the surrounding matrix phase. Therefore, an object is classified as a mass fractal with $D_{\mathrm{m}} = \tau$ when the experimentally determined value of τ is smaller than d, while for $d - 1 < \tau < d$, it is a surface fractal with dimension $D_{\mathrm{s}} = 2d - \tau$ [45,46].

3. Results and Discussions

3.1. Construction of the Multifractal Model

In constructing the Vicsek-like [35] multifractal model one starts with a square of edge size l_0 in which we inscribe a disk of radius r_0, with $0 < r_0 < l_0/2$ such as that their centers coincide (Figure 1). This is called the zero-order iteration (i.e., $m = 0$), or the initiator. We choose a Cartesian system of coordinates with the origin coinciding with the center of the square and disk, and axes parallel with the edges of the square. By replacing the initial disk with $k_1 = 4$ smaller disks of radii $r_1 = \beta_{s1} r_0$ situated in the corners of the square, and with $k_2 = 1$ disk of radius $r_2 = \beta_{s2} r_0$ situated in the center, we obtain the first iteration or generator ($m = 1$). Here, β_{s1} and β_{s2} are the scaling factors. The positions of the four corner disks are chosen in such a way that their centers are given by:

$$\boldsymbol{a}_j = \frac{1 - \beta_{s1}}{2} \{\pm l_0, \pm l_0\}, \tag{16}$$

with all combinations of the signs and with $j = 1, \cdots, 4$. The second fractal iteration ($m = 2$) is obtained by performing a similar operation on each of the $k_1 + k_2$ disks. For arbitrarily iterations m, the total number of disks is:

$$N_m = (k_1 + k_2)^m , \tag{17}$$

In the high m iteration number limit one obtains the multifractal, whose fractal dimension is given by [35]:

$$\sum_{i=1}^{2} k_i \beta_{si}^D = 1. \tag{18}$$

Note that the fractal dimension for the well-known Vicsek fractal is recovered for $\beta_{s1} = \beta_{s2} = 1/3$.

Figure 1 shows the first three iterations of the multifractal at various values of the scaling factors β_{s1} and β_{s2}. The different colors in Figure 1 represent the disks which arise at a given iteration number m. Black color denotes the disks arising at $m = 1$, orange those at $m = 2$, while green is used for the third iteration. The upper row from Figure 1 shows that for $\beta_{s1} = 0.1$ and $\beta_{s2} = 0.8$, denoted here model M1, a more heterogeneous structure is obtained when compared with the model M2 from the middle-row (i.e., for $\beta_{s1} = 0.2$ and $\beta_{s2} = 0.6$) or with the model M3 in the lower-raw, constructed using $\beta_{s1} = 0.3$ and $\beta_{s2} = 0.4$. It can also be noted that for model M3 the structure consists of very closely sized disks, and it resembles the single scale Vicsek fractal, as pointed out before.

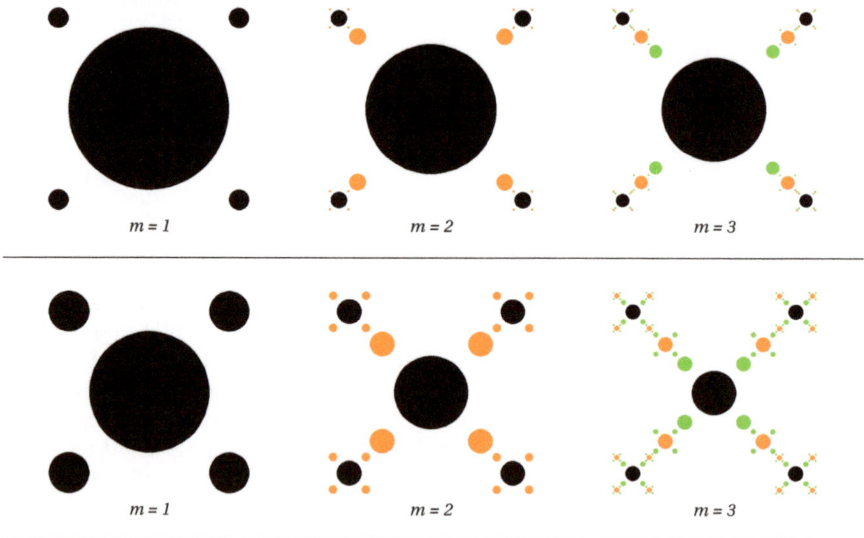

Figure 1. *Cont.*

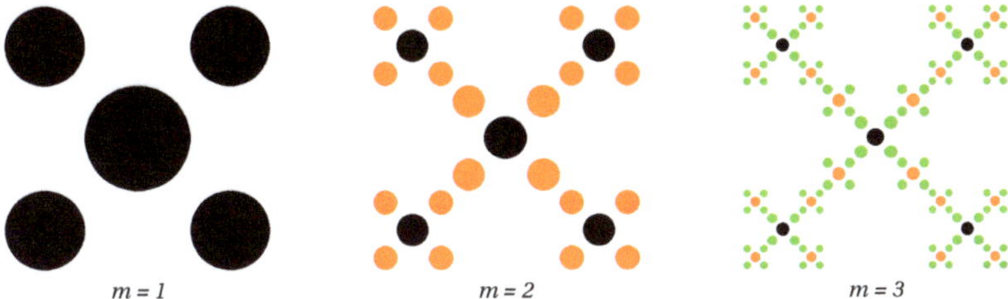

Figure 1. (Color online) First three iterations of the two-scale multifractal models. Upper row: $\beta_{s_1} = 0.1$ and $\beta_{s_2} = 0.8$ (Model M1). Note that for $m = 3$ the disks of radii $l_0\beta_{s_1}^3/2 = 0.0005l_0$ are too small to be seen in the figure (at the given size). Middle row: $\beta_{s_1} = 0.2$ and $\beta_{s_2} = 0.6$ (Model M2). Lower row: $\beta_{s_1} = 0.3$ and $\beta_{s_2} = 0.4$ (Model M3). Black, orange and green colors denote the disks generated at iterations $m = 1$, $m = 2$, and respectively at $m = 3$.

3.2. Dimension Spectra

The corresponding dimension spectra D_s for the three models M1, M2 and M3 are plotted using Equation (2) and can be seen in Figure 2 for $-10 < s < 10$. The spectrum for model M1 (black curve) clearly shows that D_s covers a broad range of values, with $0.15 \lesssim D_s \lesssim 1.85$. This can be explained by a high degree of heterogeneity, with the densest regions having the fractal dimension $\simeq 1.85$, while the most rarefied ones have dimension $\simeq 0.2$. The spectrum for model M2 (red curve) covers the much tighter range between $0.81 \lesssim D_s \lesssim 1.55$. But still, pronounced differences between regions with high and low densities are easily observed. The spectrum of model M3 (green curve) is almost a horizontal line, as expected, since the two scaling factors have close values, thus leading to an almost homogeneous fractal structure, with a fractal dimension of $D_s \simeq 1.42$. The vertical blue dotted line indicates the $s = 0$ axis. The box-counting dimensions of the three models can be determined using the intersection of the fractal dimension spectrum with this axis, so that: $D_0 \simeq 1.22$ (for model M1), $D_0 \simeq 1.31$ (for model M2) and $D_0 \simeq 1.42$ (for model M3).

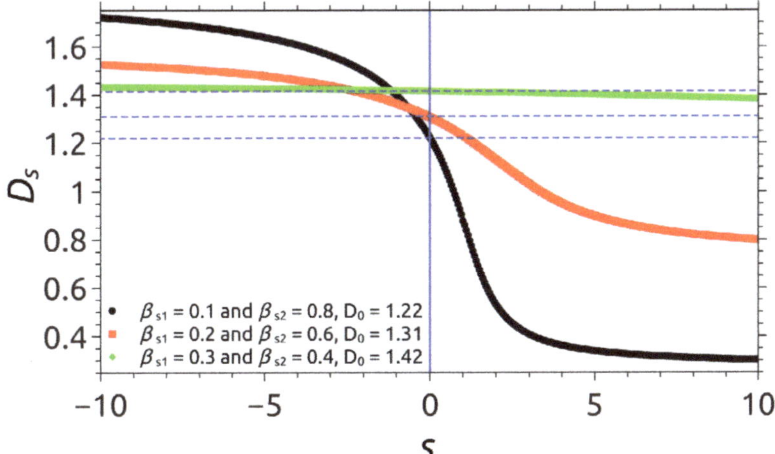

Figure 2. (Color online) Dimension spectra D_s for the three multifractal models: M1 (black), M2 (red), M3 (green). The intersection of the vertical line with each horizontal (dashed) line gives the box-counting dimension D_0.

3.3. Pair Distance Distribution Function

Figure 3 shows the real space characteristics for the same models M1, M2 and M3, at fractal iteration number $m = 4$ using the pddf function defined in Equation (13). The coefficients C_p are calculated numerically using simple combinatorial analysis. The general feature is the presence of distance-groups on a double logarithmic scale.

For $\beta_{s1} = 0.1$ and $\beta_{s2} = 0.8$ the periodicity is clearly visible (Figure 3a), the main groups being separated by gaps at $r/l_0 \simeq 4.5 \times 10^{-3}, 4.5 \times 10^{-2}$ and respectively at 4.5×10^{-1}, indicating the absence of the corresponding distances inside the fractal. The position of these groups is well described as having the periodicity $\log_{10}(1/\beta_{s1})$, and thus, they are related to the scaling factor β_{s1}. Inside each main group, other less pronounced smaller gaps can be noticed, which can be described as having a periodicity related to β_{s1}.

For models M2 and M3, when the scaling factors are relatively closer to each other, the gaps between main groups are less pronounced but still some periodicity can be seen (Figure 3b,c). However, the gaps within a single group are significantly smeared out (especially for model M3) and thus determining the scaling factor β_{s2} in these cases cannot be done with sufficient accuracy. This can be explained by the fact that when the scaling factors have close values, the position of the gaps corresponding to a scaling factor start to "interfere" with the positions of those corresponding to the other one. This leads to a more homogeneous structure, also reflected by the almost constant line in the dimension spectra of model M3 (the green curve from Figure 2).

3.4. Small-Angle Scattering Form Factor

In the proposed model, the positions of the four disks with scaling factor β_{s1} can be described by: $G_1(\boldsymbol{q}) = \cos\left(q_x l_0 \left(1 - \beta_{s1}/2\right)\right) \cos\left(q_y l_0 \left(1 - \beta_{s1}/2\right)\right)$ while the position of the disk with scaling factor β_{s2} is given by $G_2(\boldsymbol{q}) = 1$. The total number of particles at m-th iteration is given by Equation (17), with $k_1 = 4$ and $k_2 = 1$, while the total surface area is $\left(k_1\beta_{s1}^2 + k_2\beta_{s2}\right)^m$. Thus, at $m = 1$, the fractal consists of k_1 disks of radius $r_1 = \beta_{s1} r_0$ and one disk of radius $r_2 = \beta_{s2} r_0$, with $r_0 = l_0/2$. Therefore, the form factor is given by:

$$F_1(\boldsymbol{q}) = \frac{k_1\beta_{s1}^2 G_1(\boldsymbol{q}) F_0(\beta_{s1}\boldsymbol{q}) + k_2\beta_{s2} G_2(\boldsymbol{q}) F_0(\beta_{s2}^2\boldsymbol{q})}{k_1\beta_{s1}^2 + k_2\beta_{s2}^2}, \tag{19}$$

where $F_0(q) = 2J_1(q)/q$ is the form factor of the disk, and $J_1(q)$ is the Bessel function of the first kind.

At $m = 2$, while repeating the same procedure for each disk, one obtains k_2 disks of radius $\beta_{s2}^2 r_0$, k_1 disks of radii $\beta_{s1}\beta_{s2} r_0$, k_1^2 disks of radii $\beta_{s2}^1 r_0$, k_1 and so on. Thus, at an arbitrarily iteration m, we can write the corresponding form factor in terms of a recurrence relation of the form [32]:

$$F_m(\boldsymbol{q}) = \frac{k_1\beta_{s1}^2 G_1(\boldsymbol{q}) F_{m-1}(\beta_{s1}\boldsymbol{q}) + k_2\beta_{s2} G_2(\boldsymbol{q}) F_{m-1}(\beta_{s2}^2\boldsymbol{q})}{k_1\beta_{s1}^2 + k_2\beta_{s2}^2}. \tag{20}$$

Thus, at arbitrary m, the scattering intensity (Equation (10)) can be written as:

$$I_m(q)/I_m(0) = \left\langle |F_m(\boldsymbol{q})|^2 \right\rangle, \tag{21}$$

Figure 4 shows the corresponding monodisperse (black curves) and polydisperse (red curves) scattering intensity for the multifractal models M1, M2 and M3. For calculating the polydispersity we used the log-normal distribution function given by Equation (14) with the relative variance $\sigma_r = 0.2$. One observes the presence of three main regions in each case. At $ql_0 \simeq \pi$ we have a Guinier region with $I(q) \propto q^0$. At $\pi \lesssim ql_0 \lesssim 2\pi/\beta_{s2}^m$ we have a mass fractal region with $I(q) \propto q^{-D_0}$, where D_0 is the box counting dimension of the multifractal, whose value coincides with that obtained from the dimension spectra (see Figure 2) and from Equation (18). At $2\pi/\beta_{s2}^m \gtrsim ql_0$ we are beyond the mass-fractal region, reaching the Porod regime with $I(q) \propto q^{-3}$. Here, the main region of interest is the mass fractal one, since the exponent of the scattering intensity can be related to the multifractal spectra given in Figure 2.

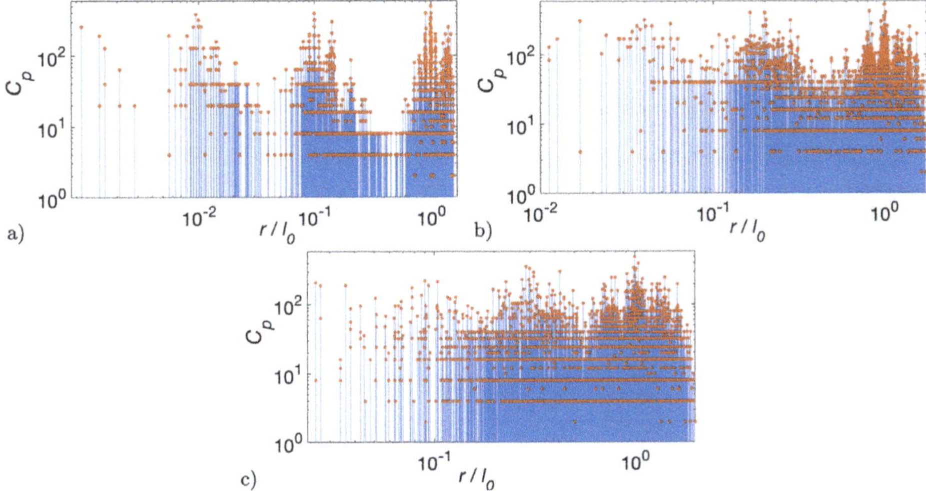

Figure 3. (Color online) The coefficients C_p (orange dots) in Equation (13) for the pair distribution function of the considered multifractal models at $m = 4$. (**a**) Model M1; (**b**) Model M2; (**c**) Model M3. For a better visualization of pddf grouping the vertical line (blue) for each distance is shown.

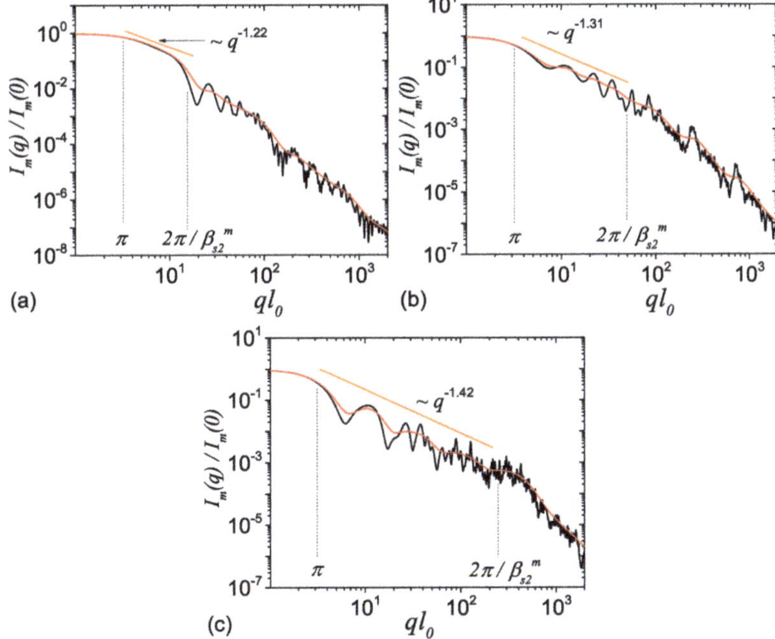

Figure 4. (Color online) Scattering form factor (Equation (21)) for monodisperse (black) and polydisperse (red) multifractal models at $m = 4$. (**a**) Model M1; (**b**) Model M2; (**c**) Model M3. Vertical lines indicate the lower and upper edges of mass fractal region.

These results show that, for the suggested model, the length of mass-fractal regions depends on the scaling factor β_{s2}, since $\beta_{s2} > \beta_{s1}$. The scattering from the three models M1, M2 and M3 illustrate three important cases. First, for the case of model M1, since $\beta_{s2} = 0.8$ is much larger than $\beta_{s1} = 0.1$, and they are related through $\beta_{s1} \equiv (1 - \beta_{s2})/2$, the upper edge of the mass fractal region is very close to its lower edge, so that no oscillations can be observed in this region (Figure 4a). However, when β_{s2} is of comparable size with β_{s1}, the length of the mass fractal region is enough large, so that a log-periodicity with the period $\log(1/\beta_{s1})$ can be observed (Figure 4a,b). In order to provide a better view of the log-periodicity, we show in Figure 5 the quantity $I(q)q^{D_0}$ vs. q. This also clearly shows the increasing complexity of the scattering curves in the fractal region, which arise due to mixing of structures of various sizes, corresponding to repeated subdivisions of the fractal with scaling factors β_{s1} and β_{s2}. In addition, for models M1 and M2 one can see that the number of most pronounced minima in the fractal region coincide with the fractal iteration number.

Note that in all cases, the corresponding polydisperse form factor smears the monodisperse curve. The degree to which the polydisperse curve is smeared-out depends on the value of the relative variance σ_r in the size distribution (Equation (14)): the higher the value of σ_r the more smooth the scattering curve. Therefore, as Figures 4 and 5 show, the periodicity and the number of fractal iteration can be recovered when the values of the relative variance are not very high.

For models M2 and M3, the relationships between the log-periodicity and the scaling factors are not so obvious as for model M1, due to superposition of maxima and minima arising from the 'mixing' of various structures of comparable sizes and of distances between them. This behavior is similar to the one observed in pddf, where separation of distance-groups is clearly visible only for model M1. Note that for 3D structures, the number of distances of a given value, is much higher than for the 2D model developed here, and thus, more pronounced minima and maxima shall be observed the SAS intensity for both the M2 and M3 models, along with more pronounced gaps in pddf.

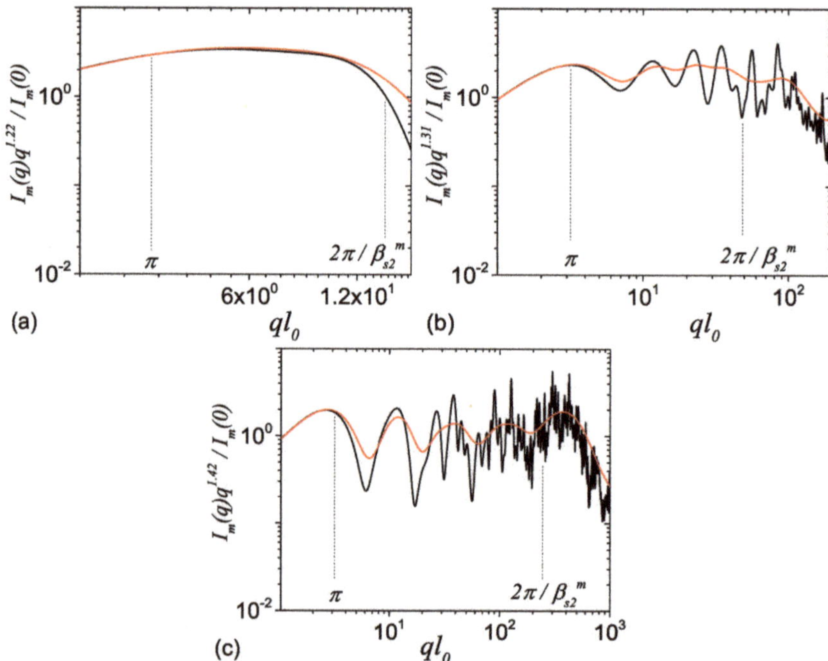

Figure 5. (Color online) The quantity $I(q)q^{D_0}$, where D_0 is the box-counting fractal dimension for monodisperse (black) and polydisperse (red) multifractal models at $m = 4$. (**a**) Model M1; (**b**) Model M2; (**c**) Model M3. Vertical lines indicate the lower and upper edges of the mass fractal region.

Therefore, a combined structural investigation involving SAXS/SANS experimental data, as well as an image analysis of multifractals, can be used to exploit the advantages provided by both reciprocal and real space. While for real space analysis the phase is not lost and thus the structure can be directly obtained, in the case of a reciprocal space analysis, the information is obtained from a macroscopic, statistically significant volume.

4. Conclusions

We developed a multifractal model that generalizes the well-known two-dimensional Vicsek fractal, with disks as basic units. The model is characterized by the presence of two-scaling factors β_{s1} and β_{s2} controlling the multifractal spectra and implicitly the box counting dimension D_0, in the range from 0 to 2.

Changes in the fractal heterogeneity are assessed using pddf and SAS intensity for several representative values of the scaling factors. The relative degree of heterogeneity is confirmed using the dimension spectra. However, depending on the relative values of the scaling factors, the changes can be more clearly visible in a particular space, that is, reciprocal or real. Thus, in order to extract additional structural information we identify three major situations, each one with its particular approach:

- If $\beta_{s1} << \beta_{s2}$, the system is highly heterogeneous and structural parameters are more clearly visible in pddf (see Figure 3a), since the mass fractal region of the scattering intensity is very short Figure 4a). The scaling factor β_{s1} is extracted from the periodicity of large groups of distances, while β_{s2} can be extracted in a relatively good approximation, from the periodicity of smaller groups found inside larger ones. The number of fractal iterations coincide with the number of large distinct groups in pddf.
- If $\beta_{s1} \lesssim \beta_{s2}$, separation of pddf in distinct groups of distance is not very clear since the values of distances arising from each of the scaling factors begin to mix with each other (see Figure 3b,c), and thus extracting exact values of the scaling factors can become a very difficult task. However, in the reciprocal space, the corresponding mass fractal region of scattering intensity is characterized by a succession of maxima and minima on a power-law decay (generalized power-law decay) and the value of the largest scaling factor can be clearly estimated from the periodicity of minima. In addition, the fractal dimension can be obtained from the scattering exponent of this power-law decay while the fractal iteration number can be obtained from the number of the minima.
- If $\beta_{s1} = \beta_{s2}$, the system reduces to a single scale fractal. Structural properties of such systems have been studied elsewhere (see Reference [30]).

Since multifractal spectra can be obtained by analyzing the images captured using various techniques, such as atomic force microscopy, scanning/transmission electron microscopy, computed tomography etc., a combined analysis of multifractal spectra together with SAS data can provide a route towards a more detailed structural analysis of multifractal structures at nano/micro-scales by exploiting the advantages provided by both real and reciprocal space analysis.

Author Contributions: Conceptualization, methodology, writing—original draft preparation, writing—review and editing, E.M.A.; investigation, formal analysis, validation, G.M.; supervision, Z.S., R.T. and D.T.

Funding: This research received no external funding. The APC was funded by Joint Institute for Nuclear Research.

Conflicts of Interest: The authors declare no conflict of interest.

References

1. Acosta, K.L.; Wilkie, W.K.; Inman, D.J. Characterizing the pyroelectric coefficient for macro-fiber composites. *Smart Mater. Struct.* **2018**, *27*, 115001. [CrossRef] [CrossRef]
2. Bica, I.; Anitas, E.M.; Lu, Q.; Choi, H.J. Effect of magnetic field intensity and γ-Fe2O3 nanoparticle additive on electrical conductivity and viscosity of magnetorheological carbonyl iron suspension-based membranes. *Smart Mater. Struct.* **2018**, *27*, 095021. [CrossRef] [CrossRef]

3. Theerthagiri, J.; Durai, G.; Karuppasamy, K.; Arunachalam, P.; Elakkiya, V.; Kuppusami, P.; Maiyalagan, T.; Kim, H.S. Recent advances in 2-D nanostructured metal nitrides, carbides, and phosphides electrodes for electrochemical supercapacitors—A brief review. *J. Ind. Eng. Chem.* **2018**, *67*, 12–27. [CrossRef] [CrossRef]
4. Kim, K.W.; Ji, S.H.; Park, B.S.; Yun, J.S. High surface area flexible zeolite fibers based on a core-shell structure by a polymer surface wet etching process. *Mater. Des.* **2018**, *158*, 98–105. [CrossRef] [CrossRef]
5. Semitekolos, D.; Kainourgios, P.; Jones, C.; Rana, A.; Koumoulos, E.P.; Charitidis, C.A. Advanced carbon fibre composites via poly methacrylic acid surface treatment; surface analysis and mechanical properties investigation. *Compos. Part B—Eng.* **2018**, *155*, 237–243. [CrossRef] [CrossRef]
6. Anagnostou, D.; Chatzigeorgiou, G.; Chemisky, Y.; Meraghni, F. Hierarchical micromechanical modeling of the viscoelastic behavior coupled to damage in SMC and SMC-hybrid composites. *Compos. Part B—Eng.* **2018**, *151*, 8–24. [CrossRef] [CrossRef]
7. Bica, I.; Anitas, E.M. Magnetic flux density effect on electrical properties and visco-elastic state of magnetoactive tissues. *Compos. Part B—Eng.* **2019**, *159*, 13–19. [CrossRef] [CrossRef]
8. El Naschie, M.S. Nanotechnology for the developing world. *Chaos Soliton Fract.* **2006**, *30*, 769. [CrossRef] [CrossRef]
9. He, J.H.; Wan, Y.Q.; Xu, L. Nano-effects, quantum-like properties in electrospun nanofibers. *Chaos Soliton Fract.* **2007**, *33*, 26–37. [CrossRef] [CrossRef]
10. Zhang, Z.; Gao, C.; Wu, Z.; Han, W.; Wang, Y.; Fu, W.; Li, X.; Xie, E. Toward efficient photoelectrochemical water-splitting by using screw-like SnO_2 nanostructures as photoanode after being decorated with CdS quantum dots. *Nano Energy* **2016**, *19*, 318–327. [CrossRef] [CrossRef]
11. Sichert, J.A.; Tong, Y.; Mutz, N.; Vollmer, M.; Fischer, S.; Milowska, K.Z.; García Cortadella, R.; Nickel, B.; Cardenas-Daw, C.; Stolarczyk, J.K.; et al. Quantum Size Effect in Organometal Halide Perovskite Nanoplatelets. *Nano Lett.* **2015**, *15*, 6521–6527. [CrossRef] [CrossRef] [PubMed]
12. Zhu, W.; Esteban, R.; Borisov, A.G.; Baumberg, J.J.; Nordlander, P.; Lezec, H.J.; Aizpurua, J.; Crozier, K.B. Quantum mechanical effects in plasmonic structures with subnanometre gaps. *Nat. Commun.* **2016**, *7*, 11495. [CrossRef] [CrossRef] [PubMed]
13. Newkome, G.R.; Wang, P.; Moorefield, C.N.; Cho, T.J.; Mohapatra, P.P.; Li, S.; Hwang, S.H.; Lukoyanova, O.; Echegoyen, L.; Palagallo, J.A.; et al. Nanoassembly of a fractal polymer: A molecular "Sierpinski hexagonal gasket". *Science* **2006**, *312*, 1782–1785. [CrossRef] [CrossRef] [PubMed]
14. Cerofolini, G.F.; Narducci, D.; Amato, P.; Romano, E. Fractal Nanotechnology. *Nanoscale Res. Lett.* **2008**, *3*, 381–385. [CrossRef] [CrossRef]
15. Berenschot, E.J.W.; Jansen, H.V.; Tas, N.R. Fabrication of 3D fractal structures using nanoscale anisotropic etching of single crystalline silicon. *J. Micromech. Microeng.* **2013**, *23*, 055024. [CrossRef] [CrossRef]
16. Li, C.; Zhang, X.; Li, N.; Wang, Y.; Yang, J.; Gu, G.; Zhang, Y.; Hou, S.; Peng, L.; Wu, K.; et al. Construction of Sierpiński Triangles up to the Fifth Order. *J. Am. Chem. Soc.* **2017**, *139*, 13749–13753. [CrossRef] [CrossRef] [PubMed]
17. Li, N.; Zhang, X.; Gu, G.C.; Wang, H.; Nieckarz, D.; Szabelski, P.; He, Y.; Wang, Y.; Lü, J.T.; Tang, H.; et al. Sierpiński-triangle fractal crystals with the C3v point group. *Chin. Chem. Lett.* **2015**, *26*, 1198–1202. [CrossRef] [CrossRef]
18. Zhang, X.; Li, R.; Li, N.; Gu, G.; Zhang, Y.; Hou, S.; Wang, Y. Sierpiński triangles formed by molecules with linear backbones on Au(111). *Chin. Chem. Lett.* **2018**, *29*, 967–969. [CrossRef] [CrossRef]
19. Fan, J.A.; Yeo, W.H.; Su, Y.; Hattori, Y.; Lee, W.; Jung, S.Y.; Zhang, Y.; Liu, Z.; Cheng, H.; Falgout, L.; et al. Fractal design concepts for stretchable electronics. *Nat. Commun.* **2014**, *5*, 3266. [CrossRef] [CrossRef]
20. Mandelbrot, B.B. *The Fractal Geometry of Nature*; W.H. Freeman: New York, NY, USA, 1982; p. 460.
21. Filoche, M.; Sapoval, B. Transfer Across Random versus Deterministic Fractal Interfaces. *Phys. Rev. Lett.* **2000**, *84*, 5776–5779. [CrossRef] [CrossRef]
22. Khoshhesab, M.M.; Li, Y. Mechanical behavior of 3D printed biomimetic Koch fractal contact and interlocking. *Extrem. Mech. Lett.* **2018**, *24*, 58–65. [CrossRef] [CrossRef]
23. Nikbakht, M. Radiative heat transfer in fractal structures. *Phys. Rev. B* **2017**, *96*, 125436. [CrossRef] [CrossRef]
24. Zhang, B.; Xiang, M.; Zhang, Q.; Zhang, Q. Preparation and characterization of bioinspired three-dimensional architecture of zirconia on ceramic surface. *Compos. Part B—Eng.* **2018**, *155*, 77–82. [CrossRef] [CrossRef]
25. Mayama, H.; Tsujii, K. Menger sponge-like fractal body created by a novel template method. *J. Chem. Phys.* **2006**, *125*, 124706. [CrossRef] [CrossRef] [PubMed]

26. Feigin, L.A.; Svergun, D.I. *Structure Analysis by Small-Angle X-ray and Neutron Scattering*; Springer: Boston, MA, USA, 1987; p. 335. [CrossRef]
27. Gille, W. *Particle and Particle Systems Characterization: Small-Angle Scattering (SAS) Applications*, 1st ed.; CRC Press: Boca Raton, FL, USA, 2013; p. 336.
28. Teixeira, J. Small-angle scattering by fractal systems. *J. Appl. Cryst.* **1988**, *21*, 781–785. [CrossRef] [CrossRef]
29. Schmidt, P.W.; Dacai, X. Calculation of the small-angle x-ray and neutron scattering from nonrandom (regular) fractals. *Phys. Rev. A* **1986**, *33*, 560–566. [CrossRef] [CrossRef]
30. Cherny, A.Y.; Anitas, E.M.; Osipov, V.A.; Kuklin, A.I. Deterministic fractals: Extracting additional information from small-angle scattering data. *Phys. Rev. E* **2011**, *84*, 036203. [CrossRef] [CrossRef] [PubMed]
31. Anitas, E.M.; Slyamov, A.M. Small-angle scattering from deterministic mass and surface fractal systems. *Proc. Rom. Acad. A* **2018**, *19*, 353–360.
32. Cherny, A.Y.; Anitas, E.M.; Osipov, V.A.; Kuklin, A.I. The structure of deterministic mass and surface fractals: Theory and methods of analyzing small-angle scattering data. *Phys. Chem. Chem. Phys.* **2019**. [CrossRef] [CrossRef]
33. Anitas, E.M.; Slyamov, A.; Todoran, R.; Szakacs, Z. Small-Angle Scattering from Nanoscale Fat Fractals. *Nanoscale Res. Lett.* **2017**, *12*, 389. [CrossRef] [CrossRef]
34. Anitas, E.M.; Slyamov, A. Structural characterization of chaos game fractals using small-angle scattering analysis. *PLoS ONE* **2017**, *12*, e0181385. [CrossRef] [CrossRef] [PubMed]
35. Vicsek, T. *Fractal Growth Phenomena*, 2nd ed.; World Scientific: Singapore, 1992; p. 528. [CrossRef]
36. Gouyet, J.F. *Physics and Fractal Structures*; Elsevier: Masson, UK, 1996; p. 234.
37. Arneodo, A.; Decoster, N.; Roux, S. A wavelet-based method for multifractal image analysis. I. Methodology and test applications on isotropic and anisotropic random rough surfaces. *Eur. Phys. J. B* **2000**, *15*, 567–600. [CrossRef] [CrossRef]
38. Decoster, N.; Roux, S.; Arnéodo, A. A wavelet-based method for multifractal image analysis. II. Applications to synthetic multifractal rough surfaces. *Eur. Phys. J. B* **2000**, *15*, 739–764. [CrossRef] [CrossRef]
39. Chhabra, A.; Jensen, R.V. Direct Determination of the f (alpha) Singularity Spectrum. *Phys. Rev. Lett.* **1989**, *62*, 1327. [CrossRef] [PubMed]
40. Kantelhardt, J.W.; Zschiegner, S.A.; Koscielny-Bunde, E.; Havlin, S.; Bunde, A.; Stanley, H. Multifractal detrended fluctuation analysis of nonstationary time series. *Phys. A* **2002**, *316*, 87–114. [CrossRef] [CrossRef]
41. Muzy, J.F.; Bacry, E.; Arneodo, A. Multifractal formalism for fractal signals: The structure-function approach versus the wavelet-transform modulus-maxima method. *Phys. Rev. E* **1993**, *47*, 875–884. [CrossRef] [CrossRef]
42. Stuhrmann, H.B. Small-angle scattering and its interplay with crystallography, contrast variation in SAXS and SANS. *Acta Cryst.* **2008**, *64*, 181–191. [CrossRef] [CrossRef]
43. Izumi, A.; Shudo, Y.; Nakao, T.; Shibayama, M. Cross-link inhomogeneity in phenolic resins at the initial stage of curing studied by 1H-pulse NMR spectroscopy and complementary SAXS/WAXS and SANS/WANS with a solvent-swelling technique. *Polymer* **2016**, *103*, 152–162. [CrossRef] [CrossRef]
44. Dierolf, M.; Menzel, A.; Thibault, P.; Schneider, P.; Kewish, C.M.; Wepf, R.; Bunk, O.; Pfeiffer, F. Ptychographic X-ray computed tomography at the nanoscale. *Nature* **2010**, *467*, 436–439. [CrossRef] [CrossRef]
45. Martin, J.E.; Hurd, A.J. Scattering from fractals. *J. Appl. Cryst.* **1987**, *20*, 61–78. [CrossRef] [CrossRef]
46. Schmidt, P.W. Small-angle scattering studies of disordered, porous and fractal systems. *J. Appl. Cryst.* **1991**, *24*, 414–435. [CrossRef] [CrossRef]

Article

Small-Angle Scattering from Fractals: Differentiating between Various Types of Structures

Eugen Mircea Anitas [1,2]

1 Joint Institute for Nuclear Research, Dubna 141980, Russia; anitas@theor.jinr.ru
2 Horia Hulubei, National Institute of Physics and Nuclear Engineering, RO-077125 Bucharest-Magurele, Romania

Received: 28 November 2019; Accepted: 26 December 2019; Published: 1 January 2020

Abstract: Small-angle scattering (SAS; X-rays, neutrons, light) is being increasingly used to better understand the structure of fractal-based materials and to describe their interaction at nano- and micro-scales. To this aim, several minimalist yet specific theoretical models which exploit the fractal symmetry have been developed to extract additional information from SAS data. Although this problem can be solved exactly for many particular fractal structures, due to the intrinsic limitations of the SAS method, the inverse scattering problem, i.e., determination of the fractal structure from the intensity curve, is ill-posed. However, fractals can be divided into various classes, not necessarily disjointed, with the most common being random, deterministic, mass, surface, pore, fat and multifractals. Each class has its own imprint on the scattering intensity, and although one cannot uniquely identify the structure of a fractal based solely on SAS data, one can differentiate between various classes to which they belong. This has important practical applications in correlating their structural properties with physical ones. The article reviews SAS from several fractal models with an emphasis on describing which information can be extracted from each class, and how this can be performed experimentally. To illustrate this procedure and to validate the theoretical models, numerical simulations based on Monte Carlo methods are performed.

Keywords: small-angle scattering; fractals; structural properties; Monte Carlo simulations; form factor; structure factor

1. Introduction

Small-angle scattering (SAS; X-rays, neutrons, light) is a widely used innovative technique for an efficient characterization of the structure of disordered materials at nano- and micro-scales [1,2]. With the recent advances of its derivative techniques, i.e., grazing-incidence [3], anomalous [4], scanning [5] or magnetic [6] SAS, the range of applications has been greatly extended. This includes now a much larger class of hierarchical materials, i.e., materials in which the structural elements themselves have a structure, such as biological macromolecules [1,7–9], polymers [10–13], composites [14–17] or cellular solids [18–20].

It is well-known that in fractal-based materials, the structural hierarchy and the geometric symmetry within a single structure play an important role in determining their bulk and surface properties. Common applications of such materials include enhancing: the lithium storage performance of CuO nanomaterials with surface fractal characteristics [21], the rheological behavior of fractal-like aggregates in polymer nanocomposites [22], the compatibility and interfacial reactivity of high-performance layered silicate/epoxy nanocomposites [23], the tensile modulus and strength in carbon nanotube fibers with mesoporous crystalline structure [24] or the electrocatalytic activity, durability and stability in Sierpinski gasket-like Pt–Ag octahedral alloy nanocrystals [25]. In addition, the fractal structure [26], also gives rise to quantum effects in various materials such as in plasmonic structures with sub-nanometer gaps [27] or in organometal halide perovskite nanoplatelets [28]. This dependence of the quantum effects and bulk properties on the fractal structure, can be attributed to the interplay between regular and chaotic dynamics of particles, confined in fractal-like regions. This interplay in classical and quantum systems leads to the emergence of dynamical symmetries, and are related to the existence of conserved quantities of the dynamics and integrability under certain conditions [29].

Therefore, understanding the influence of the fractal structure on physical properties and their evolution, can guide the preparation of advanced materials tailored for specific applications. Within this context, development of theoretical models for SAS data analysis and interpretation form the first steps in elucidating the structure-bulk properties relationship. However, the evolution of their properties, is usually determined by complex quantum mechanical models.

In SAS analysis, one of the key issues is the development of models that allow extracting the most important structural parameters from a given set of SAS measurements. Due to technological limitations, until several years ago almost all measurements were performed on random fractals, i.e., on structures where only statistical properties remain invariant at various magnifications, such as clusters generated by diffusion-limited aggregation (DLA). Thus, parameter-extraction procedures concerned obtaining only the fractal dimensions D_{s}, D_{m} or D_{p} (see below), and the inner and outer fractal cutoffs l, and respectively ξ [30–36]. While the fractal dimension is related to the self-similarity property under scale transformations [26], the two cutoffs set the range in which this self-similarity can be observed. For physical samples, the inner cutoff is defined by the size of the building blocks that make up the fractal (typically atoms or molecules). The outer cutoff is defined by the largest distance separating two points inside the fractal.

The fractal dimension is captured in the modeling process by writing the scattering intensity, such as [37]:

$$I(q) \propto q^{D_{\mathrm{s}}-2(D_{\mathrm{m}}+D_{\mathrm{p}})+2d}. \tag{1}$$

This is a simple power-law decay on a double logarithmic scale within a given q-range limited by reciprocal fractal outer and inner cutoffs q_{oc}, and respectively q_{ic}, i.e., when $q_{\mathrm{oc}} < q < q_{\mathrm{ic}}$. It is also known in the literature as the fractal region. Here, q is the module of the scattering vector, D_{s}, D_{m} and D_{p} are the surface, mass and respectively the pore fractal dimensions, and d is the Euclidean dimension of the space in which the fractal is embedded. In the case of surface fractals $d-1 < D_{\mathrm{s}} < d$, $D_{\mathrm{m}} = D_{\mathrm{p}} = d$ and the scattering intensity reduces to $I(q) \propto q^{D_{\mathrm{s}}-2d}$ [38]. For mass fractals $0 < D_{\mathrm{m}} < d$, $D_{\mathrm{m}} = D_{\mathrm{s}}$ and $D_{\mathrm{p}} = d$, and $I(q) \propto q^{-D_{\mathrm{m}}}$ [39], while for pore fractals $D_{\mathrm{p}} = D_{\mathrm{s}}$ and $D_{\mathrm{m}} = d$, and thus $I(q) \propto q^{-D_{\mathrm{p}}}$ [40]. Therefore, by using Equation (1) one can obtain the fractal dimension from the slope (τ) of the experimental scattering curve. Moreover, one can differentiate between mass and surface fractals, i.e., if the measured slope is $\tau < d$ then the sample is a mass fractal, while if $d < \tau < d+1$, the sample is a surface fractal.

To obtain the outer fractal cutoff ξ, Equation (1) is extended by adding a properly weighted exponential term which gives rise to a region (also known as Guinier region) where $I(q) \propto q^{0}$ at $q < q_{\mathrm{oc}}$. Then, the position of the transition point q_{oc} gives an estimation of fractal radius of gyration, which in turn is a measure of its outer cutoff [41,42]. The inner fractal cutoff l is obtained by addition of a second exponential term and of a power-law decay at $q > q_{\mathrm{ic}}$. Then, the position of q_{ic} is related to the radius of gyration of the smallest building block composing the fractal [41,42]. Moreover, by considering a succession of power-law decays interleaved with Guinier regions, one can describe hierarchically structured systems. This succession forms the basis of Beaucage [43,44] and Guinier–Porod [45] models, which are widely used as empirical models to analyze SAS data from random fractal and particulate systems, such as those occurring in some amphiphilic triblock polyelectrolytes [46], rocks [47], fertilizers encapsulated by starch-based superabsorbent polymers [48], silica-filled silicone rubber [49], aerosol nanoparticles [50] or micro/nano-sized TATB crystallites [51].

However, recent advances in materials science and nanotechnology allows fabrication of deterministic fractal materials with predefined structures, i.e., structures where an intrinsic pattern repeats itself *exactly* under scaling, such as: Sierpinski triangular [52–54], supramolecular [25,55], octahedral [56] or Cantor fractals [57]. In addition, theoretical developments in SAS from deterministic fractals have shown that the corresponding intensity curves are characterized by a much more complex behavior, as compared to scattering from random fractals. The main difference is that in the former case, the simple power-law decay is replaced by a generalized power-law decay, where maxima and minima superimposed on a monotonically decreasing curve [58–60]. Also, the complexity of the patterns formed by these maxima and minima increases with the magnitude of the scattering vector q. Generally, both features arise as an effect of two competing symmetries present in a deterministic fractal. First, there is dilation invariance symmetry, which is responsible for the power-law decay. Second, there is geometric symmetry, which is the source of maxima splitting as q increases. These two symmetry-effects are the defining characteristics of SAS from deterministic fractals, and are generally used to differentiate them from random fractals [58].

Although separating these effects from SAS intensity is not an easy task, some general guidelines for extracting additional structural parameters have been provided for various types of deterministic fractals. In the case of SAS from deterministic mass fractals one can extract [58]:

- The fractal dimension D_{m}, from the generalized power-law decay:
- The fractal scaling factor β_{s}, from the period of the scattering curve on a double logarithmic scale,
- The number of fractal iterations m, equal to the number of the main minima,
- The inner and outer fractal cutoffs from the beginning and the end of periodicity region, i.e., from fractal regime,
- The total number of basic objects k_m composing the fractal, from the relation $k_m = (1/\beta_{\mathrm{s}})^{mD_{\mathrm{m}}}$.

For mass fractals, the log-periodicity of the scattering curve arises from the self-similarity of *distances* between the basic objects composing the fractal. In the case of deterministic surface fractals one can extract basically the same information [60]. However, the nature of log-periodicity in the fractal region is different, and arises from the self-similarity of *sizes* of the basic objects. In the case of deterministic fat fractals, i.e., fractals in which the scaling factor is not constant, but it increases after a given number of iterations, it was shown that one can obtain the fractal dimensions and scaling factors at *each* scale [59,61]. In the case of deterministic multifractals with two scaling factors, i.e., fractals with various scaling factors at the same scale, it was shown that under certain conditions, one can also obtain *both* scaling factors from SAS curves [62].

In this survey are presented and discussed the latest theoretical advances useful for differentiating between random, deterministic mass, surface, fat and multifractals, from SAS data. Detailed "receipts" are provided that show how to extract the structural parameters about each fractal type. They are subsequently applied to numerical data generated from Monte Carlo methods. To this aim, the pair-distance distribution function (pddf), the form and structure factors are calculated. The importance of preparing samples with a high degree of monodispersity for a clear and unambiguous description of experimental SAS data, is highlighted by studying the influence on the scattering curve of a log-normal distribution of the fractal size.

The paper is structured as follows. Section 2 presents the theoretical background required for a proper understanding of the results discussed thereafter. Here, the main properties of fractals, of SAS technique and Monte Carlo simulations relevant to SAS from fractals are described. In Section 3 are introduced the main types of fractals together with representative models. Analytic expressions of the form and structure factors are derived and compared with results from Monte Carlo simulations. The main conclusions and prospects for future theoretical developments are summarized in Section 4.

2. Theoretical Background

2.1. Fractals

Mathematically, characterization of fractals requires a rigorous definition of the fractal (Hausdorff) dimension [63]. This which involves abstract concepts from measures theory [64]. For this purpose, let us consider S a subset of the n-dimensional Euclidean space, and that the set $\{\mathrm{C}_i\}$ is a cover of S with $c_i = \text{diam}(C_i) \leq s$, with $s \in S$. The Hausdorff measure $m^\alpha(S)$ of the set S is defined by taking the infimum over all possible coverings, i.e.:

$$m^\alpha(S) = \lim_{s \to 0} \inf_{\{C_i\}} \sum_i c_i^\alpha, \quad \text{with} \quad \alpha \in \mathbb{R}^+. \tag{2}$$

Then, the fractal dimension D of the set S is defined by:

$$D \equiv \inf\{\alpha : m^\alpha(S) = 0\} = \sup\{\alpha : m^\alpha(S) = +\infty\}. \tag{3}$$

This corresponds to the value of α for which the Hausdorff measure changes from zero to infinity. When $\alpha = D$, $m^\alpha(S)$ can take arbitrarily values within this range.

However, in practice is very difficult to apply Equation (3) for determination of the fractal dimension, and therefore one must resort to other methods. A common approach is to determine the variation of the fractal measure M, such as mass, area, volume, or any scalar quantity attached to the fractal support, within a sphere of dimension n and radius r centered on the fractal. For this purpose, we can write [65]:

$$M(r) = A(r)\, r^D, \tag{4}$$

where $\lim_{r\to\infty} \log A(r) / \log r \to 0$. This is known in the literature as the mass-radius relation [65].

Let us consider a fractal of size L composed of balls of size a. Then, the number of balls enclosed by the imaginary sphere of radius r with a ball in the center, is given by [65]:

$$N(r) \propto (r/a)^D \propto r^D, \tag{5}$$

with $l \lesssim r \lesssim L$. If the fractal is a line then $D = 1$, if it is a smooth surface, $D = 2$, while if the fractal is a regular Euclidean three-dimensional object, then $D = 3$. For $D < 1$, the structure reduces to a set of disconnected points.

In the case of fractals with a scaling factor β_s, one can use the property that at first iteration the fractal consists of k copies of itself, each of size $\beta_s L$, and write that [65]:

$$M(L) = kM(\beta_s L). \tag{6}$$

Then, by using Equation (4), one obtains:

$$k\beta_s^{D_m} = 1, \tag{7}$$

which can be used to obtain the fractal dimension D. In the case of fractals with multiple scaling factors β_{si} and k_i copies with $i = 1, \cdots, n$, the above equation can be extended to [65]:

$$\sum_{i=1}^{n} k_i \beta_{si}^{D_m} = 1. \tag{8}$$

For surface fractals one can write a similar relationship as in Equation (5), and write as [66]:

$$S(r) = S_0 r^{2-D_s}, \tag{9}$$

where $S(r)$ represents the area between the boundary of the (rough) surface and the envelope of all spheres of radius r centered on the boundary. Here, S_0 is a constant, which is the surface area itself for a smooth surface, i.e., when $D_s = 2$. Analogous relations as those given by Equation (7) can be written for surface fractals also. For three-dimensional fractals, when $D \to 3$ the surface is so folded that it almost fulfills all the available space, while when $D \to 2$, the surface is almost completely smooth.

The fractal dimensions obtained above are equivalent with the box-counting dimension described in Ref. [26]. This is the single dimension which characterize fractals with a single scaling factor. However, for fractals with multiple scaling factors, a more detailed description of their structure requires determination of additional fractal dimensions. This is achieved by making use of the multifractal formalism [67,68], and in particular of the moment method [69]. Although other methods have been developed for determination of the dimension spectra, the moment method is widely used for its general applicability, including image analysis.

Within this method, one considers an object S covered by a grid of boxes $B_i(l)$ of size l. Let us suppose that the measure determined by the probability of hitting the object in the box B_i is $\mu(B)$. Therefore, the number of covered boxes N at resolution l is $N \propto 1/l^2$, and the corresponding "partition function" Z_q can be written as [70]:

$$Z_q(l) = \sum_{i=1}^{N} p_i^q(l). \tag{10}$$

Here, i denotes each individual box, and $p_i = \mu(B)$ with $\sum_{i=1}^{N} p_i = 1$, are the hitting probabilities. Then, the generalized dimension spectrum is given by [70]:

$$D_q \equiv \frac{1}{1-s} \lim_{l \to 0} \frac{\ln Z_q(l)}{-\ln l}. \tag{11}$$

The function Z_q has a power-law behavior when $l \to 0$ and $N \to \infty$, so that $Z_q \propto l^{D_q(q-1)}$. Thus:

$$D_q = \lim_{l \to 0} \frac{1}{1-q} \frac{\ln \sum_{i=1}^{N} p_i^q(l)}{-\ln l}, \tag{12}$$

where the ratio $p_i \equiv N_i(l)/N$ gives the relative weight of the i-th box.

One of the main property of D_q is that is a monotonically decreasing function, with horizontal asymptotes at $\alpha_{\max} = \lim_{q \to -\infty} D_q$ and $\alpha_{\min} = \lim_{q \to \infty} D_q$. Therefore, one can describe the scaling properties of the most rarefied, and respectively of the most dense regions in the fractal in terms of $\alpha_{\max}$ and $\alpha_{\min}$, i.e., the object is heterogeneous (multifractal) if $\alpha_{\max} \neq \alpha_{\min}$, and homogeneous (simple fractal) otherwise. Also, the box-counting dimension is recovered, at $q = 0$, and is given by [65]:

$$D_0 = \lim_{l \to 0} \frac{\log N(l)}{-\log l}, \tag{13}$$

where $N(l)$ is the number of boxes in the minimal cover. At $q = 1$, one obtains D_1 which is called the information, i.e.:

$$D_1 = \lim_{l \to 0} \frac{\sum_{i=1}^{N} p_i \log p_i}{-\log l}, \tag{14}$$

and describes how the morphology increases as $l \to 0$, i.e., the lower the values of D_1, the less uniform the density. For $q = 2$, Equation (12) gives the *two-point correlation dimension* D_2, and measures the correlation between pairs of points in each box. The higher the values of D_2 the more compact the fractal. As we shall see later, we can extract only the dimension D_0 from SAS data.

2.2. Small-Angle Scattering

In SAS, a beam of particles (usually X-ray or neutrons) hit a sample, after which they are scattered elastically at various angles θ, as shown in Figure 1 [71]. While X-rays are scattered by the electron cloud, neutrons are scattered by nuclei or by the magnetic moments associated with unpaired electron spins in magnetic materials. The structure of the sample is to be determined from the distribution of scattered beam around the beam-stop.

2.2.1. General Background

Basically, one can describe a SAS process by considering that the beam propagates with the wave vector $\boldsymbol{k} = (2\pi/\lambda)\,\hat{\boldsymbol{k}}$ along an axis which coincides with the axis of the collimator with beam opening Δr. Here, λ is the wavelength. The role of the collimator is to select a monochromatic and well collimated beam, such that it provides a sharp energy spectrum, concentrated at a unique eigenvalue. A detector, placed at a large distance from the sample, records the number of particles passing through a small opening $d\Omega$ (Figure 1). Let us consider that its position is given by the vector $\boldsymbol{r} = (x, y, z)$, which is oriented along the direction $\hat{\boldsymbol{r}}$, and at an angle (θ, ϕ) with respect to the direction of the incident beam. As such, the structure of the scattering particles can be described in terms of the wave vector $\boldsymbol{k}'$. To this aim, one introduces the scattering vector $\boldsymbol{q}$ defined by [71]:

$$\boldsymbol{q} = \boldsymbol{k}' - \boldsymbol{k}, \tag{15}$$

where $|\boldsymbol{k}'| = |\boldsymbol{k}|$ = k. Thus, the magnitude of the scattering vector is given by:

$$q = \sqrt{k^2 + k'^2 - 2kk' \cos\theta} = \sqrt{2k^2\,(1 - \cos\theta)}. \tag{16}$$

Since $\cos\theta = 1 - 2\sin^2\theta/2$, and $k = 2\pi/\lambda$, the previous equation becomes:

$$q = \frac{4\pi}{\lambda}\sin\frac{\theta}{2}. \tag{17}$$

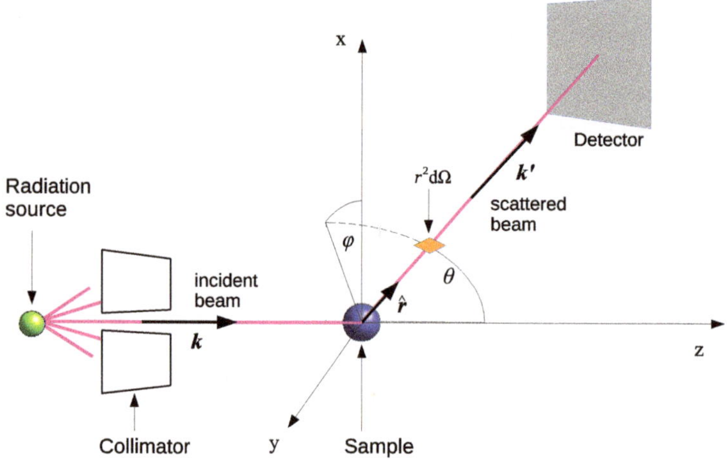

Figure 1. Overall configuration of a SAS experiment. First, a beam of particles is emitted from a radiation source, and a small fraction pass through the collimator. Then, they hit the sample, and the scattered particles around the beam-stop are recorded. The quantities $\boldsymbol{k}$ and $\boldsymbol{k}'$ denote the incident and, respectively the scattered wave vectors, Ω is the solid angle, and $\hat{\boldsymbol{r}}$ is the unit vector along the beam scattered at angles ϕ and θ [72].

Let us consider that an incident beam of neutrons or X-rays is scattered by a sample of volume V' containing a macroscopic number of objects with scattering lengths b_j. After such an event, the scattering amplitude can be written as [71]:

$$A(\boldsymbol{q}) \equiv \int_{V'} \rho_s(\boldsymbol{r}) e^{-i\boldsymbol{q}\cdot\boldsymbol{r}} \mathrm{d}^3 r, \tag{18}$$

where $\rho_s(\boldsymbol{r}) = \sum_j b_j \delta(\boldsymbol{r} - \boldsymbol{r}_j)$ is the scattering length density (SLD), $\boldsymbol{r}_j$ are the microscopic objects positions and δ is the Dirac's δ function. In SAS, multiple scattering is neglected, and the scattering intensity $I(q)$, i.e., the differential scattering cross section per unit volume, is expressed as the product between the scattering amplitude and its complex conjugate, i.e. [71]:

$$I(q) \equiv \frac{1}{V'} \frac{\mathrm{d}\sigma}{\mathrm{d}\Omega} = \frac{1}{V'} |A(\boldsymbol{q})|^2. \tag{19}$$

Since the positions r_j, and thus SLD, are fixed in time, Equation (19) can be written as [73]:

$$I(q) = \frac{1}{V'} \int_{V'} \Gamma_\eta(\boldsymbol{r}) e^{-i\boldsymbol{q}\cdot\boldsymbol{r}} \mathrm{d}^3 r, \tag{20}$$

where $\Gamma_\eta(\boldsymbol{r})$ is given by:

$$\Gamma_\eta(\boldsymbol{r}) = \int_{V'} \eta(\boldsymbol{r}') \eta(\boldsymbol{r}' + \boldsymbol{r}) \mathrm{d}^3 \mathrm{r}', \tag{21}$$

and represents the autocorrelation function of the deviations $\eta(\boldsymbol{r})$ of SLD $\rho_{\mathrm{s}}(\boldsymbol{r})$ from its mean value $\langle \rho_{\mathrm{s}} \rangle$, i.e., $\eta(\boldsymbol{r}) = \rho_{\mathrm{s}}(\boldsymbol{r}) - \langle \rho_{\mathrm{s}} \rangle$. Then, the normalized correlation function $\gamma(\boldsymbol{r})$ is defined as [74]:

$$\gamma(\boldsymbol{r}) \equiv \frac{\Gamma_\eta(\boldsymbol{r})}{\Gamma_\eta(0)}, \quad \text{with} \quad \gamma(0) = 1, \tag{22}$$

and where the normalization constant $\Gamma_\eta(0)$ is given by:

$$\Gamma_\eta(0) \equiv \int_{V'} \eta(\boldsymbol{r}') \eta(0 + \boldsymbol{r}') \mathrm{d}^3 \mathrm{r}' = V' \left\langle \eta^2 \right\rangle, \tag{23}$$

Here, $\langle \eta^2 \rangle$ is the mean square fluctuation of the density fluctuations about its mean value throughout the sample. Therefore, Equation (20) takes the form:

$$I(q) = \left\langle \eta^2 \right\rangle \int_{V'} \gamma(\boldsymbol{r}) e^{-i\boldsymbol{q}\cdot\boldsymbol{r}} \mathrm{d}^3 r. \tag{24}$$

The function $\gamma(\boldsymbol{r}) \to 0$ as $\boldsymbol{r}$ increases, there are no correlations between fluctuations of $\eta(\boldsymbol{r}')$ and $\eta(\boldsymbol{r}' + \boldsymbol{r})$ at large $\boldsymbol{r}$, and the integration over the volume V' can be replaced by an integration over an infinite region. Also, $\gamma(\boldsymbol{r}) \neq 0$ only for small values of $\boldsymbol{r}$, and the properties of $\gamma(\boldsymbol{r})$ depend only on the magnitude of $\boldsymbol{r}$. Therefore, the intensity in Equation (24) becomes:

$$I(q) = 4\pi \left\langle \eta^2 \right\rangle \int_0^\infty r^2 \gamma(r) \frac{\sin qr}{qr} \mathrm{d}r, \tag{25}$$

where the property $\langle e^{-i\boldsymbol{q}\cdot\boldsymbol{r}} \rangle = \sin(qr)/(qr)$ has been used (see also Equation (29) below). The function $\gamma(r)$ depends also on the system's geometry and gives the probability to find a volume element v_i within the system, at a distance r from another volume element v_j.

2.2.2. Two-Phase Fractal Systems

We consider here mainly two-phase system, in which particles, i.e., fractals with scattering length density (SLD) ρ_m are "frozen" in an embedding matrix with "pore" SLD ρ_m. Therefore, by subtracting the "pore" density we can consider the system as if the fractals were "frozen" in a vacuum and had the density $\Delta\rho = \rho_m - \rho_p$. This procedure can be performed since a constant shift of $\Delta\rho$ is important only when $q \to 0$, and which is beyond the instrumental q-range. Then, for practical applications the total scattering amplitude is represented as a sum amplitude of rigid fractals. Thus, the scattering intensity can be written in a general form as [71]:

$$I(q) = n|\Delta\rho|^2 V^2 P(q) S(q), \tag{26}$$

where n is the particle concentration and V is the volume of each fractal. The quantity $P(q)$ describes the spatial distribution of the scattering lengths of fractal's atoms, and is defined by [71]:

$$P(q) \equiv \left\langle |F(\boldsymbol{q})|^2 \right\rangle = \left\langle |\frac{1}{V}\int_V e^{-i\boldsymbol{q}\boldsymbol{r}} \mathrm{d}\boldsymbol{r}|^2 \right\rangle, \tag{27}$$

where $F(\boldsymbol{q})$ is the fractal form factor (see next subsection). The structure factor can be written as [71]:

$$S(q) = 1 + 4\pi n \int_0^\infty [g(r) - 1]\, r^2 \frac{\sin(qr)}{qr} \mathrm{d}r, \tag{28}$$

and contains information about the spatial arrangements of the fractals.

The quantity $\langle \cdots \rangle$ in Equation (27) denotes an ensemble averaging over all possible orientations of the fractals. Thus, if the probability of any orientation is the same, in 3D it can be calculated according to [71]:

$$\langle f(q_x, q_y, q_z) \rangle = \frac{1}{4\pi} \int_0^\pi \mathrm{d}\theta \sin\theta \int_0^{2\pi} \mathrm{d}\phi f(q, \theta, \phi), \tag{29}$$

where $q_x = q\cos\phi\sin\theta$, $q_y = q\sin\phi\sin\theta$ and $q_z = q\cos\theta$, and f is an arbitrarily function. In 2D we can write a similar expression:

$$\langle f(q_x, q_y) \rangle = \frac{1}{2\pi} \int_0^{2\pi} f(q, \phi) \mathrm{d}\phi, \tag{30}$$

where $q_x = q\cos\phi$ and $q_y = q\sin\phi$.

2.2.3. Form Factor

By following a similar procedure as the one used in obtaining Equation (25), one can write the average of the squared form factor in terms of the pair-distance distribution function $p(r) = r^2\gamma(r)$ [71], as:

$$P(q) = 4\pi \int_0^\infty p(r) \frac{\sin qr}{qr} \mathrm{d}r. \tag{31}$$

Here, $\gamma(r)$ represents the correlation function of the fractal. Then, the function $p(r)$ is related to the number of lines with lengths between r and $r + dr$ joining distinct volume elements inside the fractal. It has the properties that $p(r) = 0$ at $r = 0$ and when $r > D$, where D is the maximum distance in the particle. Therefore, for homogeneous fractals as discussed here, $p(r)$ represents a histogram of distances. However, due to the averaging, it does not contain any information about the orientations of these lines. The sine term in the last equation can be approximated by a Taylor series expansion:

$$\sin(qr) \simeq \sin a + \frac{\cos a}{1!}(qr - a) - \frac{\sin a}{2!}(qr - a)^2 - \frac{\cos a}{3!}(qr - a)^3 + \cdots. \tag{32}$$

Since SAS data are recorded near the beam-stop, i.e., close to the zero angle, we can set $a = 0$ in Equation (32). Therefore, we have:

$$\sin(qr) \simeq qr - \frac{1}{3!}(qr)^3 - \frac{1}{5!}(qr)^5 + \cdots. \tag{33}$$

Also, since q is very small, in a good approximation we can keep only the first two terms in the last equation, and thus Equation (31) becomes [71]:

$$
\begin{aligned}
P(q) &\simeq 4\pi \int_0^\infty p(r) \frac{1}{qr} \left(qr - \frac{1}{3!}(qr)^3 - \cdots \right) \mathrm{d}r \\
&\simeq 4\pi \int_0^\infty p(r)\mathrm{d}r - 4\pi \frac{1}{3!} \int_0^\infty p(r)(qr)^2 \mathrm{d}r + \cdots \\
&\simeq 4\pi \int_0^\infty p(r)\mathrm{d}r - 4\pi \frac{q^2}{3!} \frac{\int_0^\infty p(r)\mathrm{d}r}{\int_0^\infty p(r)\mathrm{d}r} \int_0^\infty p(r) r^2 \mathrm{d}r + \cdots \\
&\simeq 4\pi \int_0^\infty p(r)\mathrm{d}r \left(1 - \frac{q^2}{3} \frac{1}{2} \frac{\int_0^\infty p(r) r^2 \mathrm{d}r}{\int_0^\infty p(r)\mathrm{d}r} + \cdots \right) \\
&\simeq P(0) \left(1 - q^2 \frac{R_g^2}{3} + \cdots \right),
\end{aligned}
\tag{34}
$$

where we have the intensity at zero angle given by:

$$
P(0) = 4\pi \int_0^\infty p(r)\mathrm{d}r, \tag{35}
$$

and the radius of gyration:

$$
R_g^2 = \frac{\int_0^\infty p(r) r^2 \mathrm{d}r}{2 \int_0^\infty p(r)\mathrm{d}r}. \tag{36}
$$

As it will be seen further, some properties are discussed in terms of the scattering curve while others in terms of pddf. Generally, the symmetry of a particle is better represented in the reciprocal space, while the shape and structure are better understood in real space.

2.2.4. Structure Factor

Experimentally, the form factor appearing in Equation (27) can be measured for samples with very small concentrations of scattering particles, i.e., when the distance between them is much larger than their sizes. This is because the interference between fractals can be neglected, and the measured data contains information only about the shape and size of particles. However, at high fractal concentrations (typically above 5%), the interference effects can no longer be ignored, and they are described by a structure factor (Equation (28)). Since we deal here only with systems in which the particles are "frozen" in a matrix, the structure factor $S(q)$ is completely defined by their positions.

Please note that throughout the paper one considers a system of randomly oriented fractals whose positions are uncorrelated. This implies small concentrations, and thus the correlations *between fractals* are ignored. However, within a single fractal, the correlations between the basic objects cannot be ignored, and thus the structure factor $S(q)$ is used to describe correlations *inside a fractal*. Therefore, we shall be concerned mainly with the relative positions of the objects forming the fractal, since they are intimately connected with the fractal properties observed in the SAS intensity. Figure 2 illustrates graphically the above observations and explains the basic terms used to study the scattering properties from fractal-based samples. Please note that more generally, $S(q)$ may depend also on the interaction potential between particles, and thus it can deliver information about thermodynamic properties such as the osmotic pressure and compressibility.

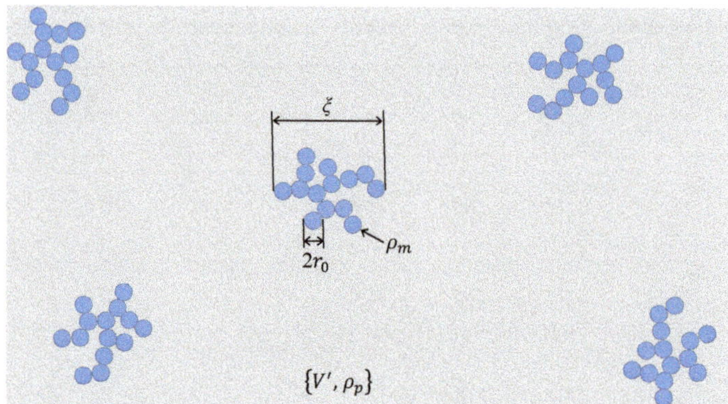

Figure 2. (Color online) Schematic representation of the sample's structure investigated in this work. The irradiated macroscopic volume V' (gray region) consists from a matrix with SLD ρ_p in which are dispersed fractals with SLD ρ_m, overall size ξ, and size of the basic objects $2r_0$. The orientations and positions of the fractals are uncorrelated. Their number is enough large such that an observable signal is recorded at detector, and enough small such that the distances between fractals are larger than their overall size ξ.

2.2.5. Polydispersity

In physical systems, the fractals usually have different sizes. We consider here that all fractals have the same shape while their size are distributed according to a distribution function $D_{\mathrm{N}}(l)$ defined such that $D_{\mathrm{N}}(l)dl$ gives the probability of finding an object whose size falls within $(l, l + \mathrm{d}l)$. Without losing generality, throughout this paper it will be used the log-normal distribution function:

$$D_{\mathrm{N}}(l) = \frac{1}{\sigma l (2\pi)^{1/2}} e^{-\frac{\left(\ln(l/l_0) + \sigma^2/2\right)^2}{2\sigma^2}}, \tag{37}$$

where $\sigma = \left(\ln(1 + \sigma_{\mathrm{r}}^2)\right)^2$ is the variance and $l_0 = \langle l \rangle_D$ is the mean value of the length, $\sigma_{\mathrm{r}} \equiv \left(\langle l^2 \rangle_D - l_0^2\right)^{1/2} / l_0$ is the *relative* variance, and $\langle \cdots \rangle_D = \int_0^\infty \cdots D_{\mathrm{N}}(l) dl$. Therefore, by averaging Equation (27) over the distribution function (37), one obtains [71]:

$$P(q) = n |\Delta\rho|^2 \int_0^\infty \left\langle |F(\boldsymbol{q})|^2 \right\rangle V^2(l) D_{\mathrm{N}}(l) \mathrm{d}l, \tag{38}$$

where $V(l)$ is the fractal volume. Analytic expressions for the form factor of each type of deterministic fractals shall be provided thereafter in their corresponding section. In the limit of strong polydispersity, i.e., when $\sigma_{\mathrm{r}} \gg 1$, the maxima and minima present in the scattering intensity of a deterministic fractal are completely smoothed and thus the intensities form random fractals can be modeled by using deterministic fractals with a high degree of size polydispersity.

Other important properties which will be used in calculating the SAS intensity are:

- $F(\boldsymbol{q}) \to F(\beta_{\mathrm{s}} \boldsymbol{q})$ when the length is scaled as $L \to \beta_{\mathrm{s}} L$,
- $F(\boldsymbol{q}) \to F(\boldsymbol{q}) e^{-i\boldsymbol{q}\cdot\boldsymbol{a}}$ when the particle is translated $\boldsymbol{r} \to \boldsymbol{r} + \boldsymbol{a}$,
- $F(\boldsymbol{q}) = \left[V_I F_I(\boldsymbol{q}) + V_{II} F_{II}(\boldsymbol{q})\right] / (V_I + V_{II})$, when the particle consists of two non-overlapping subsets I and II.

Certainly, for 2D models, all the volumes shall be replaced by the corresponding areas.

2.3. Monte Carlo Simulations

A Monte Carlo-based algorithm similar to the one described in Ref. [75] is used to generate the pddf $p(r)$. To this aim, the area occupied by a given fractal is filled with a large number of randomly generated points pairs (i, j), and all the distances $|\boldsymbol{r}_i - \boldsymbol{r}_j|$ between them are collected into a histogram with about M = 250 boxes. Such a histogram gives directly the function pddf: $p(r) = C^{-1} \sum_{(i,j)} |\boldsymbol{r}_i - \boldsymbol{r}_j| \xi^{-1} M$. Here, the coefficient C ensures that the form factor is normalized such that $P(0) = 1$. Then, for arbitrarily values of the scattering vector q, the form factor is calculated by performing numerically the Fourier transform in Equation (31). The simulated scattering curves are used to validate the analytic expressions for the form factors.

Figure 3 illustrates the results of the above steps for a simple example of a disk of radius a. At the first step, the square which circumscribe the disk is approximated by a set of random points (upper left part). Those found inside the disk are kept (blue points) and the others (gray points) are eliminated. The upper-right part of the same figure shows the corresponding pddf, while the lower part shows the corresponding scattering intensity obtained from Equation (31) (red curve). For comparison, the same figure shows the analytic curve (black) of an infinite thin circular disk of the same size, and with a known expression of the form factor given by [76]:

$$F_{\text{disk}}(qa) = \frac{2J_1(qa)}{qa}, \tag{39}$$

where $J_1(\cdot)$ is the Bessel function of the first kind and the first order.

The results show an excellent agreement between the two curves in almost the whole q-range. However, small deviations begin to arise at large values of q ($\gtrsim 40$) and are due to the intrinsic property of Monte Carlo method to approximate the disk by a finite collection of random points. The approximation can be improved by increasing the number of scattering points.

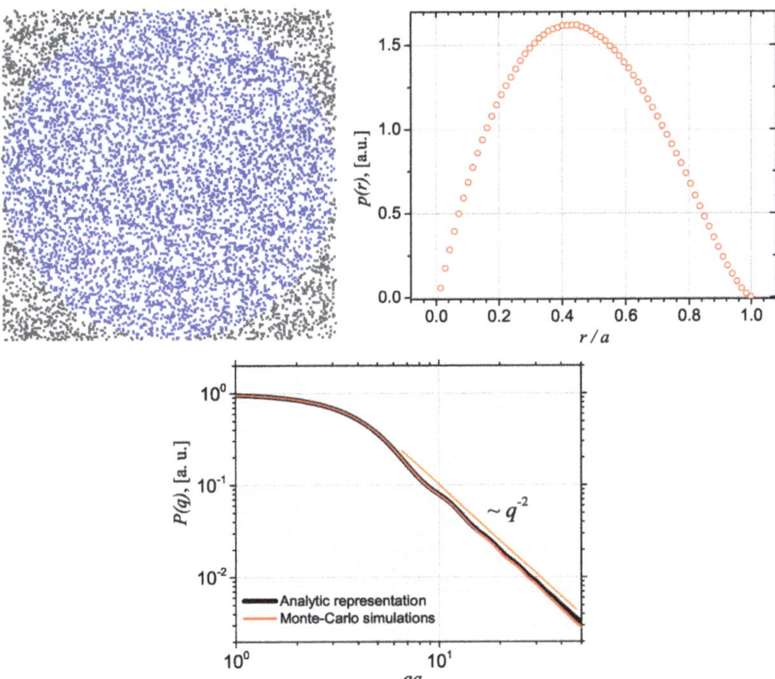

Figure 3. (Color online) Upper left: Approximating a disk of radius a with a set of randomly distributed points (blue). Upper-right: the corresponding pddf $p(r)$. Lower part: A comparison between the analytic expression (black curve) and Monte Carlo simulations (red curve) of SAS intensities.

3. Small-Angle Scattering from Fractals

In this section, analytic forms of scattering intensity shall be derived for various types of fractals, by providing and deriving explicit expressions for the form, and respectively the structure factor appearing in Equation (26). For simplicity, the fractals considered here consist either from disks or from balls, but any other geometric form can be used. While in the former case the form factor is given by Equation (39), in the later one, the form factor of a ball of radius a can be written as [71]:

$$F_{\text{ball}}(qa) = \frac{3(\sin qa - qa\cos qa)}{(qa)^3}. \tag{40}$$

3.1. Random Mass Fractals

To derive the scattering structure factor of a mass fractal, let us choose first an imaginary sphere centered on the fractal and consider a spherical layer of radius r and width $\mathrm{d}r$. Then, the number of balls (particles) within the spherical layer is:

$$\mathrm{d}N(r) = ng(r)\mathrm{d}V, \tag{41}$$

where $\mathrm{d}V = 4\pi r^2\mathrm{d}r$ is the layer volume. Thus, the total number of particles within the sphere is given by: $N(r) = n\int_0^r g(r)4\pi r^2\mathrm{d}r$ [39]. By differentiating Equation (5) and comparing with Equation (41), one obtains [39]:

$$ng(r) = \frac{D_{\text{m}}}{4\pi a^{D_{\text{m}}}} r^{D_{\text{m}}-3}, \tag{42}$$

where $a = r_0$ (see Figure 2).

This expression shows that $g(r) \to 0$ for large values of r, since the fractal dimension is $D_{\text{m}} < 3$. However, any sample is characterized by a macroscopic density at large scale. Thus, in order to describe the large-scale behavior of $g(r)$, an exponential term of the form $e^{-(r/\xi)^\beta}$, known as the cutoff function is introduced [77]. Here, ξ is the fractal outer cutoff, and it gives the distance above which the mass distribution is no longer described by the mass fractal law, and thus it coincides with the fractal size, as pointed out in Section 1. This cutoff function is faster than any power law, and it is clear that the larger the value of β, the sharper the cutoff. In the reciprocal space, an increase of β leads to the formation of a distinct jump just beyond the Guinier regime [77]. After this hump, the power-law decay $q^{-D_{\text{m}}}$ (see Equation (47)) is still visible but at different levels, depending on the values of β, and "ripples" occur for $\beta \gtrsim 5$ [77]. One of the most common cutoff functions, which provides a reasonable assumption for describing the behavior of $g(r)$ at large distances, corresponds to $\beta = 1$, and this value will be used in the following. With the introduction of the exponential cutoff, a uniform density shall be subtracted, to avoid the divergence of the structure factor, and therefore we have [39]:

$$n(g(r) - 1) = \frac{D_{\text{m}}}{4\pi a^{D_{\text{m}}}} r^{D_{\text{m}}-3} e^{-r/\xi}. \tag{43}$$

By introducing Equation (43) into Equation (28), one can write:

$$S(q) = 1 + \frac{D_{\text{m}}}{a^{D_{\text{m}}}} \int_0^\infty r^{D_{\text{m}}-1} e^{-r/\xi} \frac{\sin qr}{qr} \mathrm{d}r \tag{44}$$

$$= 1 + \frac{1}{(qa)^{D_{\text{m}}}} \frac{D_{\text{m}}\Gamma(D_m - 1)}{(1 + q^{-2}\xi^{-2})^{(D_{\text{m}}-1)/2}}, \tag{45}$$

where $\Gamma(\cdot)$ is the gamma function. Depending on the values of the scattering vector q, we distinguish several main regions in the scattering intensity in Equation (31), with the structure factor given by Equation (45) and form factor by Equation (40):

- When $q \to 0$,
$$S(q) \simeq 1 + \Gamma\left(D_{\mathrm{m}} + 1\right)\left(\xi/a\right)^{D_{\mathrm{m}}}\left(1 - q^2\xi^2 D_{\mathrm{m}}\left(D_{\mathrm{m}} + 1\right)/6\right), \tag{46}$$
and therefore the scattering intensity is dominated by $S(q)$. By comparing this with Equation (34) one can see that the outer cutoff ξ is related to the mass fractal radius of gyration R_g by [39]: $R_g^2 = D_{\mathrm{m}}\left(D_{\mathrm{m}} + 1\right)\xi^2/2$.
- When $\xi^{-1} \lesssim q \lesssim a^{-1}$,
$$S(q) \propto q^{-D_{\mathrm{m}}}, \tag{47}$$
and therefore $S(q)$ still dominates in Equation (34) since yet $P(q) \simeq 1$. The last equation shows that the scattering intensity of a mass fractal is a straight line on a double logarithmic scale. Therefore, deviations from this line at low and values of the scattering vector q provide information about the fractal overall size ξ, and respectively about the radius a of fractal basic unit.
- When $q \to \infty$,
$$S(q) \simeq 1 + \frac{D_{\mathrm{m}}\Gamma\left(D_{\mathrm{m}} - 1\right)\sin\left(\left(D_{\mathrm{m}} - 1\right)\pi/2\right)}{q^{D_{\mathrm{m}}} a^{D_{\mathrm{m}}}}, \tag{48}$$
and thus, for $qa \gg 1$, $S(q) \to 1$, while $P(q)$ dominates in the scattering intensity [73]. Therefore, scattering at large q values provides information about the fractal basic particles.

As an application of the above results, let us consider a two-dimensional (2D) DLA cluster, as a representative example of random mass fractals. The DLA is generated by particles undergoing a random walk due to Brownian motion, and which cluster into aggregates [78]. It is widely used to describe structures in which diffusion is the main means of transport, and it occurs often in dielectric breakdown, electrodeposition or Hele-Shaw flow.

The overall size of the generated DLA is $\xi \simeq 234$ nm and consists from particles of radius $a \simeq 1$ nm (Figure 4—upper row). The numerical values were chosen such that they fit into a typical range covered by SAS measurements. The corresponding pddf is shown in Figure 4—lower row (left; black points). For comparison, the pddf of an equivalent disk, i.e., a disk centered on the cluster and with radius $\xi/2$ is also shown (red points). The total number of distances used in calculating each pddf is about 32.7×10^5. When the positions of the disks forming the DLA and their shape are known, the calculations can be greatly simplified if one uses only the mass-center coordinates, since the number of points used in calculating the pddf is much lower. The corresponding Fourier transform will give the structure factor describing the relative positions of the disk inside the aggregate. By multiplying it with the disk form factor, the scattering intensity (Equation (31)) is recovered.

The results show that in the case of DLA, although the pddf has an overall similar behavior as for the disk, it is characterized by the presence of local minima and maxima, with the main maxima (at $r \simeq 75$ nm) shifted to the left. These reflects deviations of the DLA from the circular symmetry of the disk. Also, the height of the maxima increases, and indicates the number of most common distances within DLA. This peak is followed by a second one, slightly shifted to the left (at $r \simeq 110$ nm), and less pronounced, and which partially follows the disk's pddf for 100 nm $\lesssim r \lesssim$ 140 nm. This indicates that within this range the number of distances within DLA and the disk coincide. Finally, for 140 nm $\lesssim r \lesssim$ 234 nm the number of distances within DLA decreases faster, and both reach the same maximum value at $\xi \simeq 234$ nm. Please note that at $r \simeq 225$ nm the DLA's pddf is quite flat in this region, and may pose difficulties in an accurate determination of overall size ξ from SAS data, due to experimental errors [79].

Figure 4 lower row right shows the SAS intensity (red curve) from DLA obtained by Fourier transform (Equation (31)) of the corresponding pddf. The analytic expression (Equation (47)) is also shown in the fractal region (black curve). The results show a very good agreement between the two curves. The scattering exponent is 1.66 and coincides with the analytic value from mean-field theory for diffusion-limited cluster formation $(d^2 + 1)/(d + 1)$ with $d = 2$ [80] and with other computer simulations of Witten-Sander aggregates [81].

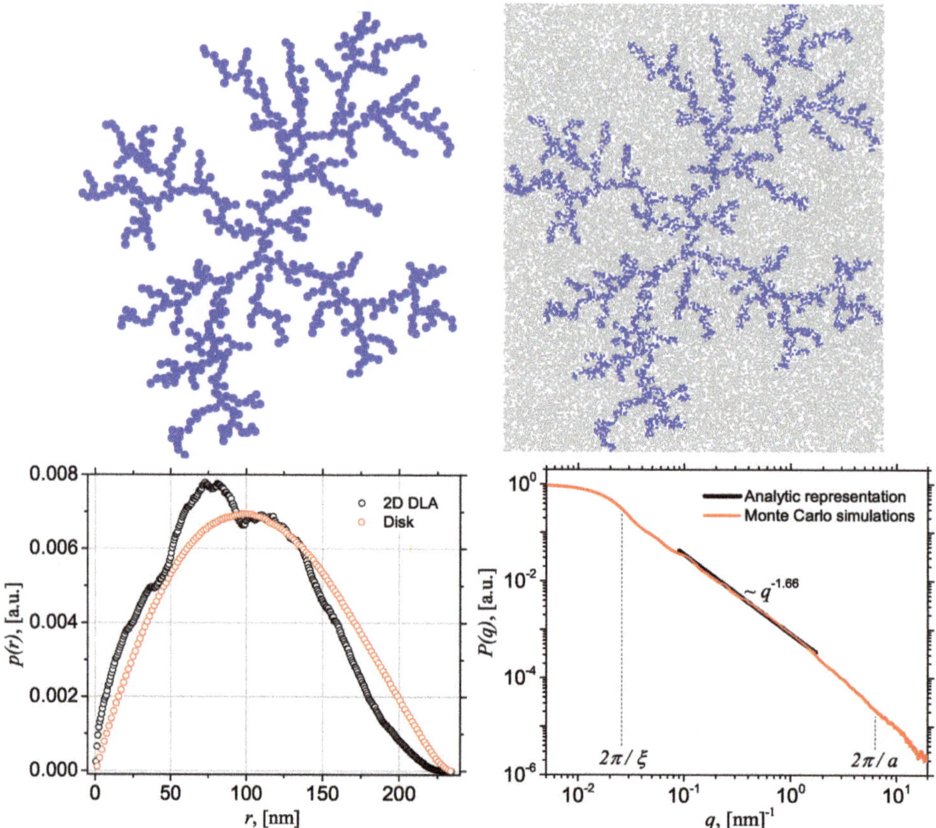

Figure 4. (Color online) Upper row: Left—2D DLA cluster of overall size $\xi \simeq 234$ nm consisting from about 500 particles of radius $a \simeq 1$ nm, Right—the corresponding approximation with a set of randomly distributed points (blue). Lower row: Left—pddf of the 2D DLA (black points) and of an equivalent disk (red; see text for additional explanations), Right—A comparison between the analytic expression of SAS intensity (black curve) with Monte Carlo simulations (red curve).

3.2. Random Surface Fractals

To derive the scattering law for surface fractals, one considers a system consisting from a random distribution of material with constant SLD ρ. By denoting with c the fraction of the volume occupied by the material, then $1 - c$ denotes the fraction of unoccupied volume, and the average density of the sample can be expressed as ρc. Therefore, the density fluctuations can be expressed as:

$$\eta(\mathbf{r}) = \begin{cases} \rho(1-c), & \mathbf{r} \in \text{occupied volume} \\ -\rho c & \mathbf{r} \notin \text{occupied volume}, \end{cases} \tag{49}$$

and the mean square volume is:

$$\left\langle \eta^2 \right\rangle = \rho^2 c(1-c). \tag{50}$$

Thus, Equation (25) becomes:

$$P(q) = 4\pi\rho^2 c(1-c) \int_0^\infty r^2 \gamma(r) \frac{\sin qr}{qr} \mathrm{d}r, \tag{51}$$

where the correlation function is defined by [38]:

$$\gamma(r) = \frac{Z(r) - c}{1 - c}. \tag{52}$$

The function $Z(r)$ represents the probability that if a point is in the occupied volume, then a second point situated at a distance r apart from the second point, belongs also to the occupied volume. Thus, from definition, $Z(r) = 1$ excepting those points found at distance r apart from the pore boundary. In the latter case, the volume of the layer with occupied regions is $V_b \equiv S_0 r^{3-D_s}$, while in the former case, the volume is $cV' - V_b$. Also, it is known that in the limit $r \to 0$, the function $Z(r)$ is given by [38]:

$$Z(r) \simeq 1 - \frac{S_0}{4cV'} r^{3-D_s}, \tag{53}$$

and therefore, by using Equation (52), the correlation function becomes [38]:

$$\gamma(r) = 1 - \frac{S_0}{4c(1-c)V'} r^{3-D_s}. \tag{54}$$

Performing two integration by parts in Equation (51) and taking into account that both $\gamma(r)$ and its derivative tend to zero at large r, the scattering intensity can be rewritten as [38]:

$$P(q) = -4\pi\delta^2 c(1-c)q^{-3} \int_0^\infty [rg\gamma''(r) + 2\gamma'(r) \sin qr] \, \mathrm{d}r. \tag{55}$$

Then, according to Erdélyi's theorem for asymptotic expansion of Fourier integrals, in the limit of large values of the scattering vector q, the intensity becomes:

$$P(q) = -\frac{4\pi\rho^2}{q^{6-D_s}} c(1-c)\Gamma(3-D_s) \sin\left[\pi\left(D_s - 1\right)/2\right] \lim_{r\to 0} \{ r^{D_s - 2} \frac{\mathrm{d}^2}{\mathrm{d}r^2} [rg(r)] \}. \tag{56}$$

Finally, by using the correlation function given by Equation (54), the scattering intensity can be rewritten as:

$$P(q) \simeq \pi S_0 \rho^2 \Gamma(5 - D_s) \sin\left[\pi\left(D_s - 1\right)/2\right] q^{6-D_s}, \tag{57}$$

which shows that for a surface fractal [38]:

$$P(q) \propto q^{-(6-D_s)}. \tag{58}$$

A similar procedure can be used to show that for $d = 2$, $P(q) \propto q^{-(4-D_s)}$. Therefore the scaling law $P(q) \propto q^{-(2d-Ds)}$ is recovered, and taking into account that $S(q) = 1$ for surface fractals discussed here, the scattering intensity in Equation (1) is recovered. Please note that when $D_s = 2$, Equation (57) reduces to Porod law, where $P(q) \simeq 2\pi\rho^2 S_0 q^{-4}$, and the density profile varies sharply over distances smaller than q^{-1}.

Since $S(q) = 1$, the function $Z(r)$ in Equation (53) gives the probability that a point at distance r in an arbitrarily direction from a given point inside the fractal will itself be in the fractal. This holds true since there is only a small probability of finding an occupied point outside a particular fractal in which the origin is chosen. As such, c is negligible and thus $1 - c \simeq 1$, Therefore, $Z(r) = \gamma(r)$, where $\gamma(r)$ represents now the correlation function of the fractal, and not of the whole sample, as in Equation (25). Therefore, Equation (51) gives:

$$P(q) = 4\pi \int_0^\infty r^2 \gamma(r) \frac{\sin qr}{qr} \mathrm{d}r, \tag{59}$$

with the SLD factor ρ^2 included in Equation (26). Thus, by using the property that $p(r) = r^2\gamma(r)$, Equation (31) is recovered.

To illustrate the above results, we consider in the following a random surface fractal model which consists from disks whose radii follow a power-law distribution. The model is shown in Figure 5 upper left row, where disks with identical radii have the same color. Specifically, one considers that the radius of the largest disk (blue), also known as the zero-th fractal iteration ($m = 0$) is r_0, and the scaling factor $\beta_s = 1/2$. Then, the radius of the disks forming the first fractal iteration (green disks) is $r_1 = \beta_s r_0$, the radius of the disks forming the second fractal iteration (black disks) is $r_2 = \beta_s^2 r_0$, and so on. From the construction one can see that each disk is tangent to at least another disk and they form an aggregate, of overall size $\xi \simeq 234$ nm. Since the number of disks at a given iteration increases by a factor of 3, then the fractal dimension of the model is $D_s \equiv \log 3/\log 2 \simeq 1.58$, and therefore in the fractal region the corresponding form factor is given by (see discussion below Equation (58)):

$$P(q) \propto q^{-(4-1.58)} = q^{-2.42}. \tag{60}$$

Figure 5 upper-right row shows the discretized version of the surface fractal model used for computing the pddf in Figure 5 lower-left row (black). For comparison, the pddf of a disk of equivalent radius is also shown in red. The total number of distances used in both cases is about 1.54×10^8. The results show that the pddf of the random surface fractal has a similar behavior as of the pddf of DLA (see Figure 4 lower-left row) in the sense that the main maximum is shifted to the left with respect to the disk's maxima, while its height increases. Thus, this reflects the asymmetry of the model as compared to the circular one. However, as opposed to the pddf of DLA, for surface fractals the maxima are much less pronounced, and generally the curve is smoother. This indicates a more uniform structure of the surface fractal, which arise from the existence of disks of different sizes.

For surface fractals, it has been recently shown that the scattering intensity can be obtained by neglecting the spatial correlations between disks, and therefore one can use the approximation of independent units, i.e. [60]:

$$P(q) \simeq \sum_{n=0}^{m} \beta_s^{n(4-D_s)} P_0 (\beta_s q) , \tag{61}$$

where $P_0(q) \equiv \langle |F_{\mathrm{disk}}(q)|^2 \rangle$, and F_{disk} is given by Equation (39). The corresponding intensity is shown in Figure 5 lower-right row (black), where one can see that the scattering exponent is in excellent agreement with the analytic value obtained above. The Monte Carlo simulations (red curve) are also in very good agreement with the theoretical results. This is to be expected, since there is an excellent agreement between theoretical and Monte Carlo-based curves of a disk (see Figure 3 lower row), while Equation (61) is just a sum of SAS intensities of disks of various radii. The edges of the fractal region are marked in Figure 5 lower row right by dotted vertical lines, and they provide information about the overall fractal size, and about the size of the smallest disk composing the fractal.

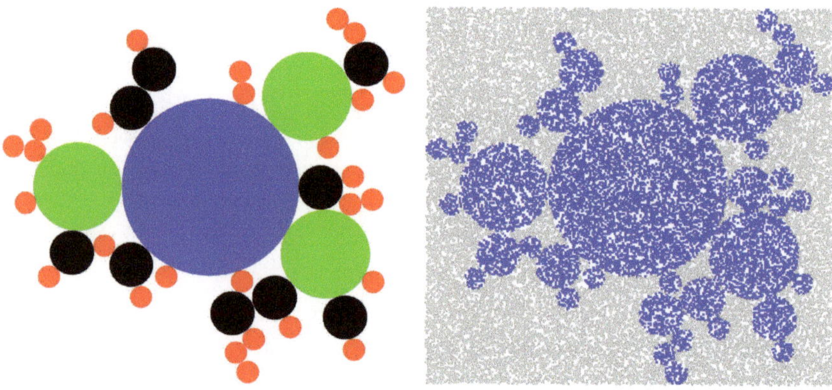

Figure 5. *Cont.*

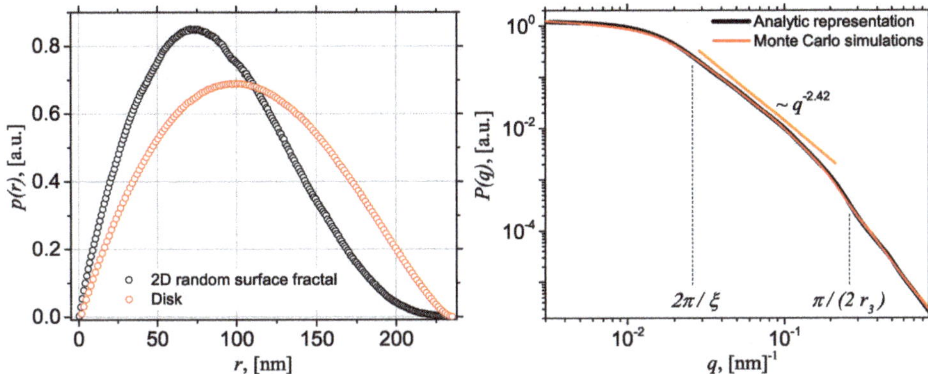

Figure 5. (Color online) Upper row: Left—random surface fractal of overall size $\xi \simeq 234$ nm consisting from disks of different radii, Right—the corresponding approximation with a set of randomly distributed points (blue). Lower row: Left—pddf of the random surface fractal (black points) and of an equivalent disk (red), Right—A comparison between the analytic expression (black curve) of SAS intensity (Equation (34)) and Monte Carlo simulations (red curve).

3.3. Deterministic Mass Factals

In the following, the scattering properties from deterministic mass fractals are illustrated on fractals similar to the well-known Cantor fractal. For this purpose, one considers first a square with edge l_0 which circumscribes a disk of radius $r_0 = l_0/2$ such that their centers coincide. One also considers a Cartesian system of coordinates whose origin lies in the square center, and with axes parallel to the square edges. At first iteration, the disk of radius r_0 is replaced by 5 smaller disks of radius $r_1 = \beta_{\mathrm{s}} r_0$, with $\beta_{\mathrm{s}} = 1/3$. One disk is situated in the origin, while the centers of the other disks are shifted from the origin by:

$$\boldsymbol{a}_j = \{0, \pm\beta_{\mathrm{t}} l_0\} \quad \text{and} \quad \boldsymbol{b}_j = \{\pm\beta_{\mathrm{t}} l_0, 0\}, \tag{62}$$

with all combinations of the signs, and $\beta_{\mathrm{t}} = (1 - \beta_{\mathrm{s}})/2$. The second iteration is obtained by performing a similar operation on each of the 5 disks of radius r_1, and so on. The "cross" mass fractal is obtained in the limit of infinite number of iterations. Figure 6 upper row shows the construction at iterations $m = 1, 2$ and 3, while in the middle row is shown the discretized version, for the same iterations. The total number of disks at m-th iteration is $k_m = 5^m$, and the radius of each disk is $r_m = \beta_{\mathrm{s}} r_0$. Therefore, in accordance with Equation (5), the fractal dimension is given by:

$$D_{\mathrm{m}} = \lim_{m\to\infty} \frac{\log k_m}{\log l_0/r_m} = \frac{\log 5}{\log 3} \simeq 1.46, \tag{63}$$

and the total area of the fractal at m-th iteration is $A_m = k_m A_0$, where $A_0 = \pi r_0^2$ is the surface area of the disk at $m = 0$.

The position of the disk centers inside the fractal can be written as:

$$G_m(\boldsymbol{q}) = \frac{1}{5}\left(1 + 2\cos q_x u_m + 2\cos q_y u_m\right), \tag{64}$$

where $u_m = l_0 \beta_{\mathrm{t}} \beta_{\mathrm{s}}^{m-1}$, and $G_0(\boldsymbol{q}) \equiv 1$. Then, for arbitrarily iteration number $m = 0, 1, 2, \cdots$, the fractal form factor is given by:

$$F_m(\boldsymbol{q}) = F_0(r_m q) \prod_{i=0}^{m} G_i(\boldsymbol{q}). \tag{65}$$

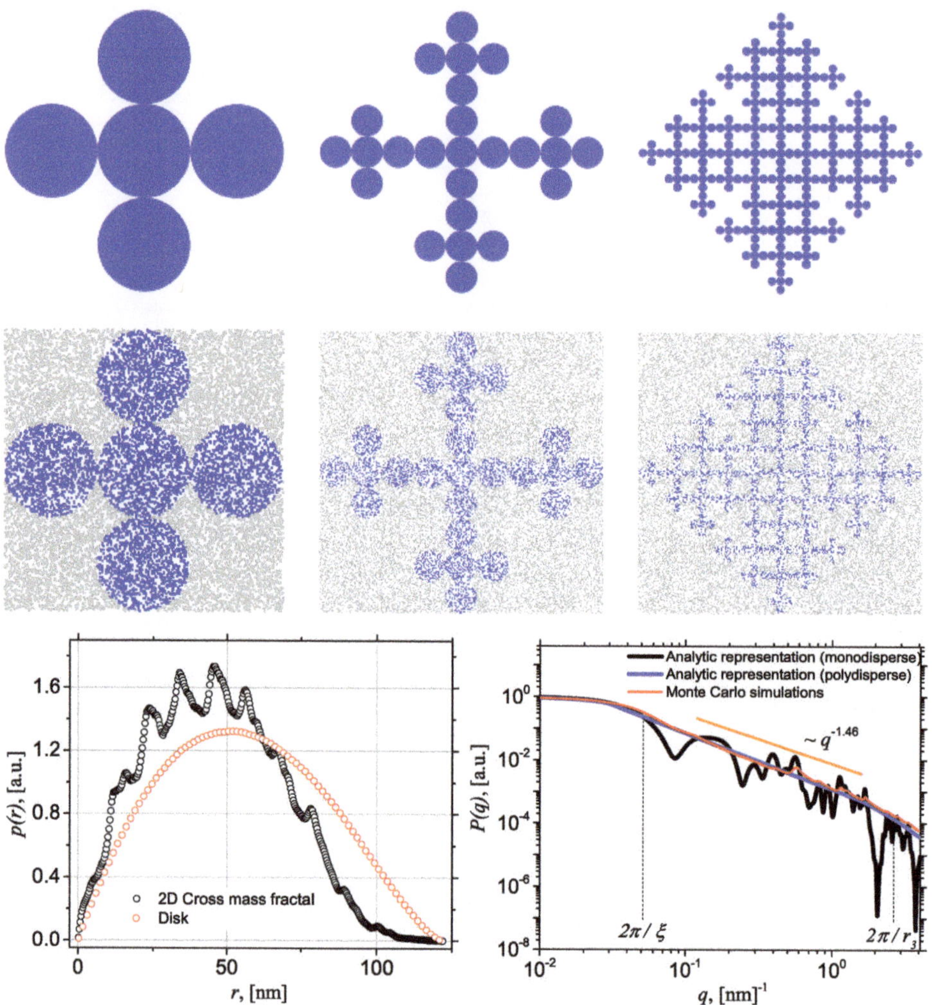

Figure 6. (Color online) Upper row: Iterations $m = 1, 2$ and 3 of the deterministic "Cross" mass fractal of overall size $\xi \simeq 125$ nm. Middle row: the corresponding approximation with a set of randomly distributed points (blue). Lower row: Left—pddf of the "Cross" mass fractal (black points) and of an equivalent disk (red points), Right—A comparison between the mono- and polydisperse analytic expression (black, and respectively blue curve) of SAS intensity given by Equation (26) with the form factor given by Equation (65), and Monte Carlo simulations (red curve). The polydisperse curve is calculated with Equation (38), and the relative variance used is $\sigma_r = 0.5$.

The pddf of the cross-mass fractal at iteration $m = 3$, overall dimension $\xi \simeq 125$ nm, is shown in Figure 6 lower row left (black). For comparison, the pddf of the equivalent disk is shown also (red). The main feature of fractal's pddf is the presence of a succession of local maxima and minima on a parabola-like function similar to that of the disk, but with the maximum shifted to smaller values of r. These features reflect both the deviations from the circular symmetry of the fractal, as well as its particular feature, in which a periodic structure arise in its construction. Please note that at $r \gtrsim 110$ nm the fractal's pddf is almost flat, thus making it difficult to estimate its overall dimension from SAS data.

The fractal form factor at $m = 3$ obtained from the Fourier transform of the pddf is shown in Figure 6 lower row right in black, while the scattering curve obtained from Monte Carlo simulations is shown in red. The analytic curve is characterized by the presence of the three main regions: Guinier, at $q \lesssim 2\pi/\xi$, fractal, at $2\pi/\xi \lesssim q \lesssim 2\pi/r_3$, where r_3 is the radius of the disk at iteration $m = 3$, and Porod, at $q \gtrsim 2\pi/r_3$. In the fractal region, the scattering curve consists from a superposition of maxima and minima on a power-law decay. This is known in the literature as a generalized power-law decays (see also Introduction section). The exponent of the power-law decay is approximately 1.46, which is agreement with the analytic value given by Equation (63). The number of main minima in the fractal region is equal to 3, and it coincides with the fractal iteration number. The periodicity of these minima on a double logarithmic scale is $\log_{10}(1/\beta_s)$, and it can be used to extract information about the fractal scaling factor. By knowing the fractal dimension D_m, iteration number m and the scaling factor β_s, the number of disks composing the mass fractal can be determined from the relation $k_m = (1/\beta_s)^{mD_m}$ [58]. When measurements are performed on an absolute scale, the Porod region can be used to extract the ratio P_m/A_m, where P_m is the total perimeter of the fractal at m-th iteration, and A_m is the total surface area (see above). For 3D fractals this would give the specific surface A_m/V_m, where V_m is the fractal volume at m-th iteration.

The corresponding polydisperse curve (blue) is calculated by using Equation (37), where the relative variance is $\sigma_r = 0.5$. In this case, the simple power-law decay, specific to scattering from random mass fractals (see Figure 4) is recovered, and the scattering exponent is preserved. Please note that generally, the oscillations in the fractal region are still visible up to $\sigma_r \lesssim 0.4$, and thus the polydisperse curve can be used to extract the fractal iteration number and scaling factor, as discussed before for the monodisperse case. This is important from an experimental point of view, since in practice it is very hard to prepare samples containing perfectly monodisperse fractals.

The Monte Carlo simulations (red) show a good agreement with the polydisperse curve, excepting the presence, in the former case, of a local bump at $q \simeq 6 \times 10^{-1}$ nm^{-1}, and of slight deviations at q close to the end of the fractal region. This agreement is to be expected, since the behavior of SAS intensity in the fractal region is dominated by the relative positions of the disks inside the fractal, as described by the generative function given by Equation (64). This implies that one can calculate the intensity based only on the center-of-mass positions of each disk [58]. Then, each of the most pronounced maxima and minima in the fractal region corresponds to the interference cluster amplitude, where each cluster represents a fractal iteration. The positions and amplitudes of these maxima and minima are related to the most common distances between the center-of-masses [58]. However, when the scattering intensity is calculated by using randomly generated points, we use only an approximation of the fractal shape and not the distances between center-of-masses. As such, the number of possible distances used is much higher, and thus the maxima and minima are significantly smeared out. As described above, the polydispersity also leads to smearing the maxima and minima. Thus, although the effects on the scattering curve are similar, their nature is different.

Please note that in building the cross-mass fractal, a single scaling factor has been used at each iteration, while the ratio between the size of the disks to the distances between them is about unity. By releasing these constraints, additional fractal classes, each one with its own "signature" on the scattering curve, can be investigated. In particular, by increasing the value of the scaling factor after a predefined number of iterations (every second, every third, etc.), one obtains a fractal with positive Lebesgue measure (also known as fat fractals). It has been recently shown that for fat fractals, the scattering curve consists from a succession of power-law decays with increasing values of the scattering exponents [59]. Such models can be used to describe the properties of hierarchically structured materials in which the fractal dimension varies with the scale (iterations with constant scaling factor). However, within each scale, the scattering behavior of regular (thin) fractals is recovered (Figure 4). Also, allowing the distances between disks to be much bigger than the disk's size, induces the appearance of a constant region (also known as a plateau or asymptotic) between the fractal and

Porod regions [60]. Therefore, the beginning and the end of this plateau can be used to estimate the length of the minimal distances between disks, and respectively their size.

3.4. Deterministic Surface Fractals

The construction process of the deterministic surface is similar to that of the cross-mass fractal, where an initial disk is repeatedly divided into a set of smaller structures, according to the same rule for each iteration. If we denote by r_0 the radius of the disk at $m = 0$ (initiator), then at the first iteration ($m = 1$; generator), there are also four disks of radius $\beta_s r_0$ whose centers are situated on the positive and negative coordinate axes, at a distance $(1 + \beta_s) r_0$ from the origin. At the m-th approximation, the surface fractal is built from the $(m-1)$-th approximation by adding disks of radii $\beta_s^m r_0$ placed at a distance $(1 + \beta_s) r_0$ from the centers of the disks of radii $\beta_s^{m-1} r_0$, in all directions of the coordinate axes and on the positions that are already not occupied by other disks (Figure (7) upper row).

At m-th iteration, the obtained Cross-surface fractal is built as a sum of mass fractals at iterations n from 0 to m. In Figure (7) upper row, each mass fractal is colored differently, and thus at $m = 3$ the surface fractal consists from mass fractals at iterations $n = 0, 1, 2, 3$. By construction, one can see that at m-th iteration, the overall size ξ of the surface fractal is very well approximated by the overall size of the mass fractal at iteration $n = m$. Please note that from the point of view of scattering properties, the main difference between cross-mass and surface fractals is that the former ones consist from disks of the same size, while the later ones consist from disks of different sizes, following a power-law distribution. Their number is given by:

$$k_m = 2 \times 3^m - 1, \tag{66}$$

and their radii are distributed in the following way: one disk of radius r_0 (blue), 4 disks of radius $r_1 = \beta_s r_0$ (green), 12 disks of radius $r_2 = \beta_s^2 r_0$ (black), and so on. Since at m-th iteration the radius is $r_m = \beta_s^m r_0$, then the surface fractal dimension is:

$$D_s = \lim_{m \to \infty} \frac{\log k_m}{\log (r_0 / r_m)} = 1. \tag{67}$$

The discrete version of the surface fractal is shown in Figure 7 middle row, and the corresponding pddf at $m = 3$ is shown in Figure 7 lower-low left (black). For comparison, the pddf of the equivalent disk is also shown (red). The figure clearly shows that the pddf of the surface fractal is much smoother than the pddf of the cross-mass fractal (Figure 6 lower row left). This is due to the presence of disks of various sizes in th former case. However, the overall behavior is similar to the one of mass fractal where the main maximum is shifted to the left of disk's maximum, while at large r, the pddf is almost completely flat. These features reflect the deviations from the circular symmetry, and which are specific to the surface fractal model.

Since the cross-surface fractal consists from a superposition of mass fractals at various iterations, by adding their amplitudes and normalizing the result to unity at $q = 0$, one can write [60]:

$$\left\langle |F_m^{(\mathrm{sf})}(q)|^2 \right\rangle = \frac{1 - h\beta_s^2}{1 - (h\beta_s^2)^{m+1}} \sum_{n=0}^{m} \left(h\beta_s^2\right)^{2n} \left\langle |F_n^{(\mathrm{mf})}(q)|^2 \right\rangle, \tag{68}$$

with $h = 3$, and where $F_n^{(\mathrm{mf})}(\cdot)$ is the mass fractal form factor at n-th iteration. Therefore, the surface fractal form factor can be written as:

$$P(q) = P(0) \left\langle |F_m^{(\mathrm{sf})}(q)|^2 \right\rangle, \tag{69}$$

where $P(0) = n|\Delta\rho|^2 A_m^2$, and A_m is the total surface area of the fractal at m-th iteration.

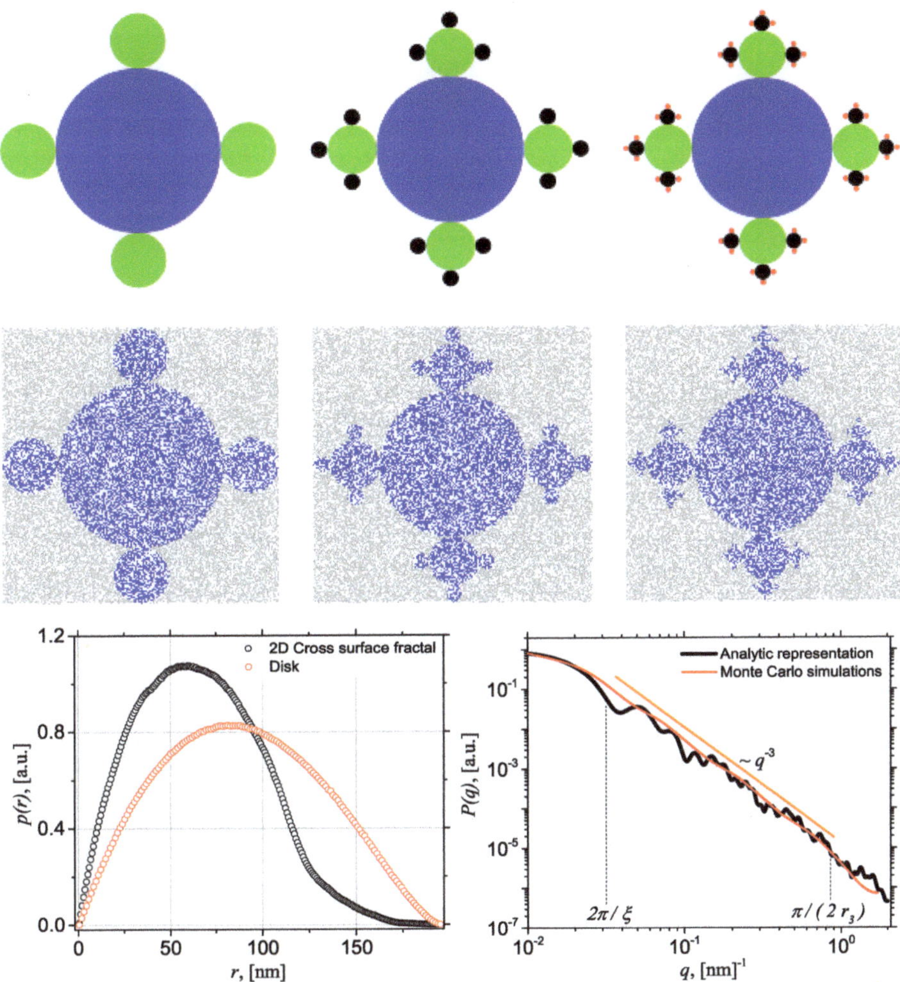

Figure 7. (Color online) Upper row: Iterations $m = 1, 2$ and 3 of the deterministic "Cross" surface fractal of overall size $\xi \simeq 195$ nm (at $m = 3$). Middle row: the corresponding approximation with a set of randomly distributed points (blue). Lower row: Left—pddf of the random surface fractal (black points) and of an equivalent disk (red points), Right—A comparison between the approximation of independent units (black curve) of SAS intensity (Equation (61)) and Monte Carlo simulations (red curve).

Please note that in some cases it is very difficult to derive an analytic expression for the mass fractal form factor in Equation (69). However, if we are not interested in the intricate behavior of the scattering curve, one can use the approximation of independent units (AIU) in Equation (61) to calculate the scattering intensity of surface fractals. Here, this approach is used, and the results are shown in Figure 7 lower row right (black). It is clear that in the fractal region i.e., for $2\pi/\xi \lesssim q \lesssim \pi/(2r_3)$, the scattering intensity decays proportional to $q^{-(4-1)} = q^{-3}$. Thus, the fractal dimension of the surface fractal is $D_s = 1$, as predicted by Equation (67). Despite using the AIU, in the fractal range one can observe the presence of three pronounced minima at $q \simeq 3.5 \times 10^{-2}, 3.5 \times 10^{-2}$ and 3.5×10^{-1} nm^{-1}. This shows that the surface fractal consists from three iterations of mass fractals. The minima positions also show an approximate log-periodicity with the scale factor $1/\beta_s$. The results are similar to those of scattering from deterministic cross-mass fractals, but the nature of the log-periodicity is different. While in the case of scattering from mass fractals, the periodicity arises from the self-similarity of distances between disks, for surface fractals, it arises from the self-similarity of disk sizes.

Monte Carlo simulations are presented in the same figure (red), and show that the scattering exponent in the fractal region of the scattering curve is recovered, but without following in detail the small oscillations present in the analytic curve. This effect is similar to the one seen in Figure 6 lower row right for mass fractals, and arise due to the discretization of the surface fractal with randomly positioned points.

3.5. Deterministic Multifractals

Single-scale fractal presented so far, lead to homogeneous structures and are the simplest cases of fractal systems. An exception is the fat fractal discussed in Section 3.3, which although have different scaling factors, they occur at different scales. However, real fractal may have far more reach scaling and self-similar properties that change from point to point. They are known as multifractals, and to model their scattering properties, one should consider structures in which various scaling factors occurs at the same scale.

For simplicity, we consider in the following a multifractal built from two scaling factors β_{s1} and β_{s2}. The initiator ($m = 0$) is a single disk, while the first iteration ($m = 1$) consists from four disks found on the coordinate axes, as in the case of deterministic cross-mass and surface fractals (see Figures 6 and 7 upper rows), each one with the scaling factor β_{s1}, and one disk in the center, with the scaling factor β_{s2}. The second iteration is obtained by repeating the same procedure for each of the five disks at $m = 1$, and so on. It is clear that when $\beta_{s1} = \beta_{s2}$, one recovers the single-scale deterministic cross-mass fractal shown in Figure 6 upper row.

Figure 8 upper row shows the cross multifractal models at iteration $m = 2$ for $\beta_{s1} = 0.15$ and $\beta_{s2} = 0.7$ (model MI, left), $\beta_{s1} = 0.20$ and $\beta_{s2} = 0.6$ (model MII, middle) and $\beta_{s1} = 0.33$ and $\beta_{s2} = 0.34$ (model MIII, right). In this figure, the disks with the same radius have the same color. Note that although for model MIII, the scaling factors are different, they are very close to each other, thus leading to disks of similar radii. Consequently, they all have the same color. The fractal dimension of each model is calculated according to Equation (8), with $k_1 = 4$ and $k_2 = 1$. Therefore, one obtains $D_\mathrm{m} \simeq 1.26$ for model MI, $D_\mathrm{m} \simeq 1.31$ for model MII and $D_\mathrm{m} \simeq 1.46$ for model MIII. The total number of particles at m-th iteration is $(k_1 + k_2)^2$, and the total surface area is $\left(k_1\beta_{s1}^2 + k_2\beta_{s2}^2\right)^m A_0$, where A_0 is the surface area of the single disk at $m = 0$.

Therefore, the multifractal form factor can be expressed through a recurrence relation between subsequent iterations. For an arbitrarily two-scale multifractal, one can write [62,82]:

$$F_m(q) = \frac{k_1\beta_{s1}^2 G_1(q) F_{m-1}(\beta_{s1}q) + k_2\beta_{s2}^2 G_2(q) F_{m-1}(\beta_{s2}q)}{k_1\beta_{s1}^2 + k_2\beta_{s2}^2}, \tag{70}$$

where the generative function $G_1(q)$ is given by Equation (64), $G_2(q) = 1$, and the form factor at $m = 0$ is given by Equation (39). Please note that the choice of the generative functions, of the initial shape, of the number of disks and their positions is arbitrarily. Thus, Equation (70) gives the form factor of a two-scale multifractal at an arbitrarily iteration m.

Figure 8 middle row shows the discretized version of the cross multifractals used to calculate the pddf (Figure 8 lower row left) of models MI (black), model MII (green) and model MIII (blue) with the overall size $\xi \simeq 100$ nm. For comparison, the pddf of the equivalent disk is also shown (red). The most pronounced differences, as compared to the disk's pddf occur in model MI, where the maxima is the left-most shifted and has the highest values. This shows that the structure corresponding to model MI is the most heterogeneous. This can be clearly seen also from their construction: while the model MI consists from disks with the highest difference in the values of their radii, for model MIII (i.e., for a single-scale fractal), the structure consists from disks of similar radii, and is more homogeneous.

In addition, the local maxima and minima present in the pddf reflect the relative values of various distances within the fractal. In particular, the main maximum at $r \simeq 25$ nm (model MI), $r \simeq 30$ nm (model MII), and $r \simeq 47$ nm (model MIII) give the most common distances inside the multifractal.

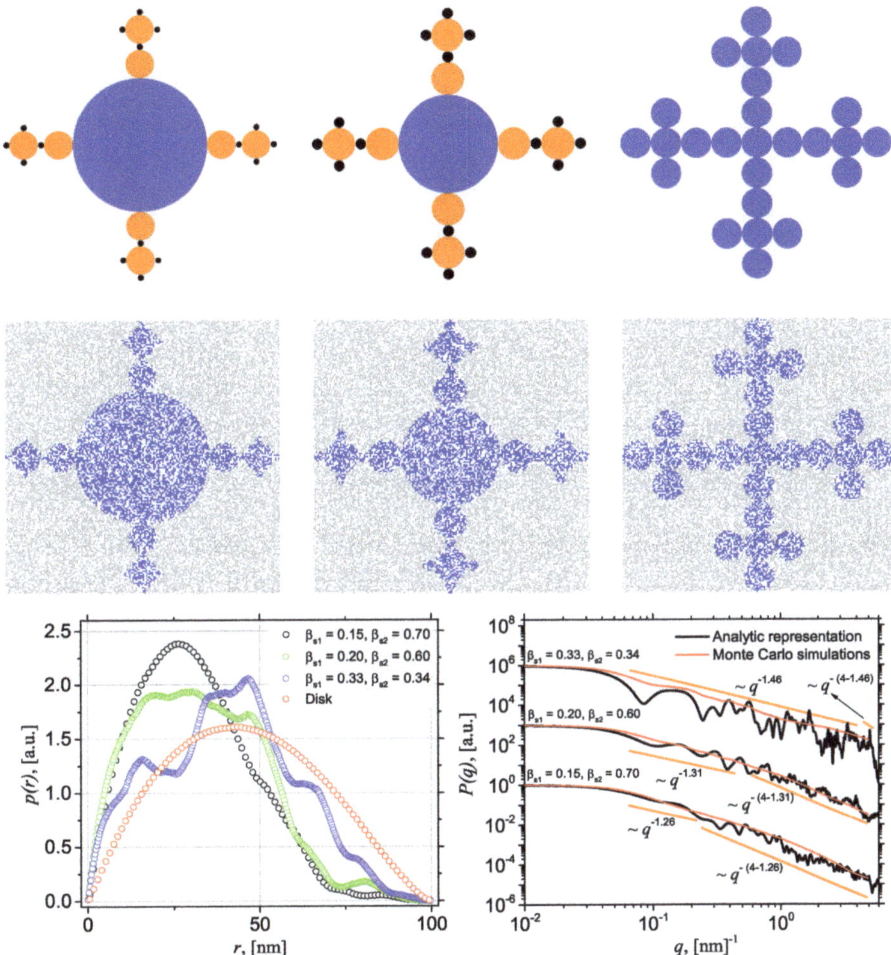

Figure 8. (Color online) Upper row: The second iteration ($m = 2$) of the deterministic "Cross" multifractal of overall size $\xi = 100$ nm at various scaling factors: Left—Model MI ($\beta_{s1} = 0.15$ and $\beta_{s2} = 0.70$), Middle—Model MII ($\beta_{s1} = 0.20$ and $\beta_{s2} = 0.60$), Right—Model MIII ($\beta_{s1} = 0.33$ and $\beta_{s2} = 0.34$). For each fractal, the disks of the same radius have the same color. Middle row: the corresponding approximation with a set of randomly distributed points (blue). Lower row: Left—pddf of the multifractal models and of an equivalent disk (red points), Right—A comparison between the analytic expression (black curve) of SAS intensity (Equation (70)) and Monte Carlo simulations (red curve). The curves for models MII and MIII are shifted vertically by a factor of 10^3, and respectively of 10^6, for clarity.

Figure 8 lower row right shows the SAS curves of the three models at iteration $m = 4$, calculated using Equation (70). The main feature of the curve is the presence of the three main regions: Guinier, fractal and Porod (not shown here) as in the case of deterministic mass and surface fractals, and which allows extraction of the main structural information. However, a specific feature of scattering from multifractals is the appearance of an additional surface fractal region between the mass fractal and Porod regions. In the mass fractal region, the scattering exponent coincides with the mass fractal dimension D_{m}, while in the surface fractal region, the exponent is $4 - D_{\mathrm{s}}$. This shows that the multifractal model consists from mass fractals in which the basic units are not simply Euclidean objects, but they are themselves complex structures, in particular surface fractals, which in their turn have disks as basic units. Please note that while the overall length of the two mass and surface fractal regions is defined by the overall fractal size, and respectively by the two scaling factors, i.e., $\pi \lesssim q \lesssim \pi/\left(\beta_{\mathrm{s1}}\beta_{\mathrm{s2}}\right)^m$, the length of each individual region depends on the value of β_{s1}, i.e., the higher the value of β_{s1} the longer the mass fractal region and the shorter the surface fractal region.

Monte Carlo simulations performed on all three models at $m = 4$ are also presented in Figure 8 lower row right (red). The results confirm the values of the scattering exponents in both regions. However, the fine structure of the scattering curve is lost in the simulated curves, for the same reason as in the case of deterministic mass and surface fractals. Therefore, experimental SAS data showing a succession of mass-to-surface fractal regions, can be attributed to multifractal (i.e., heterogeneous) structures. Thus, the procedure of extracting structural information about multifractals involves a separate analysis of both the mass, and respectively the surface fractal regions (Sections 3.3 and 3.4).

4. Conclusions

The latest advances in preparation of complex materials allow the synthesis of various types of nanomaterials based on deterministic fractals. These materials have improved physical properties such as optical, electronic, mechanical thermal ones, with important applications in surface engineering, heterogeneous catalysis, manufacturing of cloaking devices etc. It is already known that generally, the main reason for the improved properties is the exact self-similar structure of the fractal. Therefore, one of the fundamental challenges is to establish the correlations between their physical and chemical properties from one side, and the structural properties from another one. While the evolution of their properties can be assessed by complex quantum mechanical models, the nanoscale structure is generally addressed by using the SAS technique. In the latter case, one of the main reasons for using SAS is that it provides statistically significant quantities averaged over a macroscopic volume, it allows for sample deuteration (when neutrons are used) and eliminates the requirement of sample preparation, specific to other structural methods.

Here, the main structural properties of deterministic fractals are determined based on information provided by SAS technique. The focus is on differentiating between various classes of fractal, by using information from the corresponding scattering curves. For this purpose, several new theoretical models are introduced, and their scattering properties are validated against data based on Monte Carlo simulations. Table 1 summarizes the structural parameters which can be obtained from SAS curves for the main classes of fractals. Please note that in Table 1 are listed only those properties specific to each class of fractals. Besides them, by using SAS one can obtain also additional properties that are not necessarily related to a specific fractal, such as the: overall size, specific surface, molecular weight, size of the basic unit, correlation/persistence length, or the mass per unit length.

Figure 9 shows schematically the scattering curves corresponding to the fractals listed in Table 1. Each of them is compared to the curve of random mass fractals (taken as a reference due to its widespread occurrence in experimental data). In this figure, the random mass fractals have been labeled differently, depending on the context, and in accordance with the most common terminology in the literature. As such, mass fractals, random fractals, monofractals and thin fractals, all represent a single-scale random mass fractal. A representative example is given by DLA, shown in Figure 4 upper row. The results in Figure 9 show that each class of fractals has its own imprint on the scattering curve, and thus, the SAS technique can differentiate between:

- *Mass and surface fractals* (Figure 9 upper row, left). The differentiation is made through the value of the scattering exponent τ in the fractal region that is $\tau = D_{\mathrm{m}}$ for random mass fractals, and $\tau = d - D_{\mathrm{s}}$ for surface fractals. Here, d is the Euclidean dimension of the space in which the fractal is embedded.

- *Random and deterministic fractals* (Figure 9 upper row, right). The differentiation is made based on the type of power-law decay, i.e., a simple power-law decay for random fractals, and a generalized power-law decay (a complex superposition of minima and maxima on a simple power-law decay) for deterministic fractals.
- *Mono and multifractals* (Figure 9 middle row, left). The differentiation is made through the presence of one or more power-decays, either simple or generalized. For monofractals, there is a single power-law decay, while for two-scale multifractals, there is a succession of a mass fractal followed by a surface fractal. When the two scaling factors have similar values, the length of the surface fractal region is very short, and vice-versa.
- *Thin and fat fractals* (Figure 9 middle row, right). The differentiation is made in a similar way as in the previous case. However, the main difference is that the surface fractal region is replaced by another mass fractal region with the exponent smaller than the one of the first mass fractal region.
- $r \simeq 1$ *and* $r \ll 1$ fractals (Figure 9 lower row). The differentiation is made through the presence of an additional region of constant intensity between the fractal and Porod regions. For fractal in which the ratio of the size of basic units to the minimal characteristic distances between them is about unity, the length of this constant region is very short. However, for fractals with $r \ll 1$, the length of the constant region is much bigger.

Most of the above results illustrated here, are validated by Monte Carlo simulations. Although they represent important advances in structural analysis of fractal structures, there are still important open question which need to be addressed.

Experimentally, the main issues arise from the loss of information about phases, the influence of instrumental limitations, polydispersity and of incoherent scattering at high q, which may hinder the extraction of structural parameters from a SAS curve. The loss of phase information is an intrinsic drawback of the SAS method, since it gives the absolute value of the Fourier transform of the density. In the case of randomly oriented fractals, the square of the Fourier transform is also averaged over all possible orientations of vector $\boldsymbol{q}$. This operation leads to an ill-posed problem in recovering the density from SAS data. Such issue can be handled by measuring the scattering of an ensemble of aligned fractals from different angles with a position sensitive detector. The instrumental limitations arise from the finite resolution of the collimation and detection systems, as well as from the wavelength spread. For samples based on deterministic fractals, these lead to either partial or a total smearing of the scattering curves, and the fine structure of the intensity is lost. In the former case, the main maxima and minima are still visible, and all the structural parameters presented in Table 1 can be recovered. However, in the latter case, all the maxima and minima are completely smeared out. This leads to the impossibility of extracting the scaling factor(s) and the iteration number. The effect of polydispersity is similar to the one of finite instrumental resolution, i.e., the higher the polydispersity, the more smeared the SAS curve. Finally, the q-independent background of SAS curve, determined by the scattering density of irradiated nuclei with nonzero spins such as ^{1}H or ^{7}Li isotopes, may hinder the structural features arising at high q. In this case, one can use deuteration to increase the contrast and reduce the background.

Theoretically, since a given fractal dimension may represent virtually an infinite number of structures, then how one can differentiate between them? Second, multifractals are characterized by a whole spectrum of dimensions given by Equations (11) or (12). Can SAS, eventually in combination with other methods, provide other dimensions, besides D_0 (see Equation (13))? And if the multifractal consists from more than two scaling factors, which is probably one of the most common situations in physical samples, can we recover all the scaling factor from SAS curves?

From a practical point of view, partial answers to these questions begin to emerge. The progress in SAS instrumentation could eliminate the disadvantage of finite instrumental resolution, to a sufficiently high degree, without sacrificing the complex morphology of deterministic fractals and without increasing the measurement time. Similarly, recent advances in nanotechnology, materials science and chemistry allows preparation of highly monodisperse samples. Of particular importance is the use of ptychographic methods for phase recovery in coherent SAS by numerical procedures. Also, a combined use of SAXS with tensor tomography, also known as 3D scanning SAXS, allows determination of 3D orientation of nanofractals in a bulk specimen, and thus symmetries within the sample can be exploited to extract additional structural information about fractals.

Table 1. Fractal specific parameters which can be obtained from a SAS experiment. Here D_{m} and D_{s} are the mass, and respectively the surface fractal dimension, β_{s} is the mass and surface fractals scaling factor, m is the mass fractal iteration number, $\beta_{\mathrm{s}}^{(i)}$, $k_{m}^{(i)}$, $m^{(i)}$ and $D_{\mathrm{m}}^{(i)}$ are the scaling factors, the number of basic units, iteration number and respectively the fractal dimensions at i-th structural level ($i = 1, 2, \cdots$) in a fat fractal, β_{s1} and β_{s2}, with $\beta_{\mathrm{s1}} < \beta_{\mathrm{s2}}$ are the multifractal scaling factors, k_m is the number of basic units in a mass fractal, h is the characteristic minimal distance between basic units in a mass fractal, and $r \equiv l/h$, with l the size of basic units composing the mass fractal. The exponents of the scaling factors and number of units occurring for fat fractals denote an index (over the structural levels) and not a power.

Fractal Type	Parameters	Source	Fractal Power-Law Decays
Random mass fractals	D_{m}	Exponent of power-law decay	A single simple power-law decay with exponent D_{m}.
Random surface fractals	D_{s}	Exponent of power-law decay	A single simple power-law decay with exponent $d - D_{\mathrm{s}}$.
Deterministic mass fractals	D_{m}	Exponent of power-law decay	A single generalized power-law decay with exponent D_{m}.
	β_{s}	Period on the logarithmic scale	
	m	Number of periods in logarithmic scale	
	k_m	$k_m = (1/\beta_{\mathrm{s}})^{mD_{\mathrm{m}}}$	
Deterministic surface fractals	D_{s}	Exponent of power-law decay	A single generalized power-law decay with exponent D_{s}.
	β_{s}	Period on the logarithmic scale	
	m	Number of periods in logarithmic scale	
Deterministic fat fractals	$D_{\mathrm{m}}^{(i)}, D_{\mathrm{m}}^{(i)}, \cdots$	Exponents of power-law decays at each structural level	A succession of generalized power-law decay with exponents $D_{\mathrm{m1}} < D_{\mathrm{m2}} < \cdots$.
	$\beta_{\mathrm{s}}^{(i)}, \cdots$	Periods on the logarithmic scale at each structural level	
	$m^{(i)}$	Number of periods in logarithmic scale at each structural level	
	$k_{m}^{(i)}$	As for deterministic mass fractals, but at each structural level	
Deterministic multifractals with two scaling factors	$D_{\mathrm{m}}, D_{\mathrm{s}}$	Exponents of power-law decays in each fractal region	A succession of mass-to-surface fractal generalized power-law decays, with exponents D_{m}, and respectively D_{s}.
	$\beta_{\mathrm{s1}}, \beta_{\mathrm{s2}}$	Periods on the logarithmic scale from mass, and surface fractal regions	
	m	Number of periods in logarithmic scale from mass or surface fractal regions	
	k_{m1}	$k_{m1} = (1/\beta_{\mathrm{s1}})^{m1D_{\mathrm{m1}}}$	
Deterministic mass fractals with $r \gg 1$	D_{m}	Exponents of the power-law decay	A region with constant intensity occurs after the fractal region.
	β_{s}	Periods on the logarithmic scale from mass fractal region	
	m	Number of periods in logarithmic scale from mass fractal regions	
	h	End of the constant region	
	k_m	As for deterministic mass fractals	

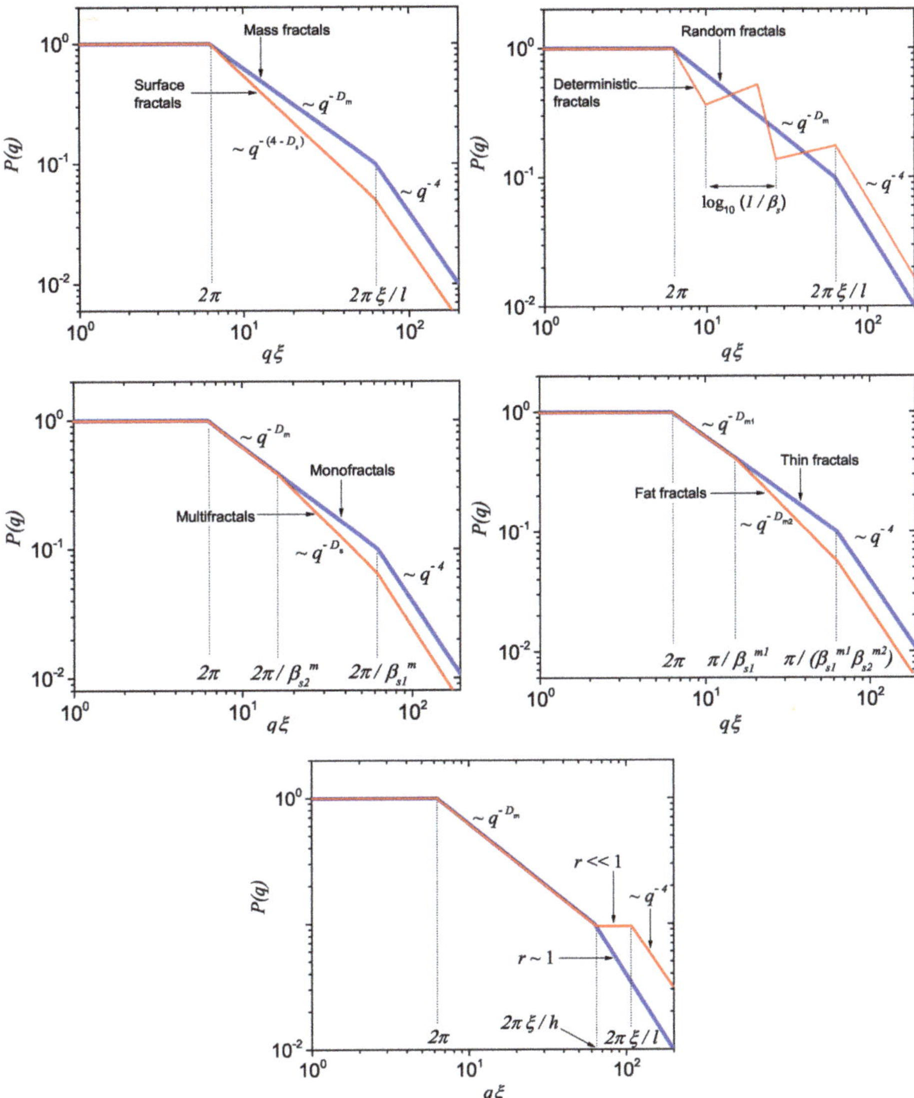

Figure 9. (Color online) Schematic representation of SAS from different classes of 2D fractals Upper row: Left—mass and surface fractals, Right—random and deterministic fractals. Middle row: Left—mono and multifractals, Right—Thin and fat fractals. Lower row: $r \ll 1$ and $r \simeq 1$ fractals (see below). Here ξ is the overall size of mass fractals, and respectively the size of the largest disk in a surface fractal, D_m (including D_{m1} and D_{m2}) and D_s are the mass and, respectively the surface fractal dimensions, l is the size of disks in a mass fractal, and respectively the size of smallest disk in a surface fractal, m (including m_1 and m_2) are the fractal iteration numbers, β_s (including β_s1 and β_s2) are the scaling factors, h is the minimal distance between the disks, and $r = l/h$.

Funding: This research received no external funding

Conflicts of Interest: The author declares no conflict of interest.

References

1. Chaudhuri, B.; Muñoz, I.G.; Qian, S.; Urban, V.S. (Eds.) *Biological Small Angle Scattering: Techniques, Strategies and Tips*, 1st ed.; Springer: Singapore, 2017; Volume 1009, p. 268; doi:10.1007/978-981-10-6038-0.
2. Glatter, O. *Scattering Methods and Their Application in Colloid And Interface Science*, 1st ed.; Elsevier: Amsterdam, The Netherlands, 2018; p. 404.
3. Jaksch, S.; Gutberlet, T.; Müller-Buschbaum, P. Grazing-incidence scattering—status and perspectives in soft matter and biophysics. *Curr. Opin. Colloid Interface Sci.* **2019**, *42*, 73–86. doi:10.1016/J.COCIS.2019.04.001.
4. Allen, A.J. Scattering methods for disordered heterogeneous materials. In *International Tables for Crystallography Volume H: Powder Diffraction*; Gilmore, C.J., Kaduk, J.A., Schnek, H., Eds.; International Union of Crystallography: New Jersey, USA 2019; Chapter 5.8; pp. 673–696; doi:10.1107/97809553602060000973.
5. Wittmeier, A.; Cassini, C.; Hemonnot, C.; Weinhausen, B.; Bernhardt, M.; Salditt, T.; Koster, S. Scanning Small-Angle-X-Ray Scattering for Imaging Biological Cells. *Microsc. Microanal.* **2018**, *24*, 336–339. doi:10.1017/S1431927618013983.
6. Mühlbauer, S.; Honecker, D.; Périgo, E.A.; Bergner, F.; Disch, S.; Heinemann, A.; Erokhin, S.; Berkov, D.; Leighton, C.; Eskildsen, M.R.; et al. Magnetic small-angle neutron scattering. *Rev. Mod. Phys.* **2019**, *91*, 015004. doi:10.1103/RevModPhys.91.015004.
7. Mahieu, E.; Gabel, F.; IUCr. Biological small-angle neutron scattering: recent results and development. *Acta Cryst. D* **2018**, *74*, 715–726. doi:10.1107/S2059798318005016.
8. Oliver, R.C.; Rolband, L.A.; Hutchinson-Lundy, A.M.; Afonin, K.A.; Krueger, J.K. Small-Angle Scattering as a Structural Probe for Nucleic Acid Nanoparticles (NANPs) in a Dynamic Solution Environment. *Nanomaterials* **2019**, *9*, 681. doi:10.3390/nano9050681.
9. Kryukova, A.E.; Shpichka, A.I.; Konarev, P.V.; Volkov, V.V.; Timashev, P.S.; Asadchikov, V.E. Shape Determination of Bovine Fibrinogen in Solution Using Small-Angle Scattering Data. *Cryst. Rep.* **2018**, *63*, 871–873. doi:10.1134/S1063774518060202.
10. Palit, S.; He, L.; Hamilton, W.A.; Yethiraj, A.; Yethiraj, A. Combining Diffusion NMR and Small-Angle Neutron Scattering Enables Precise Measurements of Polymer Chain Compression in a Crowded Environment. *Phys. Rev. Lett.* **2017**, *118*, 097801. doi:10.1103/PhysRevLett.118.097801.
11. Sakurai, S. Recent developments in polymer applications of synchrotron small-angle X-ray scattering. *Polym. Int.* **2017**, *66*, 237–249. doi:10.1002/pi.5136.
12. Horkay, F.; Basser, P.J.; Hecht, A.M.; Geissler, E. Ionic effects in semi-dilute biopolymer solutions: A small angle scattering study. *J. Chem. Phys.* **2018**, *149*, 163312. doi:10.1063/1.5028351.
13. Mortensen, K.; Annaka, M. Stretching PEO–PPO Type of Star Block Copolymer Gels: Rheology and Small-Angle Scattering. *ACS Macro Lett.* **2018**, *7*, 1438–1442. doi:10.1021/acsmacrolett.8b00792.
14. Bocz, K.; Decsov, K.E.; Farkas, A.; Vadas, D.; Bárány, T.; Wacha, A.; Bóta, A.; Marosi, G. Non-destructive characterisation of all-polypropylene composites using small angle X-ray scattering and polarized Raman spectroscopy. *Compos. Part A* **2018**, *114*, 250–257. doi:10.1016/J.COMPOSITESA.2018.08.020.
15. Wolff, M.; Saini, A.; Simonne, D.; Adlmann, F.; Nelson, A. Time Resolved Polarised Grazing Incidence Neutron Scattering from Composite Materials. *Polymers* **2019**, *11*, 445. doi:10.3390/polym11030445.
16. Bhaway, S.M.; Qiang, Z.; Xia, Y.; Xia, X.; Lee, B.; Yager, K.G.; Zhang, L.; Kisslinger, K.; Chen, Y.M.; Liu, K.; et al. Operando Grazing Incidence Small-Angle X-ray Scattering/X-ray Diffraction of Model Ordered Mesoporous Lithium-Ion Battery Anodes. *ACS Nano* **2017**, *11*, 1443–1454. doi:10.1021/acsnano.6b06708.
17. Wang, L.M.; Petracic, O.; Kentzinger, E.; Rücker, U.; Schmitz, M.; Wei, X.K.; Heggen, M.; Brückel, T. Strain and electric-field control of magnetism in supercrystalline iron oxide nanoparticle–$BaTiO_3$ composites. *Nanoscale* **2017**, *9*, 12957–12962. doi:10.1039/C7NR05097F.
18. Choi, Y.; Paik, D.; Bogdanov, S.; Valiev, E.; Borisova, P.; Murashev, M.; Em, V.; Pirogov, A. Neutron scattering on humane compact bone. *Phys. B* **2018**, *551*, 218–221. doi:10.1016/J.PHYSB.2018.01.020.
19. Busch, A.; Kampman, N.; Bertier, P.; Pipich, V.; Feoktystov, A.; Rother, G.; Harrington, J.; Leu, L.; Aertens, M.; Jacops, E. Predicting Effective Diffusion Coefficients in Mudrocks Using a Fractal Model and Small-Angle Neutron Scattering Measurements. *Water Resour. Res.* **2018**, *54*, 7076–7091. doi:10.1029/2018WR023425.
20. Penttilä, P.A.; Rautkari, L.; Österberg, M.; Schweins, R. Small-angle scattering model for efficient characterization of wood nanostructure and moisture behaviour. *J. Appl. Cryst.* **2019**, *52*, 369–377. doi:10.1107/S1600576719002012.

21. Li, A.; He, R.; Bian, Z.; Song, H.; Chen, X.; Zhou, J. Enhanced lithium storage performance of hierarchical CuO nanomaterials with surface fractal characteristics. *App. Surf. Sci.* **2018**, *443*, 382–388. doi:10.1016/J.APSUSC.2018.03.019.
22. Wang, Y.; Maurel, G.; Couty, M.; Detcheverry, F.; Merabia, S. Implicit Medium Model for Fractal Aggregate Polymer Nanocomposites: Linear Viscoelastic Properties. *Macromolecules* **2019**, *52*, 2021–2032. doi:10.1021/acs.macromol.8b02455.
23. Wei, R.; Wang, X.; Zhang, X.; Wang, S.; Zhu, S.; Du, S. A Novel Method for Manufacturing High-Performance Layered Silicate/Epoxy Nanocomposites Using an Epoxy-Diamine Adduct to Enhance Compatibility and Interfacial Reactivity. *Macromol. Mater. Eng.* **2018**, *303*, 1800065. doi:10.1002/mame.201800065.
24. Yue, H.; Reguero, V.; Senokos, E.; Monreal-Bernal, A.; Mas, B.; Fernández-Blázquez, J.; Marcilla, R.; Vilatela, J. Fractal carbon nanotube fibers with mesoporous crystalline structure. *Carbon* **2017**, *122*, 47–53. doi:10.1016/J.CARBON.2017.06.032.
25. Zhang, J.; Li, H.; Ye, J.; Cao, Z.; Chen, J.; Kuang, Q.; Zheng, J.; Xie, Z. Sierpinski gasket-like Pt–Ag octahedral alloy nanocrystals with enhanced electrocatalytic activity and stability. *Nano Energy* **2019**, *61*, 397–403. doi:10.1016/J.NANOEN.2019.04.093.
26. Mandelbrot, B.B. *The Fractal Geometry of Nature*; W.H. Freeman: New York, NY, USA 1982; p. 460.
27. Zhu, W.; Esteban, R.; Borisov, A.G.; Baumberg, J.J.; Nordlander, P.; Lezec, H.J.; Aizpurua, J.; Crozier, K.B. Quantum mechanical effects in plasmonic structures with subnanometre gaps. *Nat. Commun.* **2016**, *7*, 11495. doi:10.1038/ncomms11495.
28. Sichert, J.A.; Tong, Y.; Mutz, N.; Vollmer, M.; Fischer, S.; Milowska, K.Z.; García Cortadella, R.; Nickel, B.; Cardenas-Daw, C.; Stolarczyk, J.K.; et al. Quantum Size Effect in Organometal Halide Perovskite Nanoplatelets. *Nano Lett.* **2015**, *15*, 6521–6527. doi:10.1021/acs.nanolett.5b02985.
29. Nazmitdinov, R.G. From Chaos to Order in Mesoscopic Systems. *Phys. Part. Nucl. Lett.* **2019**, *16*, 159–169. doi:10.1134/S1547477119030154.
30. Dale W. Schaefer.; Ryan S. Justice†. How Nano Are Nanocomposites? *Macromolecules* **2007**, *40*, 8501–8517. doi:10.1021/MA070356W.
31. Yurekli, K.; Mitchell, C.A.; Krishnamoorti, R. Small-angle neutron scattering from surfactant-assisted aqueous dispersions of carbon nanotubes. *J. Am. Chem. Soc.* **2004**, *126*, 9902–9903. doi:10.1021/ja047451u.
32. Yin, W.; Dadmun, M. A New Model for the Morphology of P3HT/PCBM Organic Photovoltaics from Small-Angle Neutron Scattering: Rivers and Streams. *ACS Nano* **2011**, *5*, 4756–4768. doi:10.1021/nn200744q.
33. Stevens, D.A.; Dahn, J.R. An In Situ Small-Angle X-Ray Scattering Study of Sodium Insertion into a Nanoporous Carbon Anode Material within an Operating Electrochemical Cell. *J. Electrochem. Soc.* **2000**, *147*, 4428–4431.
34. Zabar, S.; Lesmes, U.; Katz, I.; Shimoni, E.; Bianco-Peled, H. Studying different dimensions of amylose-long chain fatty acid complexes: Molecular, nano and micro level characteristics. *Food Hydrocoll.* **2009**, *23*, 1918–1925. doi:10.1016/j.foodhyd.2009.02.004.
35. Jacques, D.A.; Trewhella, J. Small-angle scattering for structural biology-Expanding the frontier while avoiding the pitfalls. *Prot. Sci.* **2010**, *19*, 642–657. doi:10.1002/pro.351.
36. Li, T.; Senesi, A.J.; Lee, B. Small Angle X-Ray Scattering for Nanoparticle Research. *Chem. Rev.* **2016**, *116*, 11128–11180. doi:10.1021/acs.chemrev.5b00690.
37. Pfeifer, P.; Ehrburger-Dolle, F.; Rieker, T.P.; González, M.T.; Hoffman, W.P.; Molina-Sabio, M.; Rodríguez-Reinoso, F.; Schmidt, P.W.; Voss, D.J. Nearly Space-Filling Fractal Networks of Carbon Nanopores. *Phys. Rev. Lett.* **2002**. doi:10.1103/PhysRevLett.88.115502.
38. Bale, H.D.; Schmidt, P.W. Small-Angle X-Ray-Scattering Investigation of Submicroscopic Porosity with Fractal Properties. *Phys. Rev. Lett.* **1984**, *53*, 596–599. doi:10.1103/PhysRevLett.53.596.
39. Teixeira, J. Small-angle scattering by fractal systems. *J. Appl. Cryst.* **1988**, *21*, 781–785. doi:10.1107/S0021889888000263.
40. Pfeifer, P. Simple Models for the Formation of Random Rough Surfaces. In *Multiple Scattering of Waves in Random Media and Random Rough Surfaces, Proceedings of an International Symposium, University Park, PA, USA, 29 July–2 August 1985*; Varadan, V.V., Varadan, V.K., Eds.; Pennsylvania State University State College: State College, PA, USA, 1987; p. 964.
41. Martin, J.E.; Hurd, A.J. Scattering from fractals. *J. Appl. Cryst.* **1987**, *20*, 61–78. doi:10.1107/S0021889887087107.

42. Schmidt, P.W. Small-angle scattering studies of disordered, porous and fractal systems. *J. Appl. Cryst.* **1991**, *24*, 414–435. doi:10.1107/S0021889891003400.
43. Beaucage, G. Approximations Leading to a Unified Exponential/Power-Law Approach to Small-Angle Scattering. *J. Appl. Cryst.* **1995**, *28*, 717–728. doi:10.1107/S0021889895005292.
44. Beaucage, G. Small-Angle Scattering from Polymeric Mass Fractals of Arbitrary Mass-Fractal Dimension. *J. Appl. Cryst.* **1996**, *29*, 134–146. doi:10.1107/S0021889895011605.
45. Hammouda, B. A new Guinier–Porod model. *J. Appl. Cryst.* **2010**, *43*, 716–719. doi:10.1107/S0021889810015773.
46. Papagiannopoulos, A.; Meristoudi, A.; Pispas, S.; Keiderling, U. Thermal response of self-organization in an amphiphilic triblock polyelectrolyte and the influence of the globular protein lysozyme. *Eur. Polym. J.* **2018**, *99*, 49–57. doi:10.1016/j.eurpolymj.2017.12.005.
47. Anovitz, L.M.; Cole, D.R. Characterization and Analysis of Porosity and Pore Structures. *Rev. Miner. Geochem.* **2015**, *80*, 61–164. doi:10.2138/rmg.2015.80.04.
48. Qiao, D.; Liu, H.; Yu, L.; Bao, X.; Simon, G.P.; Petinakis, E.; Chen, L. Preparation and characterization of slow-release fertilizer encapsulated by starch-based superabsorbent polymer. *Carbohydr. Polym.* **2016**, *147*, 146–154. doi:10.1016/J.CARBPOL.2016.04.010.
49. Liu, D.; Song, L.; Song, H.; Chen, J.; Tian, Q.; Chen, L.; Sun, L.; Lu, A.; Huang, C.; Sun, G. Correlation between mechanical properties and microscopic structures of an optimized silica fraction in silicone rubber. *Compos. Sci. Technol.* **2018**, *165*, 373–379. doi:10.1016/j.compscitech.2018.07.024.
50. Bauer, P.; Amenitsch, H.; Baumgartner, B.; Koberl, G.; Rentenberger, C.; Winkler, P.M. In-situ aerosol nanoparticle characterization by small angle X-ray scattering at ultra-low volume fraction. *Nat. Commun.* **2019**, *10*, 1122. doi:10.1038/s41467-019-09066-4.
51. Song, P.; Tu, X.; Bai, L.; Sun, G.; Tian, Q.; Gong, J.; Zeng, G.; Chen, L.; Qiu, L. Contrast Variation Small Angle Neutron Scattering Investigation of Micro- and Nano-Sized TATB. *Materials* **2019**, *12*, 2606. doi:10.3390/ma12162606.
52. Shang, J.; Wang, Y.; Chen, M.; Dai, J.; Zhou, X.; Kuttner, J.; Hilt, G.; Shao, X.; Gottfried, J.M.; Wu, K. Assembling molecular Sierpiński triangle fractals. *Nat. Chem.* **2015**, *7*, 389–393. doi:10.1038/nchem.2211.
53. Wang, L.; Liu, R.; Gu, J.; Song, B.; Wang, H.; Jiang, X.; Zhang, K.; Han, X.; Hao, X.Q.; Bai, S.; et al. Self-Assembly of Supramolecular Fractals from Generation 1 to 5. *J. Am. Chem. Soc.* **2018**, *140*, 14087–14096. doi:10.1021/jacs.8b05530.
54. Wang, Y.; Xue, N.; Li, R.; Wu, T.; Li, N.; Hou, S.; Wang, Y. Construction and Properties of Sierpiński Triangular Fractals on Surfaces. *Chem. Phys. Chem.* **2019**, *20*, 2262–2270. doi:10.1002/cphc.201900258.
55. Newkome, G.R.; Moorefield, C.N. From 1 $\rightarrow$ 3 dendritic designs to fractal supramacromolecular constructs: understanding the pathway to the Sierpiński gasket. *Chem. Soc. Rev.* **2015**, *44*, 3954–3967. doi:10.1039/C4CS00234B.
56. Malankowska, M.; Schlautmann, S.; Berenschot, E.; Tiggelaar, R.; Pina, M.; Mallada, R.; Tas, N.; Gardeniers, H. Three-Dimensional Fractal Geometry for Gas Permeation in Microchannels. *Micromachines* **2018**, *9*, 45. doi:10.3390/mi9020045.
57. Wu, Z.; Chen, X.; Zheng, Y.; Wang, T. Fabrication of Poly(methyl methacrylate) (PMMA) Microfluidic Chip with Cantor Fractal Structure Using a CO2 Laser System. *Laser Eng.* **2019**, *43*, 121–129.
58. Cherny, A.Y.; Anitas, E.M.; Osipov, V.A.; Kuklin, A.I. Deterministic fractals: Extracting additional information from small-angle scattering data. *Phys. Rev. E* **2011**, *84*, 036203. doi:10.1103/PhysRevE.84.036203.
59. Anitas, E.M. Small-angle scattering from fat fractals. *Eur. Phys. J. B* **2014**, *87*, 139. doi:10.1140/epjb/e2014-41066-9.
60. Cherny, A.Y.; Anitas, E.M.; Osipov, V.A.; Kuklin, A.I.; IUCr. Scattering from surface fractals in terms of composing mass fractals. *J. Appl. Cryst.* **2017**, *50*, 919–931. doi:10.1107/S1600576717005696.
61. Anitas, E.M.; Slyamov, A.; Todoran, R.; Szakacs, Z. Small-Angle Scattering from Nanoscale Fat Fractals. *Nanoscale Res. Lett.* **2017**, *12*, 389. doi:10.1186/s11671-017-2147-0.
62. Anitas, E.M.; Marcelli, G.; Szakacs, Z.; Todoran, R.; Todoran, D. Structural Properties of Vicsek-like Deterministic Multifractals. *Symmetry* **2019**, *11*, 806. doi:10.3390/sym11060806.
63. Hausdorff, F. Dimension und äußeres Maß. *Math. Ann.* **1918**, *79*, 157–179. doi:10.1007/BF01457179.
64. Rogers, C.A. *Hausdorff Measures*; Cambridge University Press: Cambridge, UK, 1970; p. 179.
65. Gouyet, J.F. *Physics and Fractal Structures*; Masson: Paris, France, 1996; p. 234.
66. Roe, R.J. *Methods of X-ray and Neutron Scattering in Polymer Science* ; Oxford University Press: New York, NY, USA, 2000; p. 352.

67. Arneodo, A.; Decoster, N.; Roux, S. A wavelet-based method for multifractal image analysis. I. Methodology and test applications on isotropic and anisotropic random rough surfaces. *Eur. Phys. J. B* **2000**, *15*, 567–600. doi:10.1007/s100510051161.
68. Decoster, N.; Roux, S.; Arnéodo, A. A wavelet-based method for multifractal image analysis. II. Applications to synthetic multifractal rough surfaces. *Eur. Phys. J. B* **2000**, *15*, 739–764. doi:10.1007/s100510051179.
69. Chhabra, A.; Jensen, R.V. Direct Determination of the f (alpha) Singularity Spectrum. *Phys. Rev. Lett.* **1989**, *62*, 1327–1130.
70. Muzy, J.F.; Bacry, E.; Arneodo, A. Multifractal formalism for fractal signals: The structure-function approach versus the wavelet-transform modulus-maxima method. *Phys. Rev. E* **1993**, *47*, 875–884. doi:10.1103/PhysRevE.47.875.
71. Feigin, L.A.; Svergun, D.I. *Structure Analysis by Small-Angle X-ray and Neutron Scattering*; Springer US: Boston, MA, USA, 1987; p. 335; doi:10.1007/978-1-4757-6624-0.
72. Anitas, E.M. *Small-Angle Scattering (Neutrons, X-rays, Light) from Complex Systems*; SpringerBriefs in Physics; Springer International Publishing: Cham, Switzerland, 2019; doi:10.1007/978-3-030-26612-7.
73. Melnichenko, Y.B. *Small-Angle Scattering from Confined and Interfacial Fluids*; Springer International Publishing: Cham, Switzerland, 2016; p. 329; doi:10.1007/978-3-319-01104-2.
74. Debye, P.; Bueche, A.M. Scattering by an Inhomogeneous Solid. *J. Appl. Phys.* **1949**, *20*, 518–525. doi:10.1063/1.1698419.
75. Kaya, H. Scattering behaviour of Janus particles. *Appl. Phys. A* **2002**, *74*. doi:10.1007/s003390101115.
76. Kratky, O.; Porod, G. Diffuse small-angle scattering of X-rays in colloid systems. *J. Colloid Sci.* **1949**, *4*, 35–70. doi:10.1016/0095-8522(49)90032-X.
77. Sorensen, C.M. Light Scattering by Fractal Aggregates: A Review. *Aerosol Sci. Technol.* **2001**, *35*, 648–687. doi:10.1080/02786820117868.
78. Witten, T.A.; Sander, L.M. Diffusion-Limited Aggregation, a Kinetic Critical Phenomenon. *Phys. Rev. Lett.* **1981**, *47*, 1400–1403. doi:10.1103/PhysRevLett.47.1400.
79. Glatter, O.; May, R. Small-angle techniques. In *International Tables for Crystallography*; International Union of Crystallography: Chester, UK, 2006; pp. 89–112; doi:10.1107/97809553602060000581.
80. Muthukumar, M. Mean-Field Theory for Diffusion-Limited Cluster Formation. *Phys. Rev. Lett.* **1983**, *50*, 839–842. doi:10.1103/PhysRevLett.50.839.
81. Meakin, P. The structure of two-dimensional Witten-Sander aggregates. *J. Phys. A* **1985**, *18*, L661–L666. doi:10.1088/0305-4470/18/11/006.
82. Cherny, A.Y.; Anitas, E.M.; Osipov, V.A.; Kuklin, A.I. The structure of deterministic mass and surface fractals: theory and methods of analyzing small-angle scattering data. *Phys. Chem. Chem. Phys.* **2019**, *21*, 12748–12762. doi:10.1039/C9CP00783K.

Article

Metal–Insulator Transition in Three-Dimensional Semiconductors

Klaus Ziegler

Institut für Physik, Universität Augsburg, D-86135 Augsburg, Germany; klaus.ziegler@physik.uni-augsburg.de

Received: 28 September 2019; Accepted: 24 October 2019; Published: 1 November 2019

Abstract: We use a random gap model to describe a metal–insulator transition in three-dimensional semiconductors due to doping, and find a conventional phase transition, where the effective scattering rate is the order parameter. Spontaneous symmetry breaking results in metallic behavior, whereas the insulating regime is characterized by the absence of spontaneous symmetry breaking. The transition is continuous for the average conductivity with critical exponent equal to 1. Away from the critical point, the exponent is roughly 0.6, which may explain experimental observations of a crossover of the exponent from 1 to 0.5 by going away from the critical point.

Keywords: particle-hole symmetry; metal–insulator transition; random gap model

---

1. Introduction

The particle-hole symmetry plays a crucial role in solid state physics. In particular in semi- as well as in superconductor physics [1], this symmetry appears due to the existence of two separate bands. Recent theoretical studies of three-dimensional Weyl materials has renewed interest in the disordered driven metal–insulator transition [2–26]. It was shown recently that Anderson localization can be prevented even in the strong disorder regime when particle-hole symmetry is present [27,28]. This can be understood by the simple picture that particle-hole pairs can be created by an infinitesimal excitation energy.

Undoped semiconductors have a small gap between the valence and the conduction band, typically of the order of 0.2, ..., 1.2 eV [1]. This gap is strongly affected by doping, which allows us to engineer a variety of useful technological applications. In particular, sufficiently strong doping closes the gap such that a metallic phase appears. A classical example for this type of metal–insulator transition is doped silicon, where typical dopants are phosphorus (Si:P) or boron (Si:B) [29–33]. Disorder plays a crucial role in these materials due to the inhomogeneous distribution of the dopants. This suggested that Anderson localization must play a crucial role in these systems, where the quantum states would undergo a transition from extended to localized states for increasing disorder. This transition should be reflected in the transport properties, where extended states lead to a metal and localized states to an insulator at vanishing temperatures.

Measurements of the conductivity $\sigma(N)$ as a function of doping concentration N in Si:P at low temperatures has indeed revealed a critical behavior. Above a critical concentration N_c a power law was found

$$\sigma(N) \sim \sigma_0(N/N_c - 1)^{\mu} \quad (N \geq N_c)$$

and a vanishing conductivity for $N \leq N_c$. The exponent μ was determined as $\mu \approx 0.5$ for some experiments [29–32], whereas a crossover from $\mu \approx 0.5$ at some distance from the critical point to $\mu \approx 1$ in a vicinity very close to N_c was observed in other experiments [32,33].

Although the picture of an Anderson transition is quite appealing, an alternative description can be provided by a random gap model. The idea is that the dopants create energy levels inside the semiconductor gap. These levels are associated with states that can overlap with the states in the semiconductor bands and eventually fill the semiconductor gap by forming extended states. The effect can be described by a random distribution of local gaps. Then the locally filled gaps can be distributed over the entire system and form eventually, after sufficient doping, a conducting "network". This is associated with a second-order phase transition which will be described in this article. The transition is distinguished from the Anderson transition by the fact that the metallic phase appears at strong disorder (i.e., high dopant concentration) and the insulating phase at weak disorder. This does not rule out an Anderson transition if we increase the disorder inside the metallic regime. However, in realistic systems it is more likely to see the transition caused by the random gap than the more sophisticated Anderson transition for $N \gg N_c$.

In the following, we will discuss and analyze the insulator-metal transition due to random gap closing in a three-dimensional system. This will be based on a two-band model with particle-hole symmetry. The latter is essential for the existence of metallic states in the presence of strong disorder.

2. Model and Symmetries

We consider a two-band model with a symmetric Hamiltonian. This can be expressed in terms of Pauli matrices σ_j ($j = 0, ..., 3$). A simple case is

$$H = h_1\sigma_1 + h_3\sigma_3 \tag{1}$$

with symmetric matrices h_1, h_3 in three-dimensional (real) space. To be more specific, we can choose the Fourier components $h_1 = k/\sqrt{2m}$ with $k \equiv \sqrt{k_1^2 + k_2^2 + k_3^2}$. For a uniform gap Δ implies $h_3 = \Delta/2$ we obtain two bands with the dispersion $E_{\mathbf{k}} = \pm\sqrt{k^2/2m + \Delta^2/4}$. Subsequently we will consider a random gap h_3 with mean values $\Delta/2$ to describe the effect of an inhomogeneous distribution of dopants and rescale $k/\sqrt{2m} \to k$.

The one-particle Hamiltonian H is invariant under an Abelian chiral transformation:

$$e^{\alpha\sigma_2} H e^{\alpha\sigma_2} = H\,. \tag{2}$$

In order to reveal the relevant symmetry for transport in this system, we construct the two-body Hamiltonian

$$\hat{H} = \begin{pmatrix} H & 0 \\ 0 & H \end{pmatrix}, \tag{3}$$

where the upper block H acts on bosons and the lower block H on fermions. The reason for introducing this two-body Hamiltonian is that we can transform the distribution of the random Hamiltonian H into a distribution of the Green's function $\hat{G}(z) = (\hat{H} - z)^{-1}$ [34,35], which is often called a supersymmetric representation of the Green's function.

Next we introduce the transformation matrix

$$\hat{U} = \begin{pmatrix} 0 & \varphi\sigma_2 \\ \varphi'\sigma_2 & 0 \end{pmatrix} \tag{4}$$

and obtain the anti-commutator relation

$$\{\hat{H}, \hat{U}\}_+ = 0\,. \tag{5}$$

This implies the non-Abelian chiral symmetry

$$e^{\hat{U}} \hat{H} e^{\hat{U}} = \hat{H}\,, \tag{6}$$

which is an extension of the Abelian symmetry (2). The Green's function $\hat{G}(z)$ does not obey this symmetry for $z \neq 0$. Therefore, z plays here the role of a symmetry-breaking field. An interesting limit is $z \to 0$, which we will study in the next section.

Now we consider the case of a random gap with mean $\langle h_{3,\mathbf{r}} \rangle = \Delta/2$ and variance $\langle h_{3,\mathbf{r}} h_{3,\mathbf{r}'} \rangle - \Delta^2/4 = g\delta_{\mathbf{r},\mathbf{r}'}$ and its effect on the average conductivity at frequency ω. The conductivity is obtained from the Kubo formula as [35,36]

$$\sigma_{kk} = -\frac{e^2}{2h}\omega^2 \lim_{\epsilon \to 0} Re \left\{ \sum_{\mathbf{r}} r_k^2 Tr_2 \left[\langle G_{0\mathbf{r}}(\omega/2 + i\epsilon) G_{\mathbf{r}0}(-\omega/2 - i\epsilon) \rangle \right] \right\}, \quad G(z) = (H - z)^{-1}. \quad (7)$$

In particular, we are interested in the DC limit $\omega \to 0$. This limit restores the chiral symmetry of $\hat{H}$ in (6) for the Green's functions. However, the symmetry can be spontaneously broken now. Since it is a continuous symmetry, this creates a massless mode, which represents fluctuations on arbitrarily large length scales.

Here it should be noticed that $\sigma_2(H+z)^{-1}\sigma_2 = -(H-z)^{-1}$. This has the consequence that the product in (7) reads $G_{0\mathbf{r}}(z)G_{\mathbf{r}0}(-z) = (H-z)^{-1}_{0\mathbf{r}}(H+z)^{-1}_{\mathbf{r}0} = -(H-z)^{-1}_{0\mathbf{r}}\sigma_2(H-z)^{-1}_{\mathbf{r}0}\sigma_2$ such that elements of $\hat{G}(\omega/2 + i\epsilon)$ are sufficient to express the conductivity.

A common approximation for the average two-particle Greens function is the factorization of the average

$$\langle G_{0\mathbf{r}}(\omega/2 + i\epsilon) G_{\mathbf{r}0}(-\omega/2 - i\epsilon) \rangle \approx \langle G_{0\mathbf{r}}(\omega/2 + i\epsilon) \rangle \langle G_{\mathbf{r}0}(-\omega/2 - i\epsilon) \rangle \quad (8)$$

and a subsequent self-consistent Born approximation for the two factors. There are corrections though, which might be divergent [35,36]. The reason is that the expression on the left-hand side decays like a power law with distance r while the expression on the right-hand side decays exponentially. The power law is a consequence of the massless mode associated with the spontaneously broken non-Abelian symmetry. This problem will be discussed and solved in Section 4.

3. Self-Consistent Approximation

We start with the self-consistent Born approximation of the average one-particle Green's function

$$\langle G(z) \rangle \approx G_0(z + i\eta), \quad G_0(z) = (\langle H \rangle - z)^{-1}, \quad (9)$$

where the self-energy η is a scattering rate, which is determined by the self-consistent equation $i\eta = G_{0,0}(z + i\eta)$ [37]. This reads in our case with momentum cut-off λ

$$i\eta = \gamma(z + i\eta) \left[\lambda - \frac{\alpha}{2} \log\left(\frac{\alpha + \lambda}{\alpha - \lambda} \right) \right] \quad (\gamma = g/2\pi^2, \; \alpha = \sqrt{(z + i\eta)^2 - \Delta^2/4})$$

and for $z = 0$ this simplifies to the relation $\eta = \eta I$ with

$$I = \gamma \left[\lambda - \beta \arctan(\lambda/\beta) \right], \quad \beta = \sqrt{\eta^2 + \Delta^2/4}.$$

In this case there are two solutions of the self-consistent equation, namely $\eta = 0$ and $\eta \neq 0$ with

$$\gamma = \frac{1}{\lambda - \beta \arctan(\lambda/\beta)}. \quad (10)$$

A nonzero η reflects spontaneous symmetry breaking with respect to (6). Such a solution exists for (10) only at sufficiently large γ. Moreover, η vanishes continuously as we reduce γ. Then there is a phase boundary which separates the symmetric and the symmetry-broken regime:

$$\gamma(\Delta) = \frac{2}{2\lambda - \Delta \arctan(2\lambda/\Delta)} \tag{11}$$

which is plotted in Figure 1. The average density of states then reads

$$\begin{aligned} \rho(E) &= \frac{1}{2\pi} \lim_{\epsilon\to 0} Im\left\{Tr_2\left[\langle G_{\mathbf{rr}}(E+i\epsilon)\rangle\right]\right\} \approx \frac{1}{2\pi} \lim_{\epsilon\to 0} Im\left\{Tr_2 G_{0,0}(E+i\epsilon+i\eta)\right\} \\ &= \frac{1}{\pi} Im\left\{(E+i\eta)\left[\lambda - \frac{\alpha}{2}\log\left(\frac{\alpha+\lambda}{\alpha-\lambda}\right)\right]\right\}, \quad \alpha = \sqrt{(E+i\eta)^2 - \Delta^2/4)}\,. \end{aligned} \tag{12}$$

As a qualitative picture the average density of states is plotted for a fixed η in Figure 2.

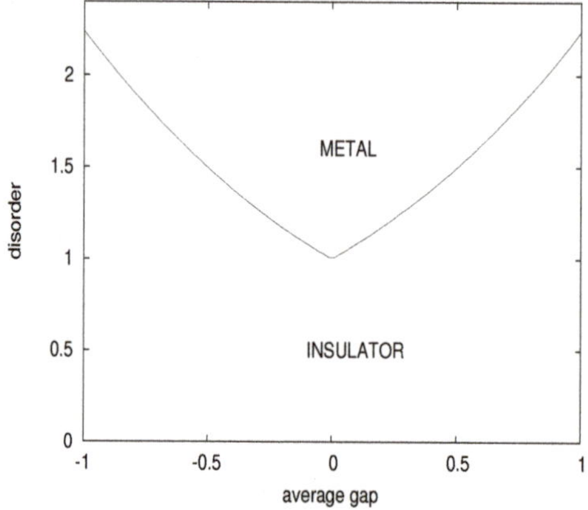

Figure 1. Phase diagram of the metal–insulator transition of the three-dimensional random gap model from Equation (11), where disorder is the parameter γ and the average gap is Δ.

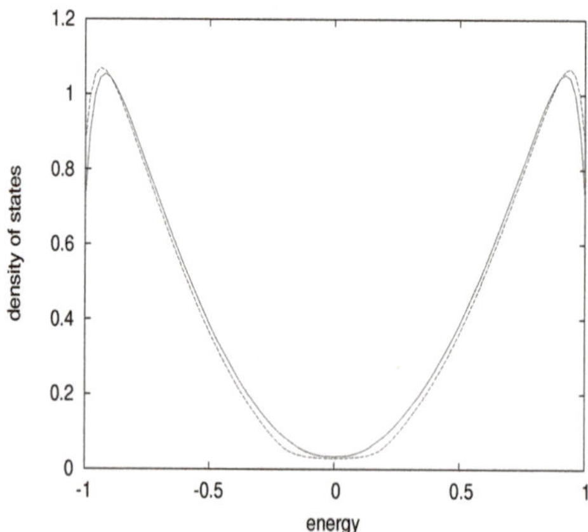

Figure 2. Average density of states of the three-dimensional random gap model for fixed $\eta = 0.04$ and average gap $\Delta = 0.4$ (full curve) and $\Delta = 0.8$ (dashed curve).

4. Scaling Relation of the Average Two-Particle Green's Function

Using the factorization of the averaged product of Green's functions in Equation (8), the conductivity in Equation (7) is approximated as [36]

$$\omega^2 \sum_{\mathbf{r}} r_k^2 Tr_2 \left[\langle G_{0\mathbf{r}}(y) G_{\mathbf{r}0}(-y)\rangle\right] \approx \omega^2 \sum_{\mathbf{r}} r_k^2 Tr_2 \left[\langle G_{0\mathbf{r}}(y)\rangle \langle G_{\mathbf{r}0}(-y)\rangle\right] \quad (y = \omega/2 + i\epsilon)\,, \tag{13}$$

where the constant prefactor $e^2/2h$ has been omitted here. This can be combined with the self-consistent Born approximation in Equation (9) to obtain

$$\omega^2 \sum_{\mathbf{r}} r_k^2 Tr_2 \left[\langle G_{0\mathbf{r}}(y)\rangle \langle G_{\mathbf{r}0}(-y)\rangle\right] \approx \omega^2 \sum_{\mathbf{r}} r_k^2 Tr_2 \left[G_{0,\mathbf{r}}(y+i\eta) G_{0,-\mathbf{r}}(-y-i\eta)\right]\,. \tag{14}$$

For the expression (7) this approximation leads to the Boltzmann (or Drude) conductivity, which reads in our specific case

$$\sigma_{kk} \approx \frac{e^2}{2h}\frac{\omega^2}{\pi^2}\int_0^\lambda \frac{\Delta^2/4 - z^2}{(\Delta^2/4 - z^2 + k^2)^3} k^2 dk \quad (z = \omega/2 + i\eta)\,. \tag{15}$$

Thus the conductivity vanishes in the DC limit $\omega \to 0$ for $\eta > 0$. The reason is that the self-consistent Born approximation creates the Green's function $G_{0,\mathbf{r}}(y + i\eta)$, which decays exponentially on the scale $1/\eta$. Consequently, the sum over the real space coordinates on the right-hand side of Equation (14) is finite.

A more careful inspection indicates that the averaged product of Green's function on the left-hand side of Equation (13) decays according to a power law as a consequence of the massless fluctuations around the spontaneous symmetry breaking solution $\eta \neq 0$ [34]. We can perform the integration with respect to these fluctuations and obtain the diffusion propagator [35]

$$Tr_2 \left[\langle G_{0\mathbf{r}}(y) G_{\mathbf{r}0}(-y)\rangle\right] \approx \frac{\eta - iy}{4}\int \frac{e^{i\mathbf{q}\cdot\mathbf{r}}}{-iy + Dq^2} d^3q \tag{16}$$

with diffusion coefficient

$$D = \frac{\eta - iy}{2}\sum_{\mathbf{r}} {r_k}^2 Tr_2 \left[G_{0,\mathbf{r}}(y+i\eta) G_{0,-\mathbf{r}}(-y-i\eta)\right]\,. \tag{17}$$

After summing over the real space coordinates we obtain the expression

$$\omega^2 \sum_{\mathbf{r}} r_k^2 Tr_2 \left[\langle G_{0\mathbf{r}}(y) G_{\mathbf{r}0}(-y)\rangle\right] = \omega^2 f(\eta/y) \sum_{\mathbf{r}} {r_k}^2 Tr_2 \left[G_{0,\mathbf{r}}(y+i\eta) G_{0,-\mathbf{r}}(-y-i\eta)\right]\,, \tag{18}$$

where the coefficient on the right-hand side is a result of the strong massless fluctuations, which didn't exist in the approximation given by Equation (14). It depends on the ratio of the order parameter of spontaneous symmetry breaking η and the symmetry-breaking field y:

$$f(\eta/y) = (1 + i\eta/y)^2\,. \tag{19}$$

This coefficient indicates that the correlations of the Green's function fluctuations are negligible only for $f(\eta/y) \approx 1$. This is the case in the absence of symmetry breaking, where $\eta = 0$ and $f(0) = 1$. This justifies the approximation by Equation (14) in the insulating regime. On the other hand, in the presence of spontaneous symmetry breaking (i.e., for $\eta > 0$) the coefficient diverges for $\omega \to 0$ and gives $\omega^2 f(\eta/y) \to -1$ in the limits $\epsilon \to 0$ and then $\omega \to 0$.

The diffusion coefficient in Equation (17) is easy to calculate and reads

$$\sum_{\mathbf{r}} r_k{}^2 Tr_2 \left[G_{0,\mathbf{r}}(y+i\eta) G_{0,-\mathbf{r}}(-y-i\eta) \right] = \frac{1}{4\pi} \frac{1}{\sqrt{\Delta^2/4 + (\eta - iy)^2}}, \tag{20}$$

which together with the scaling relation (18) gives for the conductivity of Equation (7) in the DC limit $\omega \to 0$

$$\sigma_{kk} = \frac{e^2}{4\pi h} \frac{\eta^2}{\sqrt{\Delta^2/4 + \eta^2}}. \tag{21}$$

The solution η of the self-consistent Equation (10) is inserted into σ_{kk} and the conductivity is plotted as a function of disorder strength γ in Figure 3. The conductivity vanishes linearly with decreasing disorder strength (i.e., with decreasing doping concentration). To illustrate the crossover to a power law with exponent 0.6, the calculated conductivity and the power-law fit are plotted together in Figure 4.

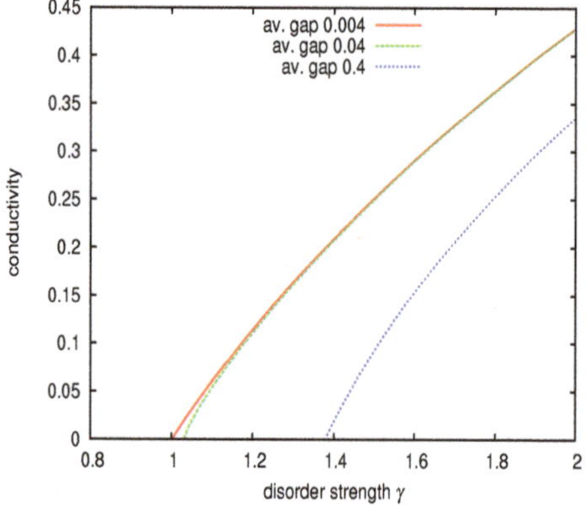

Figure 3. Conductivity as a function of disorder for an average gap $\Delta = 0.004$ (red curve), $\Delta = 0.04$ (green curve) and $\Delta = 0.4$ (blue curve). There is a metal–insulator transition at $\gamma \approx 1$, at $\gamma \approx 1.03$ and at $\gamma \approx 1.37$, respectively.

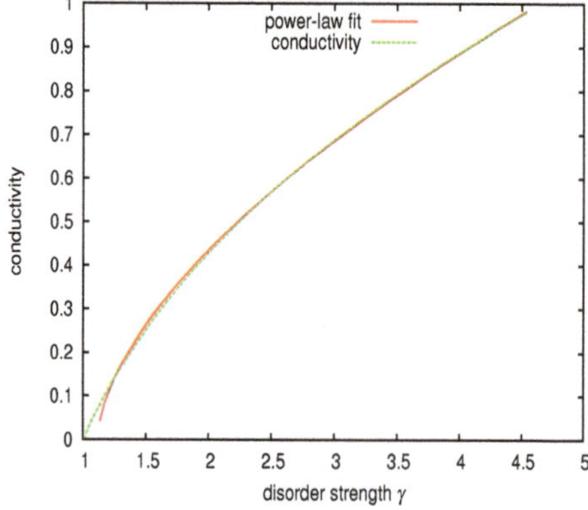

Figure 4. This plot demonstrates the crossover in the critical regime of the conductivity (green curve) through the fitting (red) curve $0.47(\gamma - \gamma_c)^{0.6}$. The latter fits the conductivity very well away from the critical point γ_c for $\Delta = 0.01$.

5. Discussion and Conclusions

Our result for the DC conductivity in Equation (21), together with the solution of the order parameter η in Equation (10), provides a simple description of a metal–insulator transition in doped three-dimensional semiconductors. The metal–insulator transition is characterized by the scattering rate η that vanishes in the insulating regime. Such a behavior is not an Anderson transition, since the latter would have a scattering rate $\eta \neq 0$ on both sides of the transition [38]. Even more important is the change of the coefficient $f(\eta/y)$: It is always 1 in the insulating regime and infinite in the metallic regime. This quantity describes the correlations of the Green's function fluctuations in the relation (18).

There is a linear behavior near the metal–insulator transition and a crossover to a non-critical power law, as depicted in Figure 4. For the linear part the slope of the conductivity is quite robust with respect to the average gap Δ (cf. Figure 3). Away from the transition point a negative curvature appears though, which can be fitted by a power law with exponent $\mu \approx 0.6$ (cf. Figure 4). The change of exponents can be related to the discussion in References [33,39] about a crossover of exponents in Si:P from $\mu \approx 1$ very close to the critical point N_c to $\mu \approx 0.5$ further away from N_c. Rosenbaum et al. have found that the conductivity close to the critical point varies from sample to sample [32]. This indicates strong conductivity fluctuations, which may also exist in our random gap model, as indicated by the strong fluctuations of the Green's functions due to the large values of $f(\eta/y)$.

As mentioned in the Introduction, a related metal–insulator transition in three-dimensional Weyl fermionic systems has attracted considerable attention recently [2–26]. Formally, this transition is very similar, although the underlying Hamiltonian is that of Weyl fermions rather than our simple semi-conductor Hamiltonian in Equation (1). This difference leads to the creation of two distinct insulating phases, characterized by the Hall conductivity $\sigma_{xy} = \mp e^2/2h$ in the lower part of the phase diagram in Figure 1 for Weyl fermions. But the role of the particle-hole symmetry, the existence of a massless mode due to spontaneous breaking of this symmetry and the role of diffusion in the metallic phase are the same in both types of models [26–28]. This indicates that metal–insulator transitions in systems with particle-hole symmetry are based on the same type of mechanism.

Funding: This research received no external funding.

Acknowledgments: This work was supported by a grant of the Julian Schwinger Foundation.

Conflicts of Interest: The author declares no conflicts of interest.

References

1. Ashcroft, N.W.; Mermin, N.D. *Solid State Physics*; Saunders College: Rochester, NY, USA, 1976.
2. Fradkin, E. Critical behavior of disordered degenerate semiconductors. I. Models, symmetries, and formalism. *Phys. Rev. B* **1986**, *33*, 3257–3262. [CrossRef]
3. Fradkin, E. Critical behavior of disordered degenerate semiconductors. II. Spectrum and transport properties in mean-field theory. *Phys. Rev. B* **1986**, *33*, 3263–3268. [CrossRef]
4. Wan, X.; Turner, A.M.; Vishwanath, A.; Savrasov, S.Y. Topological semimetal and Fermi-arc surface states in the electronic structure of pyrochlore iridates. *Phys. Rev. B* **2011**, *83*, 205101. [CrossRef]
5. Smith, J.; Banerjee, S.; Pardo, V.; Pickett, W.E. Dirac Point Degenerate with Massive Bands at a Topological Quantum Critical Point. *Phys. Rev. Lett.* **2011**, *106*, 056401. [CrossRef]
6. Burkov, A.A.; Balents, L. Weyl Semimetal in a Topological Insulator Multilayer. *Phys. Rev. Lett.* **2011**, *107*, 127205. [CrossRef]
7. Burkov, A.A.; Hook, M.D.; Balents, L. Topological nodal semimetals. *Phys. Rev. B* **2011**, *84*, 235126. [CrossRef]
8. Xu, G.; Weng, H.; Wang, Z.; Dai, X.; Fang, Z. Chern Semimetal and the Quantized Anomalous Hall Effect in $HgCr_2Se_4$. *Phys. Rev. Lett.* **2011**, *107*, 186806. [CrossRef] [PubMed]
9. Witczak-Krempa, W.; Kim, Y.B. Topological and magnetic phases of interacting electrons in the pyrochlore iridates. *Phys. Rev. B* **2012**, *85*, 045124. [CrossRef]
10. Young, S.M.; Zaheer, S.; Teo, J.C.Y.; Kane, C.L.; Mele, E.J.; Rappe, A.M. Dirac Semimetal in Three Dimensions. *Phys. Rev. Lett.* **2012**, *108*, 140405. [CrossRef] [PubMed]

11. Hosur, P.; Parameswaran, S.A.; Vishwanath, A. Charge Transport in Weyl Semimetals. *Phys. Rev. Lett.* **2012**, *108*, 046602. [CrossRef]
12. Wang, Z.; Sun, Y.; Chen, X.-Q.; Franchini, C.; Xu, G.; Weng, H.; Dai, X.; Fang, Z. Dirac semimetal and topological phase transitions in A_3Bi (A = Na, K, Rb). *Phys. Rev. B* **2012**, *85*, 195320. [CrossRef]
13. Singh, B.; Sharma, A.; Lin, H.; Hasan, M.Z.; Prasad, R.; Bansil, A. Topological electronic structure and Weyl semimetal in the $TlBiSe_2$ class of semiconductors. *Phys. Rev. B* **2012**, *86*, 115208. [CrossRef]
14. Cho, G.Y. Possible topological phases of bulk magnetically doped Bi2Se3: Turning a topological band insulator into the Weyl semimetal. *arXiv* **2012**, arXiv:1110.1939v2.
15. Halász, G.B.; Balents, L. Time-reversal invariant realization of the Weyl semimetal phase. *Phys. Rev. B* **2012**, *85*, 035103. [CrossRef]
16. Nandkishore, R.; Huse, D.A.; Sondhi, S. Rare region effects dominate weakly disordered three-dimensional Dirac points. *Phys. Rev. B* **2014**, *89*, 245110. [CrossRef]
17. Liu, C.-X.; Ye, P.; Qi, X.-L. Chiral gauge field and axial anomaly in a Weyl semimetal. *Phys. Rev. B* **2013**, *87*, 235306. [CrossRef]
18. Biswas, R.R.; Ryu, S. Diffusive transport in Weyl semimetals. *Phys. Rev. B* **2014**, *89*, 014205. [CrossRef]
19. Kobayashi, K.; Ohtsuki, T.; Imura, K.-I. Disordered Weak and Strong Topological Insulators. *Phys. Rev. Lett.* **2013**, *110*, 236803. [CrossRef]
20. Huang, Z.; Das, T.; Balatsky, A.V.; Arovas, D.P. Stability of Weyl metals under impurity scattering. *Phys. Rev. B* **2013**, *87*, 155123. [CrossRef]
21. Kobayashi, K.; Ohtsuki, T.; Imura, K.-I.; Herbut, I.F. Density of States Scaling at the Semimetal to Metal Transition in Three Dimensional Topological Insulators. *Phys. Rev. Lett.* **2014**, *112*, 016402. [CrossRef]
22. Ominato, Y.; Koshino, M. Quantum transport in a three-dimensional Weyl electron system. *Phys. Rev. B* **2014**, *89*, 054202. [CrossRef]
23. Roy, B.; Das Sarma, S. Diffusive quantum criticality in three-dimensional disordered Dirac semimetals. *Phys. Rev. B* **2014**, *90*, 241112(R). [CrossRef]
24. Sbierski, B.; Pohl, G.; Bergholtz, E.J.; Brouwer, P.W. Quantum Transport of Disordered Weyl Semimetals at the Nodal Point. *Phys. Rev. Lett.* **2014**, *113*, 026602. [CrossRef] [PubMed]
25. Syzranov, S.V.; Gurarie, V.; Radzihovy, L. Critical Transport in Weakly Disordered Semiconductors and Semimetals. *Phys. Rev. Lett.* **2015**, *114*, 166601. [CrossRef]
26. Ziegler, K. Quantum transport in 3D Weyl semimetals: Is there a metal-insulator transition? *Eur. Phys. J. B* **2016**, *89*, 268. [CrossRef]
27. Ziegler, K. Quantum transport with strong scattering: beyond the nonlinear sigma model. *J. Phys. A Math. Theor.* **2015**, *48*, 055102. [CrossRef]
28. Ziegler, K. Zero mode protection at particle-hole symmetry: A geometric interpretation. *J. Phys. A Math. Theor.* **2019**, *52*, 455101. [CrossRef]
29. Rosenbaum, T.F.; Andres, K.; Thomas, G.A.; Bhatt, R.N. Sharp Metal-Insulator Transition in a Random Solid. *Phys. Rev. Lett.* **1980**, *45*, 1723–1726. [CrossRef]
30. Rosenbaum, T.F.; Milligan, R.F.; Paalanen, M.A.; Thomas, G.A.; Bhatt, R.N.; Lin, W. Metal-insulator transition in a doped semiconductor. *Phys. Rev. B* **1983**, *27*, 7509–7523. [CrossRef]
31. Roy, A.; Sarachik, M.P. Susceptibility of Si:P across the metal-insulator transition. II. Evidence for local moments in the metallic phase. *Phys. Rev. B* **1988**, *37*, 5531–5534. [CrossRef]
32. Rosenbaum, T.F.; Thomas, G.A.; Paalanen, M.A. Critical behavior of Si:P at the metal-insulator transition. *Phys. Rev. Lett.* **1994**, *72*, 2121. [CrossRef] [PubMed]
33. Löhneysen, H. Current issues in the physics of heavily doped semiconductors at the metal-insulator transition. *Philos. Trans. R. Soc. Lond. A* **1998**, *356*, 139–156. [CrossRef]
34. Ziegler, K. Scaling behavior and universality near the quantum Hall transition. *Phys. Rev. B* **1997**, *55*, 10661–10670. [CrossRef]
35. Ziegler, K. Random-Gap Model for Graphene and Graphene Bilayers. *Phys. Rev. Lett.* **2009**, *102*, 126802. [CrossRef] [PubMed]
36. Altshuler, B.L.; Simons, B.D. *Mesoscopic Quantum Physics*; Akkermans, E., Montambaux, G., Pichard, J.-L., Zinn-Justin, J., Eds.; North-Holland: Amsterdam, The Netherlands, 1995.
37. Altland, A.; Simons, B.D. *Condensed Matter Field Theory*; Cambridge University Press: Cambridge, MA, USA, 2010.

38. Schäfer, L.; Wegner, F. Disordered system with n orbitals per site: Lagrange formulation, hyperbolic symmetry, and goldstone modes. *Z. Phys. B* **1980**, *38*, 113–126. [CrossRef]
39. Stupp, H.; Hornung, M.; Lakner, M.; Madel, O.; Löhneysen, H. Possible solution of the conductivity exponent puzzle for the metal-insulator transition in heavily doped uncompensated semiconductors. *Phys. Rev. Lett.* **1993**, *71*, 2634–2637. [CrossRef]

Article

Symmetry Breaking in Stochastic Dynamics and Turbulence

Michal Hnatič [1,2,3,*,†], Juha Honkonen [4,†] and Tomáš Lučivjanský [1,†]

1 Faculty of Science, Šafárik University, Moyzesova 16, 040 01 Košice, Slovakia; tomas.lucivjansky@upjs.sk
2 Bogolyubov Laboratory of Theoretical Physics, Joint Institute for Nuclear Research, 141980 Dubna, Russia
3 Institute of Experimental Physics, Slovak Academy of Sciences, Watsonova 47, 040 01 Košice, Slovakia
4 Department of Military Technology, National Defence University, 00861 Helsinki, Finland; juha.honkonen@helsinki.fi
* Correspondence: hnatic@saske.sk
† These authors contributed equally to this work.

Received: 19 August 2019; Accepted: 13 September 2019; Published: 23 September 2019

Abstract: Symmetries play paramount roles in dynamics of physical systems. All theories of quantum physics and microworld including the fundamental Standard Model are constructed on the basis of symmetry principles. In classical physics, the importance and weight of these principles are the same as in quantum physics: dynamics of complex nonlinear statistical systems is straightforwardly dictated by their symmetry or its breaking, as we demonstrate on the example of developed (magneto)hydrodynamic turbulence and the related theoretical models. To simplify the problem, unbounded models are commonly used. However, turbulence is a mesoscopic phenomenon and the size of the system must be taken into account. It turns out that influence of outer length of turbulence is significant and can lead to intermittency. More precisely, we analyze the connection of phenomena such as behavior of statistical correlations of observable quantities, anomalous scaling, and generation of magnetic field by hydrodynamic fluctuations with symmetries such as Galilean symmetry, isotropy, spatial parity and their violation and finite size of the system.

Keywords: stochastic dynamics; symmetry breaking; field-theoretic renormalization group

---

1. Introduction

The success of physics is to a large extent dictated by its enormous predictive power describing many natural phenomena. Ranging from microscopic distances probed at colliders facilities up to macroscopic scales observed through sophisticated astronomical devices, physics develops theories and models that describe reality to a very high precision. Once one understands basic principles, one could not stop wonder about the incredible efficiency with which fundamental physical laws are constructed. To a large extent, guiding principles in physics are based on a correct recognition of underlying symmetries.

From a historical point of view, the first nontrivial symmetry found was Galilei's discovery of equivalence of inertial frames whose direct consequence is momentum conservation. A further development in theoretical physics, most notably utilized by E. Noether in her work, uncovers the fundamental role in classical physics played by symmetries. Later on, it was realized that many physical theories can be based on a proper identification of symmetries. The prototypical and most successful example is how the Lorentz covariance of particle physics, underlying principles of quantum mechanics and local gauge symmetry, restrict the permissible form of theory to such an extent that every such attempt results in a kind of quantum field theory [1]. From a modern perspective, this is related to the observation that most quantum field theory models should be interpreted as kind of effective field models that describe physics sufficiently up to a certain energy scale. An input from experiments is still needed to impose restrictions on theory, for example, in mirror symmetry, time reversal and charge conjugation play a fundamental role in the formulation of the standard model [2]. Indications of whether a given interaction is accompanied by some symmetry are inferred by the experiment and not from theoretical reasoning alone.

Symmetry considerations are not restricted to particle physics only. They are important in other research areas as well. In particular, our aim in this article is to describe them in the context of classical physics related mainly to applications in fluid mechanics. To set the context and provide the basic framework for theoretical considerations and relations with other branches of physics, we discuss the underlying ideas more broadly.

The problems considered in this paper belong to statistical physics, which forms a cornerstone of the modern science. Since its foundation as a scientific discipline in the works of Gibbs and Boltzmann, it has evolved to a great depth both in scope and rigorousness. At present, methods primarily devoted to the study of physical systems are applied in such diverse scientific fields as chemistry, biology, economics, sociology and computer science. Such a success might be explained by the generality of the fundamental laws of statistical physics and genuine appearance of systems with great resemblance to the models studied in physics. In general, they could be characterized by a large number of entities (atoms, spins and so on) that interact with each other. It is important to realize how large this number is. For instance, merely one mole of ordinary matter under normal circumstances contains as many as 10^{23} atoms or molecules. This is an incredibly large value beyond human ability to imagine. The rigorous treatment of such a system in terms of particle dynamics (via classical Euler–Lagrange equations or Schrödinger equation in the quantum case) is apparently meaningless. Nonetheless, it is an experimental fact that under stationary boundary conditions all closed macroscopic systems tend to evolve to the equilibrium state that is characterized by a constant value of its macroscopic (coarse-grained) characteristics, e.g., temperature, volume, magnetization, etc. At mesoscopic scales of space and time, this tendency remains true, but the thermodynamic parameters are slowly varying functions of position and time. In a more precise sense, this property of nature is formulated in the second law of thermodynamics. In an isolated system, it also identifies an equilibrium with the most disordered (equivalent to the most probable) state under given external conditions.

In many instances encountered in equilibrium physics, it is possible to use an approximation in which interactions or fluctuations between microscopic constituents of the model can be neglected or treated as a small perturbation to the ideal situation of noninteracting particles. The ideal gas and van der Waals model are famous examples. Note that in the latter case attractive interaction between gas molecules are effectively taken into account by the corresponding virial term [3].

When matter is more dense, interactions between neighboring particles cannot be neglected. The fundamental property of additivity of energy and entropy [3] in the equilibrium is maintained by the assumption that the system may be considered comprising of noninteracting subsystems ("mesoscopic" elements of matter). It should be noted that, for the very possibility of considering noninteracting particles or subsystems as the ideal situation, particles are assumed to have short-range interactions. At the microscopic level, this corresponds to electrically neutral molecules or clusters of molecules. Problems related to formation and structure of these quantities [4] are beyond the scope of this article.

However, there also exist situations, in which neglecting fluctuations is not appropriate at all. The theory of critical phenomena [5,6], which deals with the second-order phase transitions in macroscopic systems, is a well-known representative. It is observed, e.g., in liquid–vapor transition, $\lambda-$transition in superfluid helium or various magnetic transitions between paramagnetic and ferromagnetic phases. A characteristic feature of these transitions is the appearance of strong fluctuations and correlations between underlying constituents (atoms or spins, subsystems comprising thereof). A parameter used for a quantitative description of correlations is the correlation length. Broadly speaking, it represents the average distance to which atoms or clusters "feel" each other or behave cooperatively. In equilibrium situations, it is of atomic order, which explains why a dilute atomic system usually can be considered consisting of effectively non-interacting atoms. In dense matter, this is rephrased to clusters of atoms.

The first attempt to tackle the problem of phase transitions was based on the use of mean field theory. Loosely speaking, it presumes that correlations between mesoscopic stochastic variables can be treated perturbatively. Thus, in the sense of the central limit theorem, deviations from the ideal Gaussian behavior of fluctuations can be constructed (in quantum field theory, this procedure is also known as the Wick theorem). In most cases, some equation of state for macroscopic quantities can be directly obtained. In situations when the system is far from the critical region, the correlation length is very small and can be effectively neglected. There is no obvious discrepancy with the experimental data. However, it turns out that this approach exhibits large quantitative differences between the theory and the experiment in the vicinity of a critical point. The problem is that here the correlations are very large. In fact, directly at the transition point, the correlation length is divergent. Therefore, perturbation theory is no longer applicable (the coupling constant is very large and by no means can be treated as a small quantity). However, the divergent correlation length reveals another important effect in the physical picture: scale invariance. In contrast to the aforementioned symmetries, this is a case of a dynamically generated (emergent) symmetry. Mathematically, it is an invariance with respect to the special class of transformations that account for a change in the scale at which a physical system is studied. A divergent correlation length means that there is no special scale and the system looks and behaves at every scale in the same way.

A well-known observation in physics is that symmetry might have enormous influence on the properties of the physical system. In terms of scale invariance, the experimentally observed power laws for the functional dependence of various thermodynamic functions can be explained and also the various relations between critical exponents can be quantitatively estimated. Another important property of the second order phase transition is *universality*. In a simple formulation, it says that the behavior of the system near its critical point is fully determined by universal quantities—dimension of space, and number of components of order parameter, symmetry constraints—that are not characteristic of one system only, but a whole class of systems and also states that universal quantities do not depend on the model-dependent parameters—coupling constants, etc. Thus, in microscopic details very different physical systems such as strongly anisotropic magnetic material (the Ising model) and liquid might have the same critical properties. One only has to know a few very general pieces of information to classify a given physical system according to its critical behavior into some universality class, wherein all the systems behave in the same way.

It is interesting to realize that there is a great intrinsic similarity between quantum models in particle physics and statistical models. A profound relation between quantum field theory and statistical mechanics is revealed through the language of path integrals [5,6]. A classical random field in this framework is completely analogous to a fluctuating quantum field. Field models are often amenable to perturbation methods. Using Feynman diagrammatic technique, terms in perturbative expansion are expressed via Feynman diagrams that are a graphical representation of certain integrals. As a rule, they contain divergences in the range of large and small scales (wavevectors). In particle physics, there are no natural restrictions on scales, therefore it is necessary to find an effective procedure to eliminate these divergences step by step in each order of a concrete perturbation scheme. In perturbation expansions of classical field theories, natural scales usually exist: at small scales, the continuum description breaks down at atomic scales (nanophysics) and there is no reason to go below this lower limit. On the other hand, real quantities of matter are of finite size. In the theoretical description, however, it is convenient and customary to extrapolate results obtained for a finite quantity of homogeneous matter to the whole space when modeling real systems. Below, we demonstrate renormalization methods in the framework of the stochastic model of developed turbulence and related applications.

The method of renormalization group (RG) was proposed in the framework of the quantum field theory in the 1950s [1,7–11]. From the practical point of view, the RG method represents an effective way to determine non-trivial asymptotic behavior of Green functions (correlation functions) in the range of large (ultraviolet) or small (infrared) wavevectors (scales). The asymptotic behavior is non-trivial if, in a given order of a perturbative calculation, the divergences in a certain range of wavevectors appear (e.g., so-called large logarithms), which compensate for the smallness of the coupling constant g. In such case, summation of all terms of a perturbation series is needed. This summation can be carried out by means of the RG approach. Technically, one obtains linear partial differential RG equations for the Green functions. The coefficient functions (RG-functions) in the differential operator (see below) are calculated at a given order of the perturbation scheme. However, the solution of the RG equation represents the sum of an infinite series. For example, if the RG-functions are calculated at the lowest non-trivial order of the perturbation theory and the corresponding RG-equation is solved, the obtained result is a sum of leading logarithms of the whole perturbation series. Moreover, if the RG-functions are calculated with an improved precision, the solution of the RG equation includes corrections to the leading logarithms.

Notwithstanding the similarity of theoretical description of quantum field theory and classical statistical models, it has to be borne in mind that there is an essential difference in the interpretation and use of the RG in statistical physics on the one hand and in particle physics on the other hand. In particle physics, we are interested in an analysis of scaling in ultraviolet (UV) regime corresponding to large momenta. On the other hand, in statistical physics, asymptotic behavior in the opposite infrared (IR) limit of small momenta is usually studied. In both statistical mechanics and hydrodynamic transport problems, the interest in the IR behavior of statistical models is determined by the property of the basic field-theoretic tool—perturbation theory—to reproduce the observed singular behavior of certain physical quantities only in the limit of an infinite (flat) space. However, this infrared limit is usually rather sensitive to the large-scale structure of the model and care has to exercised when passing to the limit. In the case of equilibrium systems, the analysis is based on the Gibbs distribution, but, in the case of steady-state stochastic systems, there is no generic tool to this end which emphasizes the role of symmetry arguments.

The final aim of the theory (either in stochastic dynamics or developed turbulence) is to find the time-space dependence of statistical correlations—mainly those that can be experimentally measured. It turns out that use of quantum field theory methods (RG included) allows deriving a linear differential equation, which contains stable solutions in the asymptotic region of large macroscopic scales.

Solutions take a form of a product of a power-like term with a nontrivial exponent and scaling function of dimensionless variables (the scaling function is not determined by the RG method). To compute critical exponents in the form of asymptotic series, one has to resort to a certain scheme (we often employ variants of dimensional renormalization). Asymptotic properties of the scaling functions are analyzed by the operator product expansion, which is another theoretical tool developed mainly by Wilson, Wegner and Kadanoff [12–14]. In the stochastic theory of fully developed turbulence, scaling functions may be singular functions of dimensionless arguments and this can drastically change the critical exponents. The results demonstrate intermittent (multifractal) behavior of statistical correlations of the random fields of concentration of advected particles. Intermittency is a typical mesoscopic phenomenon, which is quantitatively revealed in singular behavior of correlation functions of velocity fluctuations with respect to an external turbulent spatial scale L.

The RG method not only leads to a quantitative description of the behavior near critical points, it also provides a new framework in which the aforementioned scale invariance and universality are naturally explained. At present, the use of the RG method in equilibrium statistical physics is well established and represents an important theoretical tool.

Contrary to the equilibrium physics, there are only few rigorous results in the case of non-equilibrium systems. Some of their properties are reminiscent of the equilibrium systems and thus it seems natural to apply the RG method to them. However, there exist also fundamental differences between them. Non-equilibrium systems might be divided [15] into two broad classes

- Systems with a Hermitian Hamiltonian, whose stationary states are described by Gibbs–Boltzmann distribution. Note that at the beginning they happen to be in a state far from the stationary (equilibrium) state. The dynamic description of such systems is obtained directly from static formulations. Examples include the Landau–Ginzburg equation for time evolution of local magnetization, kinetic Ising model, and models A-H for various models of critical dynamics [16]. All these equations are specific realizations of a rather general Langevin equation [5].

- Systems without Hermitian Hamiltonian or without Hamiltonian description at all, which in general do not need to have a stationary state. The detailed balance condition is not satisfied for them, which implies that Einstein relation between thermal fluctuations and friction forces cannot be stated. Typical examples of such systems cover: fluid in turbulent state, irreversible chemical processes, surface growing models, etc. Other approaches to such systems have to be used via quite general stochastic differential equation, which can be considered as an extension of a Langevin equation or using a master equation [17]. The former equation is suggested for some macroscopic quantity. Neglect of microscopic degrees of freedom is replaced by an introduction of random force. Then, according to underlying physical observations, properties of random force have to be specified. The latter approach is probably more fundamental, but also more difficult to handle.

In what follows, our main interest concerns specific problems of the second type related to hydrodynamics. In hydrodynamics, dissipation of mechanical energy to heat is an essential part of the physics. Since a Hamiltonian description is not possible in this case, we are dealing with steady states of the latter class.

As we know from everyday experience, fluids can exhibit very different behaviors from very simple, e.g., laminar flow, which is very predictable, to very chaotic, as is realized in turbulent motions. Turbulence is important in the analysis of phenomena in a wide range of scales from particle collisions in accelerators [18] and circulation of human blood [19] to the flow of air and water in the atmosphere and oceans, solar wind [20] and clusters of galaxies [21]. Let us stress that all studied systems belong to open systems that need a continuous input of energy in order to maintain steady state.

Theoretical analysis of turbulence is based on the statistical analysis of solutions of the Navier–Stokes problem. Symmetry and similarity arguments have allowed infering important conclusions about the scaling behavior of velocity correlation functions in the case of very large Reynolds numbers (the famous Kolmogorov theory in the first place) [22,23]. However, a more detailed statistical description of this fully developed turbulence as well as the onset of turbulence in a laminar flow are still lacking. Notwithstanding the rapid development of experimental methods [24], one of the major problems in the study of turbulence is the deficit of high-resolution experimental data. Therefore, numerical methods have become important tools in the investigation of turbulence [24] and provide solid benchmarks for testing of analytic results.

It is well-known that weather forecasting can be done for no more than a few days. This is caused by the intrinsic instability of Navier–Stokes (NS) equations, which are believed to describe motion of viscous (non-relativistic) fluids [22]. The formidable task of finding its solution remains one of the last unsolved classical problems [23]. For classification of various fluid states, the Reynolds number Re has been introduced. It is defined as $\mathrm{Re} = VL/\nu$, where V is typical average flow velocity, L is an external scale (e.g., a dimension of an obstacle, which causes perturbation to the regular flow) and ν is kinematic viscosity of the medium. It thus expresses the ratio between inertial and friction (dissipation) forces in a given fluid. In the case of low values, $\mathrm{Re} \ll 1$, regular (laminar) flow is observed. With an increasing value of Re, very different phenomena occur ranging from the periodical ones as Kármán vortices to very chaotic irregular motion for the limit of very high values of $\mathrm{Re} \gg 1$ (in practice value $\mathrm{Re} \geq 10^6$ is large enough) [23,25]. This state of fluid is known as fully developed turbulence.

At first sight, a very complicated problem turns out to be theoretically tractable because of appearance of new symmetries (again kind of emergent symmetry)—statistical symmetries. Kolmogorov postulated hypotheses [23,26] that could explain turbulence and also predict statistical and scaling properties of various correlation and structure functions. The Kolmogorov theory can be considered a kind of theory for "ideal" turbulence in the sense that it assumes the infinite value for the Reynolds number. These hypotheses are still not proved from the first principles—in this case from the Navier–Stokes equations. It should also be borne in mind that Kolmogorov's hypotheses are stated for the case of homogeneous isotropic flat space but without any specific indication how this limit is approached when the Reynolds number grows without limit.

In contrast to mathematics, the physicist's approach to turbulence follows a different path. Instead of considering a difficult mathematical problem related to boundary and initial conditions, their effect is replaced by properly chosen random force. The Navier–Stokes equation is amended by an additive random variable, which also mimics continuous input of mechanical energy into the system. The choice of the structure and statistics of the random force is the most essential point in modeling of the large-scale effects on the scale-invariant behavior predicted by the Kolmogorov theory. To this end, random forces concentrated at large spatial scales are used. In the basic setup of the stochastic problem, rotational symmetry of force correlations is assumed, but variation of the symmetry properties of force correlations (e.g., anisotropy and reflection asymmetry) may be used to probe the effect of large-scale properties on the scaling behavior of velocity correlations. The modeled large-scale induced effects include the appearance of a set of anomalous scaling dimensions in corrections to Kolmogorov scaling (multifractality) and magnetohydrodynamic dynamo. On the other hand, the Galilei invariance and small-scale anisotropy have been shown to be stable against large-scale perturbations.

In 1977, D. Forster, D. R. Nelson and M. J. Stephen applied the RG method to calculate the correlations of velocity field [27] governed by the stochastic Navier–Stokes equation with external random forcing. This work was motivated by Wilson's momentum shell approach to RG, in which tracing out of fast degrees of freedom is supplemented with scale transformation [12,15]. Later, it was shown by C. De Dominicis and P. C. Martin [28] that in the range of small wave numbers the correlations of the velocity field manifest a scaling behavior with the celebrated Kolmogorov exponents. The stochastic NS equation was proposed to justify the Kolmogorov theory and has to be distinguished from the usual NS equations.

The essential idea of applying RG in the theory of developed turbulence consists in elimination of the direct influence of the modes with high wave numbers on observed quantities. Their influence is included in some effective variables, e.g., to the turbulent viscosity. Such an approach based on momentum shell approach was later developed further [29,30]. Let us note that in this paper we consider a different field-theoretic renormalization group technique [23,26,31,32], whose main advantage is more transparent and easier calculations.

Another interesting problem related to turbulence is the advection of some quantity [33,34] (temperature field, concentration field or tracer) by the turbulent field. In addition to the practical importance of such a problem, it is also very interesting from the theoretical point of view. It is still not clear to what extent turbulence is intermittent [23], i.e., what its fractal nature is. On the other hand, advection of a passive scalar quantity by simpler models (e.g., Kraichnan model that is described in detail below) than turbulence exhibits very strong intermittent behavior.

Naively, basic assumptions of Kraichnan-like models can be considered too crude. A typical approach to stochastic dynamics starts from an analysis of ideal systems—homogeneous in space and time, isotropic, incompressible (in the case of fluids), possessing mirror symmetry, etc. In the present review, the corresponding results for fully developed turbulence are summarized. However, real systems almost always exhibit some form of anisotropy, compressibility or violated mirror symmetry. The effect of such deviations from the ideal system on fluctuating random fields has been an object of intensive research activity, whose arguments and conclusions are described. The results have led to the general conclusion that such effects play a very important role. They can drastically change the large-scale behavior predicted by models of ideal systems.

In the original formulation of the Kraichnan model, the velocity field is assumed to be Gaussian, isotropic, incompressible and uncorrelated in time (white noise). More sophisticated models aim to incorporate effects of anisotropy, compressibility and finite correlation time [35]. Recent studies have pointed out some crucial differences between problems with vanishing and finite correlation time [33,36] and between the compressible and incompressible flows [37,38]. We employ the Kraichnan rapid-change ensemble to model the turbulent mixing [39]. Thus, we assume that velocity field is given by time-decorrelated Gaussian variable with the pair velocity function of the following form $\langle vv \rangle \propto \delta(t-t')k^{-d-\varepsilon}$, where $k = |\boldsymbol{k}|$ is the wave number and $0 < \varepsilon < 2$ is a free parameter of the theory. The physically most interesting value $\varepsilon = \frac{4}{3}$ corresponds to the realistic ("Kolmogorov") scaling behavior. This model gained popularity in the past mainly because of the insight it provides into the explanation of intermittency and anomalous scaling in turbulent flows. In the context of our study, it is worth mentioning that the Kraichnan ensemble allows a straightforward incorporation of compressibility, which appears complicated if the velocity is modeled by dynamical equations. The Kraichnan ensemble has been generalized further to the case of finite correlation time (see, e.g., [34,36,40] for the passive scalar and [41] for the passive vector fields). However, such synthetic models with non-vanishing correlation time are plagued by the lack of Galilean symmetry.

In Section 2, a short introduction to field-theoretic approach to stochastic dynamic is given and we briefly discuss the choice of the functional representation of the perturbation expansion for the solution of the Langevin equation. Section 3 is reserved for discussion of stochastic Navier–Stokes equation and basics of Kraichnan model. Section 4 is devoted to basic information of renormalization group approach and the related Section 5 to operator product expansion that is a specific method of the RG technique. In Section 6, Ward identities are used to obtain information about energy and momentum transfer in turbulent media. Section 7 is devoted to use of Ward identities in the RG analysis. Section 8 describes a dynamic restoration of initially broken Galilean symmetry. Section 9 is dedicated to a mechanism of spontaneous symmetry breaking in magnetohydrodynamic problem which is responsible for creation of magnetic dynamo. In Section 10, we discuss the effect of anisotropy and Section 11 is reserved for final remarks and comments.

2. Field-Theoretic Formulation

It is a well-known fact [5] that the failure of Landau theory of the second-order phase transition lies in the assumption of analyticity of the energy functional $F = F(\varphi)$, where $\varphi(x)$ is an order parameter configuration. The equilibrium physics of the phase transition is described by the order parameter at the minimum of the energy functional. The fluctuation theory of phase transitions takes as the fundamental quantity the random field $\varphi(x)$, whose probability density function is defined by the energy functional as the effective Hamilton function of the Gibbs distribution. The difference of the fluctuation theory from the mean field theory of the microcanonical ensemble is to take the Landau functional as the fundamental Hamiltonian of the canonical ensemble instead that of an exact microscopic model. To calculate physical quantities, one has to average over all configurations of $\varphi(x)$. Although a complete mathematical proof of the equivalence between microscopic and fluctuation model is missing, the latter approach has a very important and useful property. In contrast to the microscopic model, it is possible to use the RG method in order to analyze its behavior and to obtain quantitative predictions for critical exponents.

According to rather general arguments, many dynamic phenomena in nature exhibit a clear separation of time scales. For instance, typical time and space scales in (classical) critical systems diverge and this allows describing relevant physical quantities in terms of continuous fields. In fact, the latter corresponds simply to slow modes stemming from conservation laws or broken symmetries. On the other hand, fast degrees of freedom enter theoretical description through random noise fields. Similar reasoning applies to other systems that do not exhibit criticality. Famous examples encompass turbulence, reaction–diffusion problems, driven systems. A general class that covers such dynamical systems is known in the literature as stochastic dynamics. From the theoretical point of view, it is important that large scale properties can be properly taken into account through a formalism of Langevin-like equations.

The Langevin approach can be briefly summarized as follows. The aim is to study a slowly varying field or a set of fields φ. Employing physical insight and symmetry reasoning, it is possible to postulate a stochastic differential equation of the form

$$\partial_t \varphi = V(\varphi) + f, \tag{1}$$

where the functional $V = V(\varphi)$ is local in the time variable, i.e., V depends only on the field φ and its spatial derivatives at a given time instant. In rare cases, V can be obtained from a microscopic model through a controlled coarse-grained procedure, but mostly its construction requires nontrivial knowledge about physical properties of a given stochastic system [42]. The random force f mimics the neglected rapid degrees of freedom and it is often modeled by means of Gaussian random variables. A nonzero mean value of f can be easily absorbed in the functional V. Thus, the complete statistical information about f is captured by specification of the first two moments

$$\langle f(x) \rangle = 0, \quad \langle f(x) f(x') \rangle = \delta(t-t') D(\boldsymbol{x}, \boldsymbol{x}'), \tag{2}$$

where for brevity we have introduced the following notation $x = (t, \boldsymbol{x})$, where $\boldsymbol{x}$ is a d-dimensional vector. Let us note that, when necessary, we write the space dimension d explicitly. This permits a straightforward check of complicated expressions and also plays an important role in perturbative RG techniques such as dimensional regularization.

The crucial difference between critical and genuine non-equilibrium systems lies in the correlation function $D(\boldsymbol{x}, \boldsymbol{x}')$ in Equation (2). In critical dynamics, we know to what the system should relax. It should end up in the thermal equilibrium described by the Gibbs probability distribution $\mathrm{e}^{-\mathcal{H}/k_B T}$, which greatly restricts the form of D. On the other hand, in non-equilibrium systems, such a relation is broken and steady states (obtained in the limit of large time) are of much more complicated dynamical nature.

There exist many theoretical approaches, which can be undertaken for an investigation of the stochastic problem in Equations (1) and (2). Remarkable equivalence of stochastic problems with certain quantum field theory models offers a plethora of possibilities to use. Due to work of H.-K. Janssen [43] and C. De Dominicis [44], a given stochastic model can be cast into a path integral formulation, which is amenable to many theoretical methods such as Feynman diagrammatic technique, the RG method, and others. To provide background for later use of functional methods and sake of notation, we recall now the De Dominicis-0Janssen statement in a succinct manner [5]. First, let us rewrite potential term $V(\varphi)$ as the sum $L\varphi + n(\varphi)$, where $L\varphi$ represents the linear part in the field φ and $n(\varphi)$ contains non-linearities. Then, stochastic problem in Equation (1) is tantamount to the quantum-field-theory model with double set of fields $\phi \equiv \{\varphi, \varphi'\}$ and action functional of form [32]

$$\mathcal{S}[\phi] = \frac{1}{2}\varphi' D \varphi' + \varphi'[-\partial_t \varphi + L\varphi + n(\varphi)]. \tag{3}$$

The auxiliary prime fields φ' were put forward in [45] and are known as Martin–Siggia–Rose response fields. Hereinafter, we have employed a condensed notation, in which integrals over space-time and summations over repeated internal indices are implied. For instance, the second term in the action functional in Equation (3) is a shorthand for the expression

$$\varphi' \partial_t \varphi = \sum \int \mathrm{d}t \int \mathrm{d}^d x\, \varphi'(t, \boldsymbol{x}) \partial_t \varphi(t, \boldsymbol{x}) \equiv \int \mathrm{d}x\, \varphi_i'(x) \partial_t \varphi_i(x), \tag{4}$$

where the index i numbers different field components. In this work, we are mainly interested in stochastic models concerning the velocity field $\boldsymbol{v}$, which is a vector quantity and thus the appropriate summation over the vector (internal) index must be taken into account as well. In particular, an analogous expression to Equation (4) for the velocity field would be written as

$$\boldsymbol{v}' \partial_t \boldsymbol{v} = \int \mathrm{d}t \int \mathrm{d}^d x\, v_i'(t, \boldsymbol{x}) \partial_t v_i(t, \boldsymbol{x}) \equiv \int \mathrm{d}x\, v_i'(x) \partial_t v_i(x). \tag{5}$$

The main goal of any statistical theory is to predict behavior of various correlation and response functions. Borrowing terminology from quantum field theory we refer to them as Green functions.

These are defined as functional averages over the fields ϕ with the weight $\exp \mathcal{S}[\phi]$, where $\mathcal{S}$ is action functional in Equation (3). Statistical averaging with the weight $e^{\mathcal{S}}$ are denoted as follows

$$\langle \ldots \rangle = \int \mathcal{D}\phi \ldots e^{\mathcal{S}[\phi]}. \tag{6}$$

All Green functions are effectively encoded into generating functional $\mathcal{G}$, which takes the form of the functional integral

$$\mathcal{G}[A] = \int \mathcal{D}\phi \exp \left[\mathcal{S}[\phi] + A\phi\right], \tag{7}$$

where $A \equiv (A_\varphi, A_{\varphi'})$ is the formal source and

$$A\phi \equiv \int dx \left[A_\varphi(x)\varphi(x) + A_{\varphi'}(x)\varphi'(x)\right]. \tag{8}$$

Further, $\mathcal{D}\phi$ in Equation (7) denotes the functional measure, i.e. $\mathcal{D}\phi \equiv \mathcal{D}\varphi\mathcal{D}\varphi'$, and the expression $\int \mathcal{D}\phi \ldots$ corresponds to a functional integral over the infinite dimensional space of all possible field configurations. Taking sufficiently many derivatives of $\mathcal{G}$ with respect to the formal sources at $A = 0$ yields any permissible Green function of the theory. For example, the response function $\langle \varphi\varphi' \rangle$ can be represented by the following functional integral

$$\langle \varphi(x)\varphi'(x') \rangle = \left. \frac{\delta^2 \mathcal{G}[A]}{\delta A_\varphi(x) \delta A_{\varphi'}(x')} \right|_{A=0} = \int \mathcal{D}\phi\, \varphi(x)\varphi'(x') e^{\mathcal{S}[\phi]}. \tag{9}$$

The normalization factor, which ensures the equality $\mathcal{G}(0) = 1$, has been included into the functional measure $\mathcal{D}\phi$. The generating functional $\mathcal{G}$ for field-theoretic models might be interpreted as an analog of the partition function in equilibrium statistical physics [6]. The formal Taylor expansion of $\mathcal{G}$ reads

$$\mathcal{G}[A] = \sum_{n=0}^{\infty} \frac{1}{n!} \int dx_1 \cdots \int dx_n G_n(x_1, \ldots, x_n) A(x_1) \cdots A(x_n), \tag{10}$$

where A on the left hand side stands for either A_φ or $A_{\varphi'}$ source field, and coefficient functions correspond to full Green functions

$$G_n(x_1, \ldots, x_n) = \langle \varphi(x_1) \cdots \varphi(x_n) \rangle = \left. \frac{\delta^n \mathcal{G}[A]}{\delta A(x_1) \cdots \delta A(x_n)} \right|_{A=0}. \tag{11}$$

The formulation in Equation (3) of the stochastic problem is advantageous for the use of the powerful machinery of field-theoretic methods such as Feynman diagrammatic technique, RG method, and operator product expansion. The starting point of perturbative techniques is a separation of action $\mathcal{S}$ into a free part $\mathcal{S}_0$ and part containing nonlinearities $\mathcal{S}_{\text{int}} = \varphi' n(\varphi)$. This division is not unique, but the necessary condition is the ability to solve the free part exactly. The free part $\mathcal{S}_0$ from Equation (3) can be symmetrized in the following way

$$\mathcal{S}_0[\phi] \equiv -\frac{1}{2}\phi K \phi \equiv -\frac{1}{2} \begin{pmatrix} \varphi \\ \varphi' \end{pmatrix} \begin{pmatrix} 0 & (\partial_t - L)^T \\ \partial_t - L & -D \end{pmatrix} \begin{pmatrix} \varphi \\ \varphi' \end{pmatrix} \tag{12}$$

with the symmetric matrix K. Here, T denotes transposing, i.e., $K^T(x, x') = K(x', x)$. The inverse matrix $\Delta = K^{-1}$ defines the set of bare propagators $\Delta_{ik}(x, x') = \langle \phi_i(x)\phi_k(x') \rangle_0$, which we number as follows

$$\Delta_{12} = \Delta_{21}^T = (\partial_t - L)^{-1}, \quad \Delta_{11} = \Delta_{12} D \Delta_{21}, \quad \Delta_{22} = 0, \tag{13}$$

where $\phi_1 \equiv \varphi$, $\phi_2 \equiv \varphi'$. Generalization to a multicomponent field φ is obvious. The propagator Δ_{12} is retarded, therefore $\Delta_{21} = \Delta_{12}^T$ is advanced. The symmetric propagator $\Delta_{11} = \Delta_{11}^T$ contains both (retarded and advanced) contributions. The interaction part generates vertices with one field φ' and two or more fields φ, which are determined by the concrete form of the nonlinear terms in the action of the model. The aforementioned functional representation in Equation (3) permits construction of standard Feynman graphs for Green functions [5,6,46] by means of Wick's theorem. The lines (propagators) are derived from the quadratic (free) part $\mathcal{S}_0$, whereas the interaction part $S_{\rm int}$ gives rise to vertices. Wick's theorem (see, e.g., [5,47] for details) for the functional in Equation (7) may be compactly written in the exponential form

$$\mathcal{G}[A] = \exp\left(\frac{1}{2}\frac{\delta}{\delta\phi}\Delta\frac{\delta}{\delta\phi}\right)\exp\left(\mathcal{S}_{\rm int}[\phi] + A\phi\right)\bigg|_{\phi=0}, \tag{14}$$

where Δ is the matrix of propagators in Equation (13) and

$$\frac{\delta}{\delta\phi}\Delta\frac{\delta}{\delta\phi} \equiv \int \mathrm{d}x \int \mathrm{d}x' \frac{\delta}{\delta\phi_i(x)}\Delta_{ik}(x,x')\frac{\delta}{\delta\phi_k(x')} \tag{15}$$

is a shorthand notation for the universal differential operation and the indices i,k enumerate all fields (response field included) in the model. Expansion of both exponents in Equation (14) leads to the celebrated Feynman diagrammatic technique, which allows perturbative calculation of all Green functions of the theory.

3. Stochastic Approach to Turbulence

The stochastic approach to the Navier–Stokes (NS) equation is analogous to fluctuation theory for critical phenomena mentioned in Section 1. It can be regarded as a microscopic approach to fully developed turbulence. The crucial difference from critical phenomena is that for turbulence there is no counterpart of Hamiltonian (free-energy) functional. The stochastic NS equation neglects such physical effects as influence of the boundaries or the precise form of system's geometry (e.g., information about the way turbulence is produced), which are in the experiments responsible for creating turbulent instabilities. In a phenomenological sense, the input of energy is modeled by the proper choice of the stochastic force. The main goal of this theory is to justify Kolmogorov hypotheses [23,25]. A general proof of the equivalence between Kolmogorov hypotheses and the stochastic NS equations is still missing; nevertheless, as various studies show, it provides a nontrivial input to scaling behavior observed in turbulent flows [31,33].

The stochastic Navier–Stokes equation, which governs the dynamics of the velocity fluctuations $v_i = v_i(x), i = 1,\ldots,d$, assumes the following form

$$\partial_t \boldsymbol{v} + (\boldsymbol{v}\cdot\boldsymbol{\nabla})\boldsymbol{v} - \nu_0\boldsymbol{\nabla}^2\boldsymbol{v} + \boldsymbol{\nabla}p = \boldsymbol{f}, \tag{16}$$

where ν_0 is the molecular kinematic viscosity, $p = p(x)$ stands for pressure fluctuations, $\boldsymbol{\nabla}$ is gradient, $\boldsymbol{\nabla}^2 = \Delta = \partial^2/\partial x_i\partial x_i = \partial_i\partial_i$ is Laplace operator, and $\boldsymbol{f} = \boldsymbol{f}(x)$ represents an external random force per unit mass. For simplicity, we consider incompressible fluid with the solenoidal velocity $\boldsymbol{\nabla}\cdot\boldsymbol{v} = 0$ and unit density of fluid ($\rho = 1$). The incompressibility condition permits elimination of the pressure field from the stochastic Navier–Stokes equation (Equation (16)) and we can consider only its transverse components

$$\partial_t \boldsymbol{v} + P(\boldsymbol{v}\cdot\boldsymbol{\nabla})\boldsymbol{v} - \nu_0\boldsymbol{\nabla}^2\boldsymbol{v} = \boldsymbol{f}, \tag{17}$$

where all fields are transverse, and P denotes the transverse projection operator, which in the momentum representation takes the form

$$P_{ij}(\boldsymbol{k}) = \delta_{ij} - k_i k_j/k^2 \tag{18}$$

with $k = |\boldsymbol{k}|$ being the magnitude of the wavevector $\boldsymbol{k}$. In view of universality, it is assumed that the large-scale random force $\boldsymbol{f}$ obeys the Gaussian distribution law. Hence, only the mean value and the second moment have to be postulated. The former takes zero value ($\langle f_i \rangle = 0$) and pair correlation function is chosen in a general form

$$\langle f_i(t_1, \boldsymbol{x}_1) f_j(t_2, \boldsymbol{x}_2) \rangle \equiv D_{ij}(x_1, x_2). \tag{19}$$

It is convenient to specify the kernel function D_{ij} in frequency–momentum representation

$$D_{ij}(x_1, x_2) \equiv \delta(t_1 - t_2) d_{ij}(\boldsymbol{x}_1, \boldsymbol{x}_2) = \delta(t_1 - t_2) \int \frac{\mathrm{d}^d k}{(2\pi)^d} P_{ij}(\boldsymbol{k}) d_f(k) \mathrm{e}^{i\boldsymbol{k}\cdot(\boldsymbol{x}_1 - \boldsymbol{x}_2)}, \tag{20}$$

where d is a dimension of space. To employ the RG technique [28,31], the energy injection $d_f(k)$ is usually chosen in the power-law form

$$d_f(k) = D_0 k^{4-d-2\varepsilon} F(kL) \tag{21}$$

where L denotes outer integral scale, D_0 is the amplitude, and the scaling function $F(kL)$ possesses the unit asymptotic behavior in the range of large wave numbers $kL \gg 1$. For our purposes, it is sufficient to consider the "massless" theory for which Equation (21) becomes simply

$$d_f(k) = g_0 \nu_0^3 k^{4-d-2\varepsilon}, \tag{22}$$

with the additional feature that the corresponding integral in Equation (20) is IR regularized at $m \sim L^{-1}$. The parameter D_0 in Equation (22) is rewritten as $g_0\nu_0^3$ for dimensional and calculational reasons. The parameter g_0 plays the role of the coupling constant, $\varepsilon \geq 0$ is a free parameter of the theory. For completeness, let us note that Equation (19) takes in frequency–momentum representation the following form

$$\langle f_i(\omega, \boldsymbol{k}) f_j(\omega', \boldsymbol{k}') \rangle = (2\pi)^{d+1} \delta(\omega + \omega') \delta(\boldsymbol{k} + \boldsymbol{k}') P_{ij}(\boldsymbol{k}) d_f(k). \tag{23}$$

From the mathematical point of view, Equation (16) represents a stochastic partial differential equation, first order in time variable. This allows us to employ the machinery of Section 2. Let us explain how these formal rules are applied to the theory of developed turbulence. According to the aforementioned De Dominicis–Janssen approach, the stochastic model described by Equation (17) is tantamount to the field-theoretic model with the action

$$\mathcal{S}_{\mathrm{NS}}[\boldsymbol{v}, \boldsymbol{v}'] = \frac{1}{2} v'_i D_{ij} v'_j + \boldsymbol{v}' \cdot \left[-\partial_t \boldsymbol{v} + \nu_0 \nabla^2 \boldsymbol{v} - (\boldsymbol{v} \cdot \boldsymbol{\nabla}) \boldsymbol{v} \right], \tag{24}$$

where D_{ij} is introduced into Equation (19), the auxiliary response vector field $\boldsymbol{v}'$ is solenoidal ($\boldsymbol{\nabla} \cdot \boldsymbol{v}' = 0$) as well as the velocity field $\boldsymbol{v}$, and ν_0 is the bare (molecular) viscosity coefficient. To distinguish it from the renormalized (turbulent) viscosity ν, which is generated in the process of the renormalization, we denote it and other similar (bare) parameters by the subscript "0" . We stress that this notation is used in the whole work.

Feynman rules for the perturbation theory are constructed by means of the general operation in Equation (14)). The explicit form of the propagators is determined by the quadratic part of the action in Equation (24) and in the frequency–momentum representation they are

$$\Delta_{ij}^{vv}(\omega_k, \boldsymbol{k}) = \frac{P_{ij}(\boldsymbol{k}) d_f(k)}{\omega_k^2 + \nu_0^2 k^4}, \quad \Delta_{ij}^{vv'}(\omega_k, \boldsymbol{k}) = (\Delta_{ij}^{v'v}(\omega_k, \boldsymbol{k}))^* = \frac{P_{ij}(\boldsymbol{k})}{-i\omega_k + \nu_0 k^2}, \quad \Delta_{ij}^{v'v'}(\omega_k, \boldsymbol{k}) = 0, \tag{25}$$

where $*$ denotes complex conjugation and the transverse projector appears due to incompressibility condition. In the time–momentum representation, the corresponding expressions are

$$\Delta_{ij}^{vv}(\boldsymbol{k}, t' - t) = \frac{P_{ij}(\boldsymbol{k}) d_f(k)}{2\nu_0 k^2} \mathrm{e}^{-\nu_0 k^2 |t' - t|}, \tag{26}$$

$$\Delta_{ij}^{vv'}(\boldsymbol{k}, t' - t) = \theta\,(t' - t)\, P_{ij}(\boldsymbol{k}) \mathrm{e}^{-\nu_0 k^2 (t' - t)}, \tag{27}$$

$$\Delta_{ij}^{v'v}(\boldsymbol{k}, t' - t) = \theta\,(t - t')\, P_{ij}(\boldsymbol{k}) \mathrm{e}^{-\nu_0 k^2 (t' - t)}, \tag{28}$$

$$\Delta_{ij}^{v'v}(\boldsymbol{k}, t' - t) = 0. \tag{29}$$

Here, the step function $\theta(t)$ displays an important physical feature of the propagator $\Delta^{vv'}$—its retardation. In fact, $\Delta^{vv'}$ is the leading order contribution to the response function $\langle vv' \rangle$ of the original model in Equations (16)–(21). The propagator Δ^{vv} represents the leading contribution to the pair correlation function of the velocity field $W_{ij} = \langle v_i v_j \rangle$. With coinciding time arguments, the latter is proportional to the kinetic energy spectrum $E(k)$ in the wavevector representation. This function enters the equation of energy balance describing the transfer of the kinetic energy from the largest spatial scales to the smallest ones, where it dissipates to heat [23]. The vertex factor

$$V_m(x_1, x_2, \ldots, x_m; \phi) = \frac{\delta^m V[\phi]}{\delta\phi(x_1)\delta\phi(x_2)\ldots\delta\phi(x_m)} \tag{30}$$

is associated to each interaction vertex of a Feynman graph. In Equation (30), the dummy field ϕ is one from the set of all fields $\{\boldsymbol{v}', \boldsymbol{v}\}$. The interaction vertex in Equation (24) is cast in a more convenient form

$$-\int \mathrm{d}t \int \mathrm{d}^d x\, \boldsymbol{v}'(\boldsymbol{v} \cdot \boldsymbol{\nabla})\boldsymbol{v} = -\int \mathrm{d}t \int \mathrm{d}^d x\, v'_i v_k \partial_k v_i = (\partial_k v'_i) v_k v_i, \tag{31}$$

where the incompressibility condition $\partial_i v_i = 0$ and integration by parts have been used. The latter step requires the standard assumption of rapid enough vanishing of velocity in the limit $|\boldsymbol{x}| \to \infty$. Furthermore, the last expression in Equation (31) corresponds to the shorthand of Equation (4). Rewriting functional in Equation (31) in the symmetric form $v_i V_{ijl} v_j v_l / 2$, we derive the explicit form for the corresponding vertex factor in the Fourier representation

$$V_{ijl} = i(k_j \delta_{il} + k_l \delta_{ij}). \tag{32}$$

Here, the wavevector $\boldsymbol{k}$ is flowing in the vertex through the field v' and is denoted by slash in Figure 2. The propagators (lines) Δ and vertices V are graphically depicted in Figures 1 and 2.

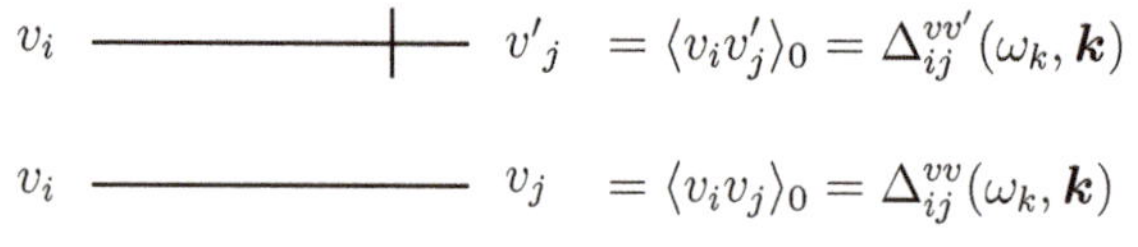

$$v_i \;\text{———┼—}\; v'_j \;=\; \langle v_i v'_j \rangle_0 = \Delta_{ij}^{vv'}(\omega_k, \boldsymbol{k})$$

$$v_i \;\text{————}\; v_j \;=\; \langle v_i v_j \rangle_0 = \Delta_{ij}^{vv}(\omega_k, \boldsymbol{k})$$

Figure 1. Nontrivial propagators for the model in Equation (24).

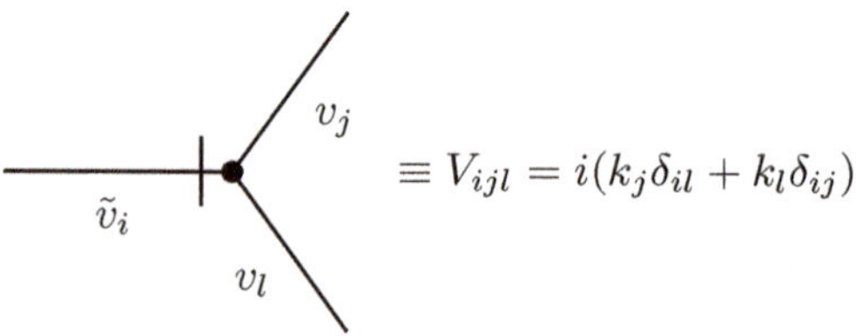

Figure 2. Interaction vertex responsible for the nonlinear interactions between velocity fluctuations in the model in Equation (24). Momentum k on the right hand side corresponds to the inflowing momentum of the auxiliary field v'.

The theoretical description of the fluid turbulence on the basis of "first principles", i.e., starting from the stochastic Navier–Stokes (NS) equation [25] remains an open problem. However, considerable progress has been made in understanding simplified model systems sharing certain essential properties with the real problem: stochastic Burgers equation [48], shell models [49] and advection by random "synthetic" velocity fields [33].

A paradigmatic model of a scalar quantity advected passively by a Gaussian random velocity field, uncorrelated in time and self-similar in space, the so-called Kraichnan's rapid-change model [39], is a famous example. The standard notation for advection problem using the Kraichnan model slightly differs from the one using stochastic Navier–Stokes ensemble. Therefore, in what follows, we give a brief overview of basic physical ideas behind the Kraichnan model and introduce the corresponding notation.

The governing equation for diffusion–advection for field θ is

$$\partial_t \theta + (\boldsymbol{v} \cdot \boldsymbol{\nabla})\theta - D_0 \boldsymbol{\nabla}^2 \theta = f^{\theta}, \tag{33}$$

where D_0 is the coefficient of molecular diffusivity and $f^{\theta} \equiv f^{\theta}(x)$ is a zero-mean Gaussian random noise with the correlation function

$$\langle f^{\theta}(x) f^{\theta}(x') \rangle = \delta(t-t') C(\boldsymbol{r}/L), \quad \boldsymbol{r} = \boldsymbol{x} - \boldsymbol{x}'. \tag{34}$$

The noise f^{θ} in Equation (33) maintains the steady-state of the system. The particular form of the correlator is not relevant, however. The sole condition which must be satisfied by the function $C(\boldsymbol{r}/L)$ is that it must fall off rapidly at $r \equiv |\boldsymbol{r}| \gg L$. Here, L is an integral scale related to the stirring.

In accordance with the generalized Kraichnan model [34,50] with finite correlation time taken into account, we assume velocity field generated by a simple linear stochastic equation

$$\partial_t v_i + R v_i = f_i^{v}, \tag{35}$$

where $R \equiv R(x)$ is a linear operation to be specified below and $f_i^{v} \equiv f_i^{v}(x)$ is a zero-mean random stirring force with the correlator

$$\langle f_i^{v}(x) f_j^{v}(x') \rangle \equiv D_{ij}^{f}(x;x') = \frac{1}{(2\pi)^{d+1}} \int \mathrm{d}\omega \int \mathrm{d}^d k \, P_{ij}(\boldsymbol{k}) D^{f}(\omega_k, \boldsymbol{k}) \mathrm{e}^{-i(t-t')+i\boldsymbol{k}\cdot(\boldsymbol{x}-\boldsymbol{x}')}. \tag{36}$$

It should be noted that, in the SDE in Equation (33), the multiplicative noise due to random velocity is not a white noise in time as in the original Kraichnan model. Therefore, there is no need to specify the interpretation of the SDE. However, in the analysis, the white-noise limit is considered and it should recalled that in this limit the results correspond to the Stratonovich interpretation of the SDE in Equation (33).

The correlator D^f is chosen [34–36] in the following form

$$D^{f}(\omega_k, \boldsymbol{k}) = g_0 \nu_0^3 (k^2 + m^2)^{2-d/2-\varepsilon/2-\eta/2}, \tag{37}$$

with the wavenumber representation of the function $R(x)$:

$$R(\boldsymbol{k}) = u_0 \nu_0 (k^2 + m^2)^{1-\eta/2}. \tag{38}$$

The positive amplitude factors g_0 and u_0 are the coupling constants of the model. Furthermore, g_0 can be regarded as a formally small parameter of the perturbation theory. The positive exponents ε and η ($\varepsilon = \mathcal{O}(\eta)$) are RG expansion parameters. They are analogous to expansion parameter $\varepsilon = 4 - d$ in the φ^4- theory. Now, the expansion is carried out in the (ε, η)-plane around the origin $\varepsilon = \eta = 0$.

Note the presence of two scales in the problem—integral scale L introduced in Equation (34) and momentum scale m, which has appeared in Equation (38). Clearly, they have different physical origins. However, these two scales can be related to each other and for technical purposes [35] it is reasonable to choose $L = 1/m$. When not explicitly stated, this relation is always assumed.

In the limit $k \gg m$ the functions in Equations (37) and (38) take on a simple powerlike form

$$D^f(\omega_k, \mathbf{k}) = g_0 \nu_0^3 k^{4-d-\varepsilon-\eta}, \quad R(k) = u_0 \nu_0 k^{2-\eta}, \tag{39}$$

which is convenient for actual calculations. The needed IR regularization will be given by restrictions on the region of integrations.

From Equations (35), (36), and (39), the statistics of the velocity field $\boldsymbol{v}$ can be determined. It obeys Gaussian distribution with zero mean and correlator

$$\langle v_i(t, \boldsymbol{x}) v_j(0, \mathbf{0}) \rangle = \int \frac{d\omega_k}{2\pi} \int \frac{d^d k}{(2\pi)^d} D_v(\omega_k, \boldsymbol{k}) e^{-i\omega_k t + i\boldsymbol{k}\cdot\boldsymbol{x}}, \tag{40}$$

where the kernel function $D_v(\omega_k, \boldsymbol{k})$ is assumed in the form

$$D_v(\omega_k, \boldsymbol{k}) = P_{ij}(\boldsymbol{k}) \frac{g_{10} u_{10} D_0^3 k^{4-d-\varepsilon-\eta}}{\omega_k^2 + u_{10}^2 D_0^2 (k^{2-\eta})^2}. \tag{41}$$

The correlator in Equation (41) is directly connected to the energy spectrum via the frequency integral [34,51–55]

$$E(k) \simeq k^{d-1} \int d\omega D^v(\omega_k, k) \simeq \frac{g_0 \nu_0^2}{u_0} k^{1-\varepsilon}. \tag{42}$$

Hence, the coupling constant g_0 and the exponent ε characterize the equal-time velocity correlator or, similarly, energy spectrum. Further, the parameter u_0 and the exponent η are related to the frequency $\omega_k \simeq u_0 \nu_0 k^{2-\eta}$ (or to the function $R(k)$, the reciprocal of the correlation time at the wave number k) which describes the mode with wave number k [34,51–57]. Let us note that in the chosen notation the value $\varepsilon = 8/3$ corresponds to the well-known Kolmogorov "five-thirds law" for the spatial scaling behavior of the velocity field, and the value $\eta = 4/3$ corresponds to the Kolmogorov frequency. A straightforward dimensional analysis reveals that the parameters (charges) g_0 and u_0 are connected to the ultraviolet (UV) momentum scale Λ (of the same order of magnitude as the inverse Kolmogorov length) by the relations

$$g_0 \simeq \Lambda^{\varepsilon+\eta}, \quad u_0 \simeq \Lambda^{\eta}. \tag{43}$$

In Ref. [50], it was demonstrated that the linear model in Equation (35) (and consequently the Gaussian model in Equation (40) as well) is not invariant under Galilean transformation and, therefore, it effectively neglects important effect of the self-advection of turbulent eddies. As a result of these so-called "sweeping effects" the different time correlations of the velocity are not self-similar and exhibit strong dependence on the integral scale [58,58–60]. However, the results presented in Ref. [50] lead to the conclusion that the Gaussian model describe the passive advection reasonably well in the appropriate frame of reference, in which the mean velocity field vanishes. An additional argument to support the model in Equation (40) is that we are mainly interested in the equal-time, Galilean invariant quantities (e.g., structure functions), which are not affected by the sweeping. Therefore, their absence in the Gaussian model in Equation (40) is not relevant [34,36,40].

The kernel function in Equation (41) is written in a very general form and allows studying various special limits, in which the numerical analysis of the resulting equations is simplified and which provide a deeper physical insight. Possible limiting cases are

- The rapid-change model corresponding to the limit $u_{10} \to \infty, g'_{10} \equiv g_{10}/u_{10} = const$. Then, the kernel function becomes

$$D_v(\omega_k, \boldsymbol{k}) \propto g'_{10} D_0 k^{-d-\varepsilon+\eta}. \tag{44}$$

The velocity correlator is obviously $\delta-$correlated in the time variable.

- The frozen velocity field arising in the limit $u_{10} \to 0$, in which the kernel function corresponds to

$$D_v(\omega_k, \boldsymbol{k}) \propto g_0 D_0^2 \pi \delta(\omega_k) k^{2-d-\varepsilon}. \tag{45}$$

- The purely potential velocity field obtained in the limit $\alpha \to \infty$ with αg_{10} =constant. This case is similar to the model of random walks in a random environment with long-range correlations [61,62].
- The turbulent advection, for which $\varepsilon = 2\eta = 8/3$. This choice mimics properties of the fully developed turbulence and yields well-known Kolmogorov scaling [23].

Using Equation (3), the stochastic problem in Equations (33)–(36) can be recast into the equivalent field theoretic model of the doubled set of fields $\phi \equiv \{\theta, \theta', \boldsymbol{v}, \boldsymbol{v}'\}$ with the action functional

$$\mathcal{S}[\phi] = \frac{1}{2}\boldsymbol{v}' D^f \boldsymbol{v}' + \theta' \left[-\partial_t \theta - (\boldsymbol{v}\cdot\boldsymbol{\nabla})\theta + \nu_0 \boldsymbol{\nabla}^2 \theta\right] + \boldsymbol{v}' \cdot \left[-\partial_t \boldsymbol{v} - R\boldsymbol{v}\right], \tag{46}$$

where D_{ij}^f is defined in Equation (36), and as usual θ' and $\boldsymbol{v}'$ are auxiliary response fields.

Generating functional of full Green functions $\mathcal{G}[A]$ is defined by Equation (7), where now a linear form $A\varphi$ is defined as

$$A\phi = A_\theta \theta + A_{\theta'}\theta' + Av_i v_i + Av_i' v_i'. \tag{47}$$

Following the argument in [34], we set $A_i^{v'} = 0$ in Equation (47) and carry out the explicit Gaussian integration over the auxiliary vector field, $\boldsymbol{v}'$ because we are not interested in the Green functions containing the auxiliary field $\boldsymbol{v}'$. After the integration, we are left with the field-theoretic model with the action functional

$$\mathcal{S}[\phi] = -\frac{1}{2}\boldsymbol{v}(D^v)^{-1}\boldsymbol{v} + \theta'[-\partial_t\theta - (\boldsymbol{v}\cdot\boldsymbol{\nabla})\theta + \nu_0\boldsymbol{\nabla}^2\theta], \tag{48}$$

where the second term represents the De Dominicis–Janssen action for the stochastic problem in Equation (33) at fixed velocity field $\boldsymbol{v}$. The first term describes the Gaussian averaging over $\boldsymbol{v}$ specified by the correlator D^v. The latter explicitly reads

$$\mathcal{S}_{\rm vel}[\boldsymbol{v}] = \frac{1}{2}\int \mathrm{d}x_1 \int \mathrm{d}x_2\, \boldsymbol{v}_i(x_1) D_{ij}^{-1}(x_1 - x_2)\boldsymbol{v}_j(x_2). \tag{49}$$

The action in Equation (48) is written in a form that is suitable for a straightforward application of the field-theoretic perturbative analysis with the use of the standard Feynman diagrammatic technique. From the quadratic part of the action, we derive the matrix of bare propagators. The wavenumber frequency representations of relevant propagators are: (a) the bare propagator $\langle\theta\theta'\rangle_0$ defined as follows

$$\langle\theta\theta'\rangle_0 = \langle\theta'\theta\rangle_0^* = \frac{1}{-i\omega + \nu_0 k^2}; \tag{50}$$

and (b) the bare propagator for the velocity field $\langle vv\rangle_0$ that reads

$$\langle v_i v_j\rangle_0 = P_{ij}(\boldsymbol{k}) D_v(\omega, \boldsymbol{k}). \tag{51}$$

The triple (interaction) vertex $-\theta' v_j \partial_j$ can be rewritten in $\theta = \theta' v_j V_j \theta$, where momentum $\boldsymbol{k}$ is flowing into the vertex via the response field θ'. A graphical representation of the perturbation elements for a Kraichnan-like model is schematically depicted in Figure 3.

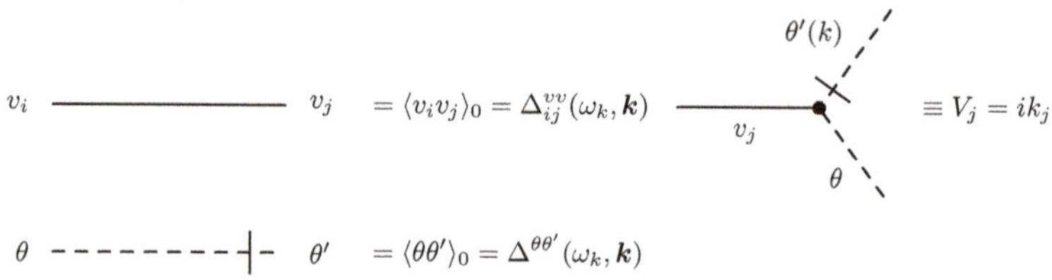

$$v_i \;\text{———}\; v_j \quad = \langle v_i v_j\rangle_0 = \Delta^{vv}_{ij}(\omega_k, \boldsymbol{k}) \qquad \text{vertex: } v_j,\ \theta'(k),\ \theta \quad \equiv V_j = ik_j$$

$$\theta \;\text{- - - - - - - -+-}\; \theta' \quad = \langle \theta\theta'\rangle_0 = \Delta^{\theta\theta'}(\omega_k, \boldsymbol{k})$$

Figure 3. Feynman rules for the model in Equation (48).

Taking as a example the Kraichnan model, let us briefly describe what kind of quantities might be studied by functional techniques. From experimental and theoretical point of view, the main focus is in the behavior of the equal-time structure functions

$$S_N(r) \equiv \langle [\theta(t, \boldsymbol{x}+\boldsymbol{r}) - \theta(t, \boldsymbol{x})]^N \rangle \tag{52}$$

in the inertial range, specified by the inequalities $l \sim 1/\Lambda \ll r \equiv |\boldsymbol{r}| \ll L = 1/m$ (l is an internal length). Brackets $\langle \cdots \rangle$ denote the functional average over fields $\phi = \{\theta, \theta', \boldsymbol{v}\}$ with the weight functional $\exp \mathcal{S}[\Phi]$ from Equation (48). In the isotropic case, the odd functions S_{2N+1} vanish identically, while for even functions S_{2N} a simple dimensional argument dictates the following form

$$S_{2N}(r) = \nu_0^{-N} r^{2N} R_{2N}(r/l, r/L, g_0, u_0), \tag{53}$$

where R_{2N} are scaling functions of purely dimensionless variables. In principle, functions R_{2N} can be calculated by means of the usual perturbation theory (i.e., as series in g_0). However, this is not a reasonable way to study the inertial-range behavior: the reason is that the coefficients are singular in the limits $r/l \to \infty$ and/or $r/L \to 0$ and compensate for the smallness of g_0. To obtain correct IR behavior the entire series have to be summed. Such a summation procedure can be effectively done by the use of the field theoretic RG and OPE (see Sections 4 and 5).

The RG analysis can be divided into two stages. During the first stage, the multiplicative renormalizability of the model is proved and the differential RG equations for its correlation (structure) functions are derived. The asymptotic behavior of functions similar to the one in Equation (52) for $r/l \gg 1$ and any fixed r/L is governed by IR stable fixed points (see next section) of the RG equations and assumes the form

$$S_{2N}(r) = \nu_0^{-N} r^{2n} (r/l)^{-\gamma_N} R_{2N}(r/L), \quad r/l \gg 1 \tag{54}$$

with so far unknown scaling functions $R_{2N}(r/L)$. Whenever γ_N is a nonlinear function of N, we refer to such case as anomalous scaling or multiscaling.

Let us remind that the quantity $\Delta[S_{2N}] \equiv -2N + \gamma_N$ is called the critical dimension. The exponent γ_N, the difference between the critical dimension $\Delta[S_{2N}]$ and the canonical dimension $-2N$, is known as the anomalous dimension. In the present case, the latter takes a simple form: $\gamma_N = n\epsilon$. For any function $R_N(r/L)$, the representation in Equation (54) implies scaling behavior in the IR region ($r/l \gg 1$, r/L fixed) with definite critical dimensions of all IR relevant parameters, $\Delta[S_{2N}] = -2N + N\epsilon$, $\Delta_r = -1$, $\Delta_{L^{-1}} = 1$ and fixed irrelevant parameters ν_0 and l.

In the second stage, the small r/L behavior of the functions $R_{2N}(r/L)$ is analyzed in the general representation in Equation (54) employing the OPE technique (Section 5). It predicts that, in the limit $r/L \to 0$, the functions $R_{2N}(r/L)$ have the asymptotic forms

$$R_{2N}(r/L) = \sum_F C_F(r/L)\,(r/L)^{\Delta_N}, \tag{55}$$

where C_F are coefficients regular in the variable r/L. In general, the summation is performed over specific renormalized composite operators F with critical dimensions Δ_n. Kraichnan model exhibits nontrivial scaling behavior as some of anomalous exponents Δ_N are negative and singular behavior on L is present. Such situation never occurs in critical phenomena [5,6] where corresponding exponents are positive and lead only to subleading corrections.

More elaborated discussion on anomalous scaling can be found in Section 10, which is devoted to generalization of Kraichnan model. Namely, assumption of isotropy is abandoned and effect of anisotropy is taken into account.

4. Renormalization Group Analysis

Let us briefly summarize main ideas of the quantum-field theory of renormalization and RG technique; a detailed account can be found in monographs [5,6,15,46].

Feynman graphs of Green functions are a convenient graphical representation of perturbation theory. Quantum field theory models are well-known for appearance of UV divergences in loop diagrams. This results from an integration at large momenta. Therefore, it is necessary to find an effective procedure to eliminate these divergences step by step in a controlled manner. Finite diagrams (free of UV divergences) are brought by an iterative renormalization procedure. The inherent ambiguity of this removal of divergences may be used to establish connection between values of Green functions at different scales without having explicit solutions for them. This is the idea of the method of renormalization group (RG) proposed in the framework of high energy physics long time ago [1,7–11].

In statistical physics, RG allows one to extract relevant information about large-scale behavior from the mutual correlation between IR behavior of Green functions and UV divergences at critical dimension [5,15]. Thus, in statistical theories, the RG method can be understood as an effective way to determine non-trivial asymptotic behavior of Green functions in the range of small (infrared) wavevectors (scales).

There is simple criterion how to determine the true asymptotic range in the framework of the RG. One of the RG-functions is the β-function, which is the coefficient of the operation ∂_g in the RG equation. The β-function is calculated perturbatively as infinite series of powers of the coupling constant g. Non-trivial asymptotic behavior is governed by *RG fixed points* g^*, which are roots of the β function (solutions of equation $\beta(g) = 0$). A fixed point can be IR or UV stable depending on the behavior of the β-function in the vicinity of g^*. Of course, physical theories contain usually many charges and these considerations have to be properly generalized [5,46].

The field theoretic RG is based on non-trivial techniques of UV renormalization. The basic procedure lies in a perturbative calculation of the RG-functions in the framework of a prescribed scheme of regularization [5,6]. To find and analyze UV divergences in a specific field-theoretic model counting of canonical scaling dimensions of fields and parameters of the model is used. The essence of such a power counting is closely connected with the existence of a scale invariance in the model.

For models considered in this paper, it is advantageous to calculate Feynman diagrams in a formal scheme [5] without UV-cut-off Λ. Then, UV-divergences manifest themselves as poles in a dimensionless parameter ε that measures deviation from a logarithmic theory, i.e., a theory in which all coupling constants become dimensionless. The procedure of multiplicative renormalization removing UV-divergences (in the present case, poles in a parameter ε) is the following: the original action $\mathcal{S}[\phi, e_0]$ is declared to be unrenormalized; its parameters e_0 (the letter e_0 stands for the whole set of parameters; for instance, coupling constants, deviation from criticality, viscosity, etc.) are the bare parameters, and they are assumed to be functions of the new renormalized parameters e. The new renormalized action is the functional $\mathcal{S}_R[\phi] = \mathcal{S}[\phi Z_\phi, e Z_e]$ with certain (to be determined perturbatively such that the Green functions generated by the renormalized action are UV finite, i.e., regular in ε) renormalization constants of fields Z_ϕ (one per each independent component of the field) and parameters Z_e. In unrenormalized full Green functions $G_N = \langle \phi \dots \phi \rangle$, the functional averaging $\langle \dots \rangle$ is performed with the weight functional $\exp \mathcal{S}[\phi]$, whereas, in renormalized functions G_N^R, with the renormalized weight functional $\exp \mathcal{S}_R[\phi]$. The relation between the functionals $\mathcal{S}[\phi]$ and $\mathcal{S}_R[\phi]$ leads to the relation between the corresponding Green functions $G_N^R = Z_\phi^{-N} G_N$, where by definition

$G_N = G_N(e_0, \varepsilon\ldots)$ (ellipsis denotes other arguments such as coordinates or wavenumbers), and, by convention, the quantities G_N^R and Z_ϕ are expressed in terms of the parameters e. The correspondence $e_0 \leftrightarrow e$ within perturbation theory is assumed to be one-to-one, therefore either of the sets e_0, e can be taken as the independent variables.

For translationally invariant theories, it is much more convenient to deal not with the full Green functions G_N, but with their connected parts W_N. Their generating functional being through the relation

$$\mathcal{W}[A] \equiv \ln \mathcal{G}[A]. \tag{56}$$

A further simplification is possible through 1-irreducible functions Γ_N (also called one particle irreducible functions or vertex functions). The generating functional for the latter is defined by the functional Legendre transform [47]

$$\Gamma[\alpha] \equiv \mathcal{W}[A] - A\alpha, \tag{57}$$

where

$$A\alpha = \int \mathrm{d}x\, (A_\varphi(x)\alpha_\varphi(x) + A_{\varphi'}(x)\alpha_{\varphi'}(x)), \quad \alpha_\varphi = \frac{\delta\mathcal{W}[A]}{\delta A_\varphi(x)}, \quad \alpha_{\varphi'} = \frac{\delta\mathcal{W}[A]}{\delta A_{\varphi'}(x)}. \tag{58}$$

To simplify notation in practical calculations, it is convenient to relabel α-variables back to the original fields ϕ. This allows us to rewrite the first relation in Equation (57) compactly as

$$\Gamma[\phi] = \mathcal{S}[\phi] + \tilde{\Gamma}[\phi], \tag{59}$$

where $\tilde{\Gamma}[\phi]$ is the sum of all one particle irreducible (1PI) loop diagrams [5].

Statements of RG theory are readily summarized at the level of corresponding Green functions. For connected and 1PI Green functions, they read

$$W_N^R(e, \varepsilon, \ldots) = Z_\phi^{-N}(e, \varepsilon) W_N(e_0(e, \varepsilon), \varepsilon, \ldots), \tag{60}$$

$$\Gamma_N^R(e, \varepsilon, \ldots) = Z_\phi^{N}(e, \varepsilon) \Gamma_N(e_0(e, \varepsilon), \varepsilon, \ldots), \tag{61}$$

where the functions $e_0(e, \varepsilon)$, $Z_\phi^N(e, \varepsilon)$ can be chosen arbitrarily, which implies an arbitrary choice of normalization of the fields and parameters e at given e_0. In the present text, we also interchangeably use the following notation for the connected Green functions

$$W_{\phi_1 \ldots \phi_N} \equiv \langle \phi_1 \ldots \phi_N \rangle_{\text{conn.}}, \tag{62}$$

and for the 1PI Green functions according to the aforementioned relabeling $\alpha \to \phi$

$$\Gamma_{\phi_1 \ldots \phi_N} \equiv \langle \phi_1 \ldots \phi_N \rangle_{\text{1-ir}}. \tag{63}$$

The crucial statement of the theory of renormalization is that for the multiplicatively renormalizable models these functions can be chosen to provide UV finiteness of Green functions as $\varepsilon \to 0$. With this choice, all UV divergences (poles in ε) contained in the functions $e_0(e, \varepsilon)$, $Z_\phi^N(e, \varepsilon)$ are absent in renormalized Green functions $W_N^R(e, \varepsilon)$. We note that the UV finiteness in this sense of any one set of Green functions (full, connected, and 1-irreducible) automatically leads to the UV-finiteness of any other. The RG equations are written for the renormalized functions W_N^R which differ from the original unrenormalized functions W_N only by normalization, and, therefore, can be used equally well to analyze the critical scaling. Let us recall an elementary derivation of the RG equations [5,46]. The requirement of elimination of divergences does not uniquely determine the functions $e_0(e, \varepsilon)$ and $Z_\phi(e, \varepsilon)$. An arbitrariness remains which allows introducing an additional dimensional parameter scale setting parameter (renormalization mass) μ in these functions (and via them also into W_N^R)

$$W_N^R(e, \mu, \varepsilon, \ldots) = Z_\phi^{-N}(e, \mu, \varepsilon) W_N(e_0(e, \mu, \varepsilon), \varepsilon, \ldots). \tag{64}$$

A change of μ at fixed e_0 leads to a change of e, Z_ϕ and W_R for unchanged $W_N(e_0, \varepsilon, \dots)$. We denote by $\tilde{\mathcal{D}}_\mu$ the differential operator $\mu\partial_\mu$ for fixed e_0 and apply it to both sides of the equation $Z_\phi W_N^R = W_N$ with it. This yields the basic RG differential equation

$$\left[\mu\partial_\mu + \sum_e \tilde{\mathcal{D}}_\mu e \partial_e + N\gamma_\phi\right] W_N^R(e, \mu, \varepsilon, \dots) = 0, \; \gamma_\phi \equiv \tilde{\mathcal{D}}_\mu \ln Z_\phi \tag{65}$$

where the operator $\tilde{\mathcal{D}}_\mu$ is expressed in the variables μ, e. The coefficients $\tilde{\mathcal{D}}_\mu e$ and γ_ϕ are called the RG functions and are calculated in terms of various renormalization constants Z. Coupling constants (charges) g are those parameters e, on which the renormalization constants $Z = Z(g)$ depend. Logarithmic derivatives of charges in Equation (65) are β functions

$$\beta_g = \tilde{\mathcal{D}}_\mu g \, . \tag{66}$$

All the RG-functions are UV-finite, i.e., have no poles in ε, which is a consequence of the functions W_N^R being UV-finite in Equation (65).

For models considered in the present work, the analysis of divergences should be augmented by the following considerations:

- For any dynamic model in Equation (1), all 1PI Green functions containing only the original fields ϕ are proportional to the closed loops of step functions, hence they vanish, and thus do not generate counterterms.
- If for some reason several external momenta or frequencies occur as an overall factor in all the Feynman diagrams of a particular Green function, the real degree of divergence δ' is less than $\delta \equiv d_\Gamma(\varepsilon = 0)$ by the corresponding number of units.
- Sometimes the divergences formally allowed by dimensionality are absent due to symmetry restrictions, for instance, the Galilean invariance of the fully developed turbulence [31] restricts the form of possible counterterms.
- Nonlocal terms of the model are not renormalized.

In principle, these general considerations permit determining all superficially divergent functions and to explicitly obtain the corresponding counter-terms for any dynamic model.

The most convenient scheme for analytic calculations is the scheme of minimal subtractions (MS) proposed in [63], in which all the renormalization constants Z in the perturbation theory are of the form

$$Z^{\rm MS}(g, \varepsilon) = 1 + \sum_{n=1}^{\infty} g^n \sum_{k=1}^{n} \varepsilon^{-k} c_{n,k} \, . \tag{67}$$

In the dimensional renormalization the contribution to the coefficient of g^n in Equation (67) may be expressed as a Laurent series in ε. In the MS scheme, only the singular part of the Laurent expansion of each coefficient is retained. In any other renormalization scheme, the renormalization constant is of the form

$$Z(g, \varepsilon) = 1 + \sum_{n=1}^{\infty} g^n \sum_{k=-n}^{\infty} \varepsilon^k c_{n,-k} \, , \tag{68}$$

where the regular part of each coefficient $\sum_{k=0}^{\infty} \varepsilon^k c_{n,-k}$ is, by and large, an arbitrary regular function of ε at the origin.

5. Composite Operators and Operator Product Expansion

In this section, we recall the basic information about renormalization and critical exponents (dimensions) of composite operators, i.e., local products of the basic fields of the model and their derivatives. In the models we are interested in, they are constructed from the velocity field v, scalar field θ or magnetic field $\boldsymbol{b}$ at the single space-time point $x \equiv (t, \boldsymbol{x})$. Examples are $\boldsymbol{v}^n$, $\boldsymbol{b}^n$, θ^n, $\partial_t \boldsymbol{v}^n$, $\boldsymbol{v}\Delta\boldsymbol{v}$, $(\boldsymbol{\nabla}\theta \cdot \boldsymbol{\nabla}\theta)^n$ and so on.

A theoretical analysis of composite operators and their renormalization is important at least for two reasons. First, their critical dimensions and correlation functions can be measured experimentally and for some operators such data are available [64,65]. For instance, in the fully developed turbulence, the mean of the energy dissipation is proportional to the statistical average of the composite operator $\boldsymbol{v}\Delta\boldsymbol{v}$. This quantity enters the equation of energy balance and contributes to the redistribution of the energy of the turbulent motion and its dissipation. Moreover, strong statistical fluctuations of the operator of energy dissipation seem to account for deviations from Kolmogorov's exponents and lead to the intermittency (multifractality) of the turbulent processes [23]. Second, the general solution of the RG equation (Equation (65)) contains an unknown scaling function depending on dimensionless effective variables (coupling constants, viscosity, etc.). This function can be calculated in the framework of usual perturbation scheme in an expansion parameter but, as mentioned above, in certain asymptotic ranges of scales, this calculation fails. Both experimental and theoretical reasons in theory of turbulence motivate us to study behavior of correlation functions with respect to outer (integral) scale L. Let us elucidate this issue in some detail. As an example we consider pair correlation function for velocity fluctuations $W_2 = \langle \boldsymbol{v}\boldsymbol{v} \rangle$ for field-theoretic model in Equation (24). There is no field renormalization in this model [32], therefore the Green function W_{2R} coincide with the unrenormalized function W_2. The only difference lies in a choice of variable and perturbation theory (expansion either in charge g or g_0, respectively). In renormalized variable correlation, function W_2 depends on $\boldsymbol{k}, g, \nu, \mu$ and L. From a dimensional consideration, we directly see that W_2 can be represented in the form

$$W_2 = \nu^2 k^{2-d} R(s, g, u) \,, \quad s = k/\mu \,, \quad u = kL \,, \tag{69}$$

where R is a function of dimensionless parameters and for brevity we have not explicitly written the transverse projection operator. The correlation function W_2 satisfies a general RG equation with $\gamma_\phi = 0$, which is a direct consequence of absence of renormalization of velocity field $\boldsymbol{v}$, and reads

$$D_{RG} W_2 = 0 \,, \quad D_{RG} = \mu\partial_\mu + \beta(g)\partial_g - \gamma_\nu(g)\nu\partial_\nu \,. \tag{70}$$

The solution of this equation can be found using the method of characteristics and presented in the form

$$W_2 = \bar{\nu}^2 k^{2-d} R(1, \bar{g}, \bar{u}) \,, \quad \bar{u} = u \,, \tag{71}$$

where $\bar{g}$, $\bar{\nu}$ are invariant variables, i.e., the first integrals of Equation (70). Using standard RG considerations, the invariant viscosity [31] takes the following form

$$\bar{\nu} = \nu \exp\left[\int_{\bar{g}}^{g} \mathrm{d}x \, \frac{\gamma_\nu(x)}{\beta(x)} \right] = \left(\frac{g\nu^3}{\bar{g}s^{2\varepsilon}} \right)^{1/3} = \left(\frac{D_0}{\bar{g}k^{2\varepsilon}} \right)^{1/3} . \tag{72}$$

As the parameter s approaches zero, invariant charge $\bar{g}$ approaches IR fixed point g_* and $\bar{\nu} \to \nu_* = (D_0/g_*)^{1/3} k^{-2\varepsilon/3}$. Hence, at fixed point g_* (far from dissipation scales $k \ll \mu \sim l^{-1}$), the single-time correlation function of velocity field takes the scaling form

$$W_2 = (D_0/g_*)^{2/3} k^{2-d-4\varepsilon/3} R(1, g_*, kL) \,. \tag{73}$$

Setting $\varepsilon = 2$ gives kinetic energy spectrum $E(k) = W_2 k^{d-1}$ that behaves as a power-law function of wavevector k. This coincides with Kolmogorov's prediction $-5/3$ for the exponent. The remaining scaling function R is not determined yet and in general it is possible to employ perturbation theory and obtain infinite series in parameter ε. In particular, in the theory of turbulence, the main interest is in the scaling function $R(1, g^*, kL)$ in the inertial interval $kL \gg 1$. In the theory of critical phenomena, the asymptotic form of scaling functions for $kL \gg 1$ (formally, $L \to \infty$) is studied using Wilson's operator product expansion (OPE) (see, e.g., [6,66]). The analog of L in turbulence is played by the correlation length r_c in critical phenomena. It turns out that this technique can be used also in the theory of turbulence and in simplified (toy) models associated with the genuine turbulence (see, e.g., [5,31,33,67,68]).

The generating functional of the correlation functions of the field ϕ with one insertion of the composite operator $F(\phi)$ has the form (compare with the generating functional in Equation (7) for the usual correlation functions of ϕ)

$$\mathcal{G}[A,F] = \int \mathcal{D}\varphi\, F(\varphi) \exp\left[\mathcal{S}[\phi] + A\phi\right]. \tag{74}$$

Since the arguments of the fields in the operator F coincide (giving rise to new closed loops in the Feynman diagrams), correlation functions with these operators contain new UV divergences, which have to removed by an additional renormalization procedure (see, e.g., [5,6,66]). The standard RG equations yield the IR scaling of the renormalized correlation functions with definite critical dimensions $\Delta_F \equiv \Delta[F]$ of a set of basis operators F. Due to the renormalization, $\Delta[F]$ is not the sum of critical dimensions of the fields and derivatives in F. A detailed analysis of the renormalization procedure of composite operators for the stochastic NS problem can be found in the review [67], and below we restrict ourselves to the necessary information only.

As a rule, composite operators are mixed during the renormalization procedure, i.e., an UV finite renormalized operator F^R (correlation functions with one insertion of F_R do not possess UV divergences) takes the form $F^R = F+$ counterterms, in which "counterterms" stands for a linear combination of the operator F itself and other unrenormalized operators mixing the the operator F. Let $F \equiv \{F_\alpha\}$ denote a closed set of operators mixing only with each other under renormalization. For this set, the matrix of renormalization constants $Z_F \equiv \{Z_{\alpha\beta}\}$ and the matrix of anomalous dimensions $\gamma_F \equiv \{\gamma_{\alpha\beta}\}$ are defined by

$$F_\alpha = \sum_\beta Z_{\alpha\beta} F^R_\beta, \qquad \gamma_F = Z_F^{-1} \tilde{\mathcal{D}}_\mu Z_F. \tag{75}$$

The subsequent matrix of critical dimensions $\Delta_F \equiv \{\Delta_{\alpha\beta}\}$ reads

$$\Delta[F] \equiv \Delta_F = d^k_F + \Delta_\omega d^\omega_F + \gamma^*_F, \tag{76}$$

in which d^k_F, d^ω_F, and d_F denote diagonal matrices of canonical dimensions of the operators of the closed set (the diagonal element corresponding to a particular operator F is equal to the sum of canonical dimensions of all fields, their derivatives and renormalized parameters in F) and $\gamma^*_F \equiv \gamma_F(g^*)$ is the matrix in Equation (75) at the fixed point.

Critical dimensions of the set $F \equiv \{F_\alpha\}$ correspond to the eigenvalues of the matrix Δ_F. The basis operators possessing definite critical dimensions are linear combinations of the renormalized operators

$$\bar{F}^R_\alpha = \sum_\beta U_{\alpha\beta} F^R_\beta, \tag{77}$$

where the matrix $U_F = \{U_{\alpha\beta}\}$ is such that the matrix $\Delta'_F = U_F \Delta_F U_F^{-1}$ is diagonal.

Counterterms generated by a given operator F are determined by all possible 1PI Green functions with one insertion of operator F and an arbitrary number of primary fields ϕ,

$$\Gamma_{N;F} = \langle F(t,\boldsymbol{x})\phi(t,\boldsymbol{x}_1)\ldots\phi(t,\boldsymbol{x}_N)\rangle. \tag{78}$$

The total canonical dimension (the formal degree of divergence) for these functions is given by

$$d_\Gamma = d_F - N_\Phi d_\Phi, \tag{79}$$

where the sum is taken over all types of field arguments. For d_Γ is a nonnegative integer.

According to the OPE, the single-time product $F_1(t,x_1)F_2(t,x_2)$ of two renormalized operators at $x \equiv (x_1+x_2)/2 = const.$, and $r \equiv x_1 - x_2 \to 0$ can be represented as follows

$$F_1(t,x_1)F_2(t,x_2) = \sum_\alpha A_\alpha(r)\bar{F}^R_\alpha(t,x). \tag{80}$$

Here, the functions A_α are the Wilson coefficients regular in L, whereas $\bar{F}^R_\alpha$ are all possible renormalized local composite operators of the type in Equation (77) allowed by symmetry arguments, with specific critical dimensions $\Delta_{\bar{F}^R_\alpha}$.

The renormalized correlator $\langle F_1(t,x_1)F_2(t,x_2)\rangle$ is obtained by averaging Equation (80) with the weight $\exp S_R$, quantities $\langle \bar{F}^R_\alpha\rangle \propto L^{-d_\alpha} f_\alpha(g, L\mu)$ involving dimensionless (scaling) functions $f_\alpha(g, L\mu)$ appear on the right hand side. Their asymptotic behavior for $L\mu \to 0$ is found from the corresponding RG equations (see [34] for the case of Kraichnan model) and has the form

$$\langle \bar{F}^R_\alpha\rangle \propto L^{\Delta_{\bar{F}^R_\alpha}}. \tag{81}$$

From the operator product expansion in Equation (80), we therefore get

$$\langle F_1(t,x_1)F_2(t,x_2)\rangle = \sum_{\bar{F}^R} C_{\bar{F}^R}(r/L)^{\Delta_{\bar{F}^R}}, \quad r/L \to 0, \tag{82}$$

where the quantities $C_{\bar{F}^R}$ generated by the Wilson coefficients A_α in Equation (80) are regular in L, the summation is carried out over all possible composite renormalized basis operators $\bar{F}^R$ allowed by the symmetry of the left side, and $\Delta_{\bar{F}^R}$ are their critical dimensions. The leading contributions for $r/L \to 0$ are those with the least dimension $\Delta_{\bar{F}^R}$. In the theory of critical phenomena, it is observed that all the nontrivial composite operators have positive critical dimensions $\Delta_{\bar{F}^R} > 0$ for small ε and the most important term in Equation (82) corresponds to the simplest operator $\bar{F}^R = 1$ with $\Delta_{\bar{F}^R} = 0$, i.e., the function $R(r/L)$ is finite as $L \equiv r_c \to 0$ (see [6]). However, as has been noted in [68] in the model of developed turbulence composite operators with *negative* critical dimensions exist and are responsible for possible singular behavior of the scaling functions such as N-point correlation functions $W_N = \langle \varphi \dots \varphi\rangle$ as $r/L \to 0$. We call operators with $\Delta_{\bar{F}^R} < 0$—if they exist—dangerous [67]. This is motivated by the fact that they correspond to contributions to Equation (82) which diverge for $r/L \to 0$. The scaling functions in Equation (82) decomposed in dangerous operators exhibit anomalous scaling behavior which is a manifestation of a nontrivial multifractal (intermittent) nature of the statistical fluctuations of the random fields under consideration and globally all the physical system.

Dangerous composite operators in the stochastic model of turbulence occur only for finite values of the RG expansion parameter ε. Let us note that within the ε expansion it is not possible to determine whether or not a given operator is dangerous, if only its critical dimension is not found exactly employing the Schwinger equations, etc. or the Galilean symmetry (see [67,69]). Furthermore, dangerous operators appear in the operator product expansion in the form of infinite families with the spectrum of critical dimensions unbounded from below. Therefore, for a proper analysis of the large L behavior, a summation of their contributions is called for.

6. Schwinger Equations and Conservation Laws

Useful information about composite operators can be gained even without an actual calculation of Feynman diagrams. Exploiting invariance properties of functional integrals provides nontrivial relations known as Schwinger equations [47]. One of the simplest symmetries is the translation invariance of action in Equation (3). It is invariance with respect to a shift $\phi \to \phi + \omega$, where $\omega = \omega(x)$ is a suitably chosen function that vanishes sufficiently fast, i.e., $\omega(\infty) = 0$. Such translations do not change the integration measure and as a result the quantity $\int \mathcal{D}\varphi F[\phi+\omega]$ does not depend on ω for any functional $F[\phi]$. We then easily derive that the first variation with respect to ω yields a formal relation

$$\int \mathcal{D}\phi \frac{\delta F[\phi]}{\delta \phi} = 0 \tag{83}$$

written in the notation of Equation (7). The following relation is of particular importance

$$\int \mathcal{D}\phi \, \frac{\delta}{\delta \phi} \exp[\mathcal{S}[\phi] + A\phi] = 0. \tag{84}$$

Performing variational derivatives gives us

$$\int \mathcal{D}\phi \left[\frac{\delta \mathcal{S}[\phi]}{\delta \phi} + A(x) \right] \mathrm{e}^{\mathcal{S}[\phi] + A\phi} = 0. \tag{85}$$

Multiplication by field ϕ_α inside the functional integral is tantamount to a differentiation with respect to the corresponding source field A. This observation allows us to rewrite Equation (85)

$$\left[\left. \frac{\delta \mathcal{S}[\phi]}{\delta \phi} \right|_{\phi \to \delta/\delta A} + A(x) \right] \mathcal{G}(A) = 0. \tag{86}$$

Substituting $\mathcal{G} = \mathrm{e}^{\mathcal{W}}$, we obtain the corresponding Schwinger equation for $\mathcal{W}[A]$ where from we can derive the equation for $\Gamma[\alpha]$. All these equations are of finite order (for polynomial action) in functional derivatives, and each of them is tantamount to an infinite chain of connected equations for the Green functions—the expansion coefficients of the corresponding functionals [47].

In the following discussion, we need one additional relation that corresponds to the Schwinger equation

$$\int \mathcal{D}\phi \, \frac{\delta}{\delta \varphi'} \Big(\varphi(x) \exp[\mathcal{S}[\phi] + A\phi] \Big) = 0, \tag{87}$$

where ϕ stands for either the fluctuating field or the corresponding response field.

As discussed in Section 5, composite operators are related to experimentally measurable physical quantities. We illustrate this claim on an example of stochastic hydrodynamics summarized in the field-theoretic action in Equation (24). Our aim is to elucidate transfer of energy in a stationary of turbulent state. The latter condition ensures that time derivatives of averaged quantities are identically zero.

Let us derive equations describing energy and momentum transfer in turbulent flows. To obtain an equation expressing momentum conservation, we employ the first equation in Equation (87), where ϕ consists of altogether two d-dimensional vector fields $\{\boldsymbol{v}, \boldsymbol{v}'\}$. First, for ϕ, we choose a response field $\boldsymbol{v}'$, and we get

$$\int \mathcal{D}\phi \, \frac{\delta}{\delta v_i'(x)} \exp[\mathcal{S}_{\rm NS}[\phi] + A\phi] = \int \mathcal{D}\phi \left[\frac{\delta \mathcal{S}_{\rm NS}[\phi]}{\delta v_i'(x)} + A_{v_i'} \right] \exp[\mathcal{S}_{\rm NS}[\phi] + A\phi] = 0. \tag{88}$$

Performing indicated derivative, we obtain differential equation

$$\langle A_{v_i'} + D_{ij} v_j' - \partial_t v_i + \nu_0 \Delta v_i - (\boldsymbol{v} \cdot \boldsymbol{\nabla}) v_i - \partial_i p \rangle = 0. \tag{89}$$

Due to transversality of response field $\boldsymbol{v}'$, a nonlocal term has appeared, which corresponds to the pressure fluctuations

$$p = -\frac{\partial_l \partial_s}{\Delta} (v_l v_s). \tag{90}$$

To derive an equation describing energy transfer, we utilize the second Schwinger equation (Equation (87)). In particular, letting $\varphi' \to \boldsymbol{v}'$ and $\varphi \to \boldsymbol{v}$ in Equation (87) yields

$$\int \mathcal{D}\phi \, \frac{\delta}{\delta v_i'(x)} \left(v_i(x) \, \mathrm{e}^{\mathcal{S}_{\mathrm{NS}}[\phi]+A\phi} \right) = 0. \tag{91}$$

In an analogous manner to Equation (89), we get

$$\langle v_i A_{v_i'} + v_i D_{ij} v_j' - v_i \partial_t v_i + \nu_0 v_i \boldsymbol{\nabla}^2 v_i - v_i (v_j \partial_j) v_i - (v_j \partial_j) p \rangle = 0 \tag{92}$$

written in a component notation. Note that all quantities have been normalized to unit mass, i.e., density has been set to unity ($\rho = 1$). Equations (89) and (92) represent conservation laws for momentum and energy. They can be further rewritten into a physically more transparent form

$$\langle \partial_t v_i + \partial_j \Pi_{ij} \rangle = \langle D_{ij} v_j' \rangle + A_{v_i'}, \tag{93}$$

$$\langle \partial_t E + \partial_i S_i \rangle = \langle -\mathcal{E} + v_i D_{ij} v_j' + v_i A_{v_i'} \rangle, \tag{94}$$

where v_i might be interpreted as momentum density, $E \equiv \boldsymbol{v}^2/2$ is energy density, Π_{ik} is tensor of momentum transfer, S_i is vector of energy flow, and $\mathcal{E}$ is a rate of energy dissipation. Direct comparison of Equations (89) and (92) with Equations (93) and (94) yields explicit expressions

$$\Pi_{ij} = p\delta_{ij} + v_i v_j - \nu_0(\partial_i v_j + \partial_j v_i), \tag{95}$$

$$S_i = p v_i - \nu_0 v_j (\partial_i v_j + \partial_j v_i) + \frac{1}{2} v_i \boldsymbol{v}^2, \tag{96}$$

$$\mathcal{E} = \frac{1}{2}\nu_0(\partial_i v_j + \partial_j v_i)^2. \tag{97}$$

We recognize Equation (93) as a stochastic Navier–Stokes equation stirred by random force $D_{ij}v_j'$ and regular force $A_{v_i'}$.

Functional averaging of Equations (93) and (94) according to the prescription in Equation (6) with weigh functional $\exp \mathcal{S}_{\mathrm{NS}}[\phi]$ leads to the balance equation for energy and momentum. Assuming vanishing external force $A_{v_i'}$, we obtain the following equation for time derivative of energy

$$\partial_t \langle E \rangle = - \langle \mathcal{E} \rangle - \partial_i \langle S_i \rangle + \left\langle v_i D_{ij} v_j' \right\rangle. \tag{98}$$

It is clear that for homogeneous and isotropic flows at zero external force the mean value $\langle F(x) \rangle$ of an arbitrary composite operator $F(x) = F(t, \boldsymbol{x})$ could not depend on the position $\boldsymbol{x}$. Hence, it is constant and consequently all spatial derivatives are identically zero. From Equation (98), we then get for steady state

$$\overline{\mathcal{E}} = \int \mathrm{d}x \, D_{ij}(x, x') \left\langle v_i(x) v_j'(x') \right\rangle, \tag{99}$$

where we have introduced the following abbreviation for mean energy dissipation

$$\overline{\mathcal{E}} \equiv \langle \mathcal{E} \rangle. \tag{100}$$

Let us recall (see Equation (19)) that pair correlation for random force can be written as $D_{ij}(x, x') = \delta(t - t') d_{ij}(\boldsymbol{x}, \boldsymbol{x}')$. Insertion of this relation into Equation (100) and integrating over time variable t yields

$$\overline{\mathcal{E}} = \int \mathrm{d}^d x \, d_{ij}(\boldsymbol{x}, \boldsymbol{x}') \left\langle v_i(t, \boldsymbol{x}) v_j'(t, \boldsymbol{x}') \right\rangle |_{t=t'}. \tag{101}$$

To the lowest order in perturbation theory, the retarded response function $\left\langle v_i(t, \boldsymbol{x}) v_j'(t, \boldsymbol{x}') \right\rangle_0$ is δ-correlated in spatial variable. This property holds also for the full response function $\left\langle v_i(x) v_j'(x') \right\rangle$ (what follows from a straightforward analysis of Feynman graphs) and therefore

$$\left\langle v_i(x)v_j'(x')\right\rangle\Big|_{t=t'} = \frac{1}{2}\delta(\boldsymbol{x}-\boldsymbol{x}')\left(\delta_{ij}-\frac{\partial_i\partial_j}{\Delta}\right). \tag{102}$$

Insertion of this relation into Equation (101), recalling Equation (20) and integrating through spatial variable $\boldsymbol{x}'$, we derive

$$\overline{\mathcal{E}} = \frac{1}{2}\int \frac{\mathrm{d}^d k}{(2\pi)^d} d_f(k) P_{ij}(\boldsymbol{k}) P_{ij}(\boldsymbol{k}). \tag{103}$$

Summation over internal indices i and j corresponds to a calculation for a trace of transverse projection operator $P_{ii}(\boldsymbol{k})$, which equals $d-1$. Thus, we finally arrive at the expression

$$\overline{\mathcal{E}} = \frac{d-1}{2(2\pi)^d}\int \mathrm{d}^d k\, d_f(k). \tag{104}$$

This relation reflects an important property of stationary homogeneous turbulence. It expresses the expected fact that, to achieve a stationary state, it is necessary to inject energy into a system in a continuous fashion. In the stochastic approach, this is done through a random force, which compensates energetic losses due to friction processes. These losses are expressed through mean energy dissipation $\overline{\mathcal{E}}$.

7. Ward Identities and Galilean Invariance

We say that a given theory possesses a symmetry, if the corresponding action functional of the theory is zero under the action generating this symmetry. Ward–Takahashi identities mathematically express inherent symmetry of a given field-theoretic action. In stochastic models of turbulence, they correspond to the well-known Galilean invariance. As a rule, these identities provide nontrivial relations between various Green functions of theory and consequently between renormalization constants. Moreover, they are also relevant for an analysis of composite operators.

As pointed out in Section 4, certain divergences present in Feynman diagrams of the perturbative expansion of a Green function might cancel each other, so that the given Green function is in fact UV finite. This compensation might be caused by the inherent fact that the underlying stochastic model describing developed turbulence is invariant with respect to the Galilean transformations. Of course, such and similar mechanisms are quite general in physics. As a further example, we can mention absence of potential UV divergences in quantum electrodynamics, or even quantum chromodynamics describing strong interactions between quarks and gluons. The underlying symmetry in these cases is gauge symmetry.

Now, we show that Galilean invariance in the model in Equation (24) ensures that UV divergences in triple 1PI function

$$\langle v_i{}'(x_1)v_j(x_2)v_k(x_3)\rangle_{1\mathrm{PI}} \equiv \Gamma_{i,jk}(x_1;x_2,x_3) \tag{105}$$

are actually absent. The essential idea consists in a derivation of certain Ward identity, which takes form of an differential equation relating the triple 1PI Green function $\Gamma_{i;sl}$ with the pair 1PI Green function $\Gamma_{i;j}$, which is an abbreviation for

$$\langle v_i'(x_1)v_j(x_2)\rangle_{1\mathrm{PI}} \equiv \Gamma_{i,j}(x_1;x_2). \tag{106}$$

These relations stem from invariance property of generating functional with respect to the Galilean transformations. Then, an absence of certain types of UV divergences in $\Gamma_{i,j}$ leads to absence of divergences in $\Gamma_{i,jk}$.

Let us consider a generalized Galilean transformation of fields $\phi \equiv \{\boldsymbol{v}', \boldsymbol{v}\}$ defined as follows

$$\phi \to \phi_w : \boldsymbol{v}_w(x) = \boldsymbol{v}(x_w) - \boldsymbol{w}(t), \quad \boldsymbol{v}_w'(x) = \boldsymbol{v}'(x_w), \tag{107}$$

where

$$\boldsymbol{x}_w \equiv \boldsymbol{x} + \boldsymbol{u}(t); \quad t = t'; \quad \boldsymbol{u} = \int_{-\infty}^{t} \mathrm{d}t'\, \boldsymbol{w}(t') = \int_{-\infty}^{\infty} \mathrm{d}t'\, \theta(t-t')\boldsymbol{w}(t'). \tag{108}$$

Here, $\boldsymbol{w}(t)$ is some velocity vector describing the Galilean transformation. The spatial vector $\boldsymbol{u}(t)$ is responsible for a shift of spatial coordinate $\boldsymbol{x}$. The transformations in Equation (107) are a generalization of the standard Galilean transformations [70], in which velocity $\boldsymbol{w}$ is constant in time variable. This is brought about by functional integration in which functional space has to be restricted by appropriate conditions. Here, it is required that velocity and response fields $\boldsymbol{v}$ and $\boldsymbol{v}'$ vanish sufficiently quickly in the limit $|t| \to \infty$. Of course, arbitrary symmetry transformation of the model must comply with this property.

Insertion of Equation (107) into the action in Equation (24) yields the following relation for the transformed action

$$\mathcal{S}_{\rm NS}[\phi_w] = \mathcal{S}_{\rm NS}[\phi] + \boldsymbol{v}' \cdot \partial_t \boldsymbol{w} = \mathcal{S}_{\rm NS}[\phi] - (\partial_t \boldsymbol{v}') \cdot \boldsymbol{w}, \tag{109}$$

where in the last equation we have transformed time derivative using partial integration. In this derivation, the following relations for variational derivatives have been utilized

$$\delta_w \boldsymbol{v}(x) = (\boldsymbol{u} \cdot \boldsymbol{\nabla})\boldsymbol{v}(x) - \boldsymbol{w}, \tag{110}$$

$$\delta_w \boldsymbol{v}'(x) = (\boldsymbol{u} \cdot \boldsymbol{\nabla})\boldsymbol{v}'(x), \tag{111}$$

$$\delta_w \partial_t \boldsymbol{v}(x) = (\boldsymbol{u} \cdot \boldsymbol{\nabla})\partial_t \boldsymbol{v}(x) + (\boldsymbol{w} \cdot \boldsymbol{\nabla})\boldsymbol{v}(x) - \partial_t \boldsymbol{w}, \tag{112}$$

which can be directly obtained from Equation (107). In infinitesimal form, Equation (109) takes form

$$\delta_w \mathcal{S}_{\rm NS}[\phi] = -(\partial_t \boldsymbol{v}') \cdot \boldsymbol{w}, \tag{113}$$

where

$$\delta_w \mathcal{S}_{\rm NS}[\phi] \equiv \mathcal{S}_{\rm NS}[\phi_w] - \mathcal{S}_{\rm NS}[\phi] = \boldsymbol{w} \cdot \left. \frac{\delta \mathcal{S}_{\rm NS}[\phi_w]}{\delta \boldsymbol{w}} \right|_{\boldsymbol{w}=0}. \tag{114}$$

The implicit assumption in derivation of Equation (113) is smallness of velocity $\boldsymbol{w}$.

In a compact form, the requirement of Galilean invariance for the model in Equation (24) is equivalent to the condition

$$\mathcal{G}[A] = \mathcal{G}[A_w] \tag{115}$$

or in an infinitesimal form

$$\delta_w \mathcal{G}[A] = 0, \qquad \boldsymbol{w} \cdot \left. \frac{\delta \mathcal{G}[A_w]}{\delta \boldsymbol{w}} \right|_{\boldsymbol{w}=0} = 0. \tag{116}$$

The Ward identities are useful not only for Green functions of basic fields $\boldsymbol{v}, \boldsymbol{v}'$, but for composite operators as well (see Section 5). This motivates an introduction of generalized generating functional that includes composite operators F. It can be presented in the following form

$$\mathcal{G}[A, a] = \int \mathcal{D}\phi \, \exp\left[\mathcal{S}_{\rm NS}[\phi] + A\phi + aF(\phi)\right], \tag{117}$$

where $aF(\phi)$ stands for $aF(\phi) \equiv \sum_i^N \int {\rm d}x \, a_i(x) F_i(\phi, x)$. Sources $a_i(x)$ correspond to N composite operators $F_i(\phi, x)$. In principle, the functional in Equation (117) generates not only all possible Green functions of basic fields, but also full Green functions consisting of arbitrary inclusion of composite operators and fields. In this regard, it is useful to compare this generating functional in Equation (117) with the functional in Equation (74) that generates Green functions containing one inclusion of composite operator F.

The functional measure is obviously translationally invariant. Hence, equality $\mathcal{D}\phi = \mathcal{D}\phi_v$ is satisfied and Equation (116) can be rewritten in the following way

$$\int \mathcal{D}\phi \, \delta_w {\rm e}^{\mathcal{S}_{\rm NS}(\phi_w) + A\phi_w + aF_w} = 0, \tag{118}$$

or

$$\int \mathcal{D}\phi \, \{\delta_w \mathcal{S}_{\rm NS}[\phi] + A\delta_w\phi + a\delta_w F\} \, {\rm e}^{\mathcal{S}_{\rm NS}[\phi]+A\phi+aF} = 0. \tag{119}$$

As source fields A and a are independent, the choice $a = 0$ yields Ward identities for Green functions of basic fields, whereas $a \neq 0$ leads to additional Ward identities for Green functions containing contributions from composite operators. Further, we concentrate on derivation of Ward identity for Green function containing solely basic fields $\phi = \{\boldsymbol{v}, \boldsymbol{v}'\}$, which we derive from Equation (119) inserting $a = 0$. We have

$$\langle\langle -\boldsymbol{w}\cdot\partial_t \boldsymbol{v}' + A\delta_w\phi \rangle\rangle = 0, \tag{120}$$

where double brackets $\langle\langle\cdots\rangle\rangle$ correspond to the functional averaging with respect to the weight functional $\exp[\mathcal{S}[\phi] + A\phi]$

$$\langle\langle\cdots\rangle\rangle = \frac{\int \mathcal{D}\phi \, \ldots \exp\left[\mathcal{S}_{\rm NS}[\phi] + A\phi\right]}{\int \mathcal{D}\phi \, \exp\left[\mathcal{S}_{\rm NS}[\phi] + A\phi\right]}. \tag{121}$$

Using the relations in Equations (108)–(111), we rewrite Equation (120) in the component notation

$$\int {\rm d}x \, w_i(t) \left\langle\left\langle -\partial_t v_i'(x) + \int_{-\infty}^{t} {\rm d}t' \left[A_j(x')\partial_i v_j(x') + A_j'(x')\partial_i v_j'(x')\right] - A_i(x) \right\rangle\right\rangle = 0, \tag{122}$$

where for brevity we have denoted $x' \equiv (t', \boldsymbol{x})$, $\partial_i = \partial/\partial x_i$, $A_j(x) \equiv A_{v_j}(x)$ and $A_j'(x) \equiv A_{v_j'}(x)$. As Galilean velocity $\boldsymbol{w}$ is arbitrary, Equation (122) can be further simplified to

$$\int {\rm d}^d x \left\langle\left\langle -\partial_t v_i'(x) + \int_{-\infty}^{\infty} {\rm d}t' \, \theta(t-t') \left[A_j(x')\partial_i v_j(x') + A_j'(x')\partial_i v_j'(x')\right] - A_i(x) \right\rangle\right\rangle = 0. \tag{123}$$

Every term in Equation (123) containing fields ϕ might be obtained by taking an appropriate number of derivatives of $\exp[A\phi]$ with respect to source A. This means that field ϕ can be effectively replaced by variational derivative $\delta/\delta A$

$$\phi \quad \text{in} \quad \langle\langle\ldots\rangle\rangle \Leftrightarrow \frac{\delta}{\delta A}. \tag{124}$$

Hence, we get

$$\int {\rm d}^d x \left\langle\left\langle -\partial_t \tfrac{\delta}{\delta A_i'(x)} + \int_{-\infty}^{\infty} {\rm d}t' \, \theta(t-t') \left[A_j(x') \tfrac{\partial}{\partial x_i} \tfrac{\delta}{\delta A_j(x')} + A_j'(x') \tfrac{\partial}{\partial x_i} \tfrac{\delta}{\delta A_j'(x')}\right] - A_i(x) \right\rangle\right\rangle = 0. \tag{125}$$

In this formulation, the whole expression in brackets does not depend on integration over fields ϕ and the equation can be further rewritten in terms of generating functional

$$\int {\rm d}^d x \left\{ -\partial_t \tfrac{\delta}{\delta A_i'(x)} + \int_{-\infty}^{\infty} {\rm d}t' \, \theta(t-t') \left[A_j(x') \tfrac{\partial}{\partial x_i} \tfrac{\delta}{\delta A_j(x')} + A_j'(x') \tfrac{\partial}{\partial x_i} \tfrac{\delta}{\delta A_j'(x')}\right] - A_i(x) \right\} \mathcal{G}(A) = 0. \tag{126}$$

Substituting $\mathcal{G} = {\rm e}^{\mathcal{W}}$, we rewrite Equation (126) into equation for generating functional $\mathcal{W}$ for connected Green functions

$$\int {\rm d}^d x \left\{ -\partial_t \tfrac{\delta \mathcal{W}}{\delta A_i'(x)} + \int_{-\infty}^{\infty} {\rm d}t' \, \theta(t-t') \left[A_j(x') \tfrac{\partial}{\partial x_i} \tfrac{\delta \mathcal{W}}{\delta A_j(x')} + A_j'(x') \tfrac{\partial}{\partial x_i} \tfrac{\delta \mathcal{W}}{\delta A_j'(x')}\right] - A_i(x) \right\} = 0. \tag{127}$$

Finally, by means of the functional Legendre transformations in Equation (57), we rewrite Equation (127) in terms of generating functional for 1PI (vertex) functions Γ

$$\int \mathrm{d}^d x \left\{ -\partial_t \alpha'_i(x) + \int_{-\infty}^{\infty} \mathrm{d}t'\, \theta(t-t') \left[\frac{\delta\Gamma(\alpha)}{\delta\alpha_j(x')} \frac{\partial \alpha_{v_j}(x')}{\partial x_i} + \frac{\delta\Gamma(\alpha)}{\delta\alpha'_j(x')} \frac{\partial \alpha_{v'_j}(x')}{\partial x_i} \right] - \frac{\delta\Gamma(\alpha)}{\delta\alpha_i(x)} \right\} = 0, \quad (128)$$

where

$$\alpha_i \equiv \alpha_{v_i}, \quad \alpha'_i \equiv \alpha_{v'_i}. \quad (129)$$

Generating functional Γ can be represented in form of formal Taylor series with respect to sources α, in which coefficients by powers α^n $(n = 0, 1, 2, \ldots, \infty)$ are 1PI Green functions. The first few terms are

$$\Gamma(\alpha) = \alpha_{v'} \Gamma_{v'v} \alpha_v + \frac{1}{2} \alpha_{v'} \Gamma_{v'vv} \alpha_v^2 + \cdots, \quad (130)$$

where, for instance, the second term actually stands for an expression

$$\alpha_{v'} \Gamma_{v'vv} \alpha_v^2 \equiv \int \mathrm{d}x_1 \int \mathrm{d}x_2 \int \mathrm{d}x_3\, \alpha_{v'_i}(x_1) \Gamma_{i,jk}(x_1; x_2, x_3) \alpha_{v_j}(x_2) \alpha_{v_k}(x_3). \quad (131)$$

Insertion of this expansion into Equation (128), we get an infinite system of the Ward identities that relates various 1PI Green functions. For the present discussion, the most relevant is the Ward identity between pair (two-point) and triple (three-point) Green functions. The substitution of the expansion in Equation (130) into Equation (128) yields a formal polynomial in variables α_i and α'_j. A comparison of terms proportional to term $\alpha_i \alpha'_j$ gives the following relation

$$\int \mathrm{d}^d x\, \Gamma_{i,jk}(x_1; x_2, x) + \left[\theta(t-t_1) \frac{\partial}{\partial x_{1k}} + \theta(t-t_2) \frac{\partial}{\partial x_{2k}} \right] \Gamma_{i,s}(x_1; x_2) = 0. \quad (132)$$

Due to translation, invariance, $\Gamma_{i,j}(x_1; x_2) = \Gamma_{i,j}(x_1 - x_2; 0)$, therefore

$$\frac{\partial}{\partial x_{2k}} = -\frac{\partial}{\partial x_{1k}}, \quad k = 1, \ldots, d. \quad (133)$$

Using this relation and integrating over time, the variable t (132) yields

$$\iint \mathrm{d}^d x\, \mathrm{d}t\, \Gamma_{i,jk}(x_1; x_2, x) + \int \mathrm{d}t\, [\theta(t-t_1) - \theta(t-t_2)] \frac{\partial}{\partial x_{1k}} \Gamma_{i,j}(x_1; x_2) = 0. \quad (134)$$

Let us integrate the second term in this equation over time t ($\Gamma_{i,j}(x_1; x_2)$ does not depend on time variable t, only on t_1 and t_2). It is straightforward to show that

$$\int_{-\infty}^{\infty} \mathrm{d}t\, [\theta(t-t_1) - \theta(t-t_2)] = \int_{t_1}^{\infty} \mathrm{d}t - \int_{t_2}^{\infty} \mathrm{d}t = \int_{t_1}^{t_2} \mathrm{d}t = t_2 - t_1. \quad (135)$$

Employing this relation in Equation (134), we finally obtain the required Ward identity in the coordinate representation

$$\int \mathrm{d}x\, \Gamma_{i,jk}(x_1; x_2, x) + (t_2 - t_1) \frac{\partial \Gamma_{i,j}(x_1; x_2)}{\partial x_{1k}} = 0. \quad (136)$$

For an analysis of UV divergences, it is useful to rewrite the Ward identity into a frequency–momentum (Fourier) representation. Let us determine Fourier transforms of pair and triple Green functions. They read

$$\Gamma_{i,j}(x_1;x_2) = \frac{1}{(2\pi)^{2d+2}} \int \mathrm{d}p_1 \int \mathrm{d}p_2\, \Gamma_{i,j}(p_1;p_2) \mathrm{e}^{i(\boldsymbol{p}_1\cdot \boldsymbol{x}_1 + \boldsymbol{p}_2\cdot \boldsymbol{x}_2) - i(\omega_1 t_1 + \omega_2 t_2)}, \tag{137}$$

where $p_i = (\omega_i, \boldsymbol{p}_i)$. Translation invariance in time and space variables dictates

$$\Gamma_{i;j}(p_1, p_2) = (2\pi)^{d+1} \delta(\omega_1 + \omega_2) \delta^{(d)}(\boldsymbol{p}_1 + \boldsymbol{p}_2) \Gamma_{i,j}(p_1). \tag{138}$$

This allows us to simplify Equation (137) into

$$\Gamma_{i,j}(x_1;x_2) = \frac{1}{(2\pi)^{d+1}} \int \mathrm{d}p\, \Gamma_{i;j}(p) \mathrm{e}^{i\boldsymbol{p}\cdot(\boldsymbol{x}_1 - \boldsymbol{p}_2\cdot \boldsymbol{x}_2) - i\omega_p(t_1 - t_2)}. \tag{139}$$

A similar consideration applies also for the triple Green function $\Gamma_{i,jk}$ and leads to the Fourier representation

$$\Gamma_{i,jk}(x_1;x_2,x_3) = \iiint \frac{\mathrm{d}p_1 \mathrm{d}p_2 \mathrm{d}p_3}{(2\pi)^{2(d+1)}} \delta(\sum_i \omega_i) \delta^{(d)}(\sum_i \boldsymbol{p}_i) \Gamma_{i,jk}(p_1;p_2,p_3) \mathrm{e}^{i\sum_{i=1}^{3}(p_i x_i - \omega_i t_i)}. \tag{140}$$

Hence, we derive

$$\int \mathrm{d}x\, \Gamma_{i,jk}(x_1;x_2,x) = \frac{1}{(2\pi)^{d+1}} \int \mathrm{d}p\, \Gamma_{i,jk}(p;-p,0) \mathrm{e}^{ip(x_1 - x_2) - i\omega_p(t_1 - t_2)}. \tag{141}$$

Insertion of the relations in Equations (139) and (141) into Equation (136) leads to the Ward identity in form

$$\Gamma_{i,jk}(p;-p,0) = p_l \frac{\partial}{\partial \omega_p} \Gamma_{i,j}(p). \tag{142}$$

This relation can be conveniently represented in a graphical form as follows

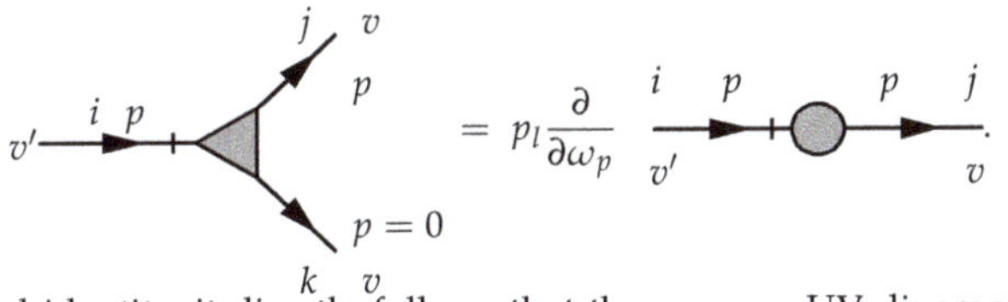

$j\ \ v$, p; v' $i\ p$; $p = 0$, $k\ \ v$ $\quad = p_l \dfrac{\partial}{\partial \omega_p}\quad$ $i\ p$, v'; $p\ j$, v. (143)

From the Ward identity, it directly follows that there are no UV divergences in triple Green functions. In fact, on the right hand side of Equation (143), we have the Green function $\Gamma_{i,j}$, which contains only divergent structures that are proportional to $\boldsymbol{p}^2$ (divergences proportional to frequency ω_p). Therefore, taking derivative with respect to frequency ω_p ensures that the divergent part on right hand side of Equation (143) vanishes. Hence, the right hand side of Equation (143) is UV finite and therefore also the left hand side of the given equation is UV finite. This concludes the proof of UV finiteness of the triple one-time 1PI Green function $\Gamma_{i,jk}$.

8. Symmetry Restoration

In the previous Section 7, we show that the Galilean symmetry in Equations (107)–(108) restricts appearances of UV divergences in the perturbation theory. In this regard, a natural question arises: What can violation of the Galilean symmetry lead to? We address this issue in the context of the stochastic Navier–Stokes model in Equation (24). Let us note that the following exposition closely follows that in [71], where all necessary technical details can be found. Violation can be achieved by various means. In particular, it can be achieved by modification of time behavior of the force correlator in Equation (20). Let us imagine that there exists microscopic finite correlation time behaving according to Ornstein-Uhlenbeck process [17,72].

Without loss of generality, let us consider a generalization of Equation (23) as

$$\langle f_i(\omega,\boldsymbol{k}) f_j(\omega',\boldsymbol{k}')\rangle = (2\pi)^{d+1}\delta(\omega+\omega')\delta(\boldsymbol{k}+\boldsymbol{k}')P_{ij}(\boldsymbol{k})d_f(\omega,\boldsymbol{k}), \tag{144}$$

where the kernel function d_f now assumes the following form

$$d_f(\omega_k,\boldsymbol{k}) = D_0 P_{ij}(\boldsymbol{k})\frac{k^{8-d-(y+2\eta)}}{\omega_k^2+\nu_0^2 u_0^2 k^{4-2\eta}}. \tag{145}$$

Exponent η is related to the dispersion law $\omega_k \propto k^{-2+\eta}$. From dimensional considerations, parameter D_0 can be represented as

$$D_0 = g_0\nu_0^5 u_0^2, \tag{146}$$

which can be interpreted as a defining relation for charge g_0. Quite general form of Equation (145) allows studying two special cases: the limit $u_0 \to 0$ fully corresponds to the time-independent correlation of random force. On the other hand, the limit $u_0 \to \infty$ yields the previously analyzed Galilean model in Equation (20).

The propagator in the frequency–momentum representation takes the form

$$\langle v_i v_j\rangle_0 = D_0\frac{k^{8-d-(y+2\eta)}}{\omega^2+\nu_0^2u_0^2k^{4-2\eta}}\frac{P_{ij}(\boldsymbol{k})}{\omega^2+\nu_0^2k^4}, \tag{147}$$

According to the theoretical consideration discussed in Sections 2 and 4, it can be shown [71] that the model corresponding to Equation (24) with Equation (145) is multiplicatively renormalizable in which two renormalization constants Z_1 and Z_2 have to be added. The renormalized action functional is

$$\mathcal{S}_{\mathrm{NS}}[\phi] = \frac{1}{2}v_i' D_{ik} v_k' + v_k'\left[-\partial_t - Z_1(v_i\partial_i) + Z_2\nu\partial^2\right]v_k, \tag{148}$$

where g, ν, and u are the renormalized counterparts of the bare (original) parameters; the correlation function D_v is written in terms of the renormalized parameters with the use of the relation $g_0\nu_0^5u_0^2 = g\mu^{y+2\eta}\nu^5u^2$; and the reference scale μ is a new free parameter of the renormalized model. The action of the renormalized model in Equation (148) can be constructed from the original action in Equation (24) by renormalization of fields $v \to Z_v v$, $v' \to Z_{v'}v'$ and parameters

$$g_0 = g\mu^y Z_g, \quad u_0 = u\mu^\eta Z_u, \quad \nu_0 = \nu Z_\nu. \tag{149}$$

The renormalization constants in Equations (148) and (149) are related as follows

$$Z_\nu = Z_2, \quad Z_u = Z_2^{-1}, \quad Z_g = Z_1^2 Z_2^{-3}, \quad Z_v = Z_{v'}^{-1} = Z_1. \tag{150}$$

The leading one-loop calculation [71] of Feynman diagrams is relatively straightforward and ensuing anomalous dimensions are

$$\gamma_1 = g\frac{1}{d(d+2)}\frac{u}{(u+1)^3}, \quad \gamma_2 = g\frac{1}{d(d+2)}\frac{u^3d(d-1)+3u^2d(d-1)+2u(d^2-d+2)}{4(u+1)^3}, \tag{151}$$

where for brevity the following redefinition of the charge g

$$g\frac{S_d}{(2\pi)^d} \equiv g\overline{S_d} \to g. \tag{152}$$

has been used. Further, from the relations in Equation (150), we get β-functions

$$\beta_g = g(-y-2\gamma_1+3\gamma_2), \quad \beta_u = u(-\eta+\gamma_2). \tag{153}$$

A straightforward analysis of the β functions reveals three IR fixed points: the trivial fixed point (zero values of charges' coordinates) and two nontrivial fixed points. The trivial (Gaussian) fixed point, at which all nonlinearities are IR irrelevant, is

$$g^* = 0, \quad u^* = 0 \tag{154}$$

and is IR attractive when both y and η are negative.

A further fixed point $\{g^*, u^*\}$ with the coordinates

$$u^* = \frac{-3 + \sqrt{1 - \frac{16(\alpha-1)}{d(d-1)(3\alpha-1)}}}{2}, \quad g^* = d(d+2)\frac{(u^*+1)^3}{u^*}\frac{3\alpha - 1}{2}y, \tag{155}$$

is actually of a saddle-point type for, which one eigenvalue of the matrix Ω is positive and the other negative.

The most interesting fixed point is actually obtained in the limit $u^* \to \infty$, from Equations (145) and (147). This case corresponds to an earlier studied model with random force uncorrelated in time [32]. Therefore, we might expect to find the known fixed point of this model. Introducing a new variable $x = 1/u$, one obtains a new β function of the following form

$$\beta_x = \widetilde{\mathcal{D}}_\mu x = -\frac{1}{u^2}\beta_u = x\left[\eta - g\frac{d-1}{4(d+2)}\right]. \tag{156}$$

Actual calculations reveal the existence of the fixed point with the coordinates

$$x^* = 0, \quad g^* = \frac{4(d+2)}{3(d-1)}y, \tag{157}$$

which coincides with that in Ref. [32] and is IR attractive in the region restricted by inequalities $y > 0$ and $\eta > y/3$.

The main conclusion from these considerations is that at the only nontrivial IR attractive fixed point correlation time vanishes. In other words, the Galilean symmetry, broken by the colored random force in Equation (145), is reestablished in the IR limit.

9. Parity Breaking in Magnetohydrodynamic Turbulence

In addition to the passive scalar problems introduced in Section 3, there exists a broad class of problems related to vector admixtures [23,35]. Due to the presence of more degrees of freedom, it is natural to expect that observed behavior can be richer than in scalar case. Among many, much attraction has been gained by a model of magnetohydrodynamics (MHD) [73–76]. The interplay between the velocity field and the magnetic field is crucial in explaining many phenomena—magnetic dynamo, convective processes, galaxy formation, etc. [77–81]—therefore it is clear that it has to be taken into account properly.

The full stochastic MHD problem is rather complicated, therefore for many purposes for a theoretical description it has been proposed to use a simplified model—the kinematic Kazantsev–Kraichnan model [35]. Its basic assumption is that the vector field of magnetic induction $\boldsymbol{B}$ (hereinafter, referred to as vector) is passively advected by the turbulent flow, but reaction on the velocity field by the magnetic field is neglected. The most notable point of criticism to the Kazantsev–Kraichnan model is the assumption about the velocity field using a simple Gaussian statistical ensemble. From a more fundamental point of view, the velocity field would be generated dynamically. Therefore, we assume here that the velocity field is brought about by the stochastic Navier–Stokes equation expounded in Section 3.

The introduction of magnetic field in the Kazantsev–Kraichnan model comes from a physical approximation called magnetohydrodynamic limit [74,75]. It is assumed that the matter is electrically neutral at large scales and that the mean free path is much shorter than the corresponding Debye length. This assures that the electric displacement field accounting for the overall motion of charged particles can be entirely neglected. The system is then completely described in terms of the variables of density, pressure, and mean velocity field. From the RG point of view the displacement field is IR irrelevant. Hence, there is no need to retain it. Faraday's law $\partial_t \boldsymbol{B} = -\boldsymbol{\nabla} \times \boldsymbol{E}$ together with the generalized Ohm's law $\boldsymbol{J} = \sigma(\boldsymbol{E} + \boldsymbol{v} \times \boldsymbol{B})$ give rise to the advection–diffusion equation $\partial_t \boldsymbol{B} - \boldsymbol{\nabla} \times (\boldsymbol{v} \times \boldsymbol{B}) = \kappa_0 \boldsymbol{\nabla}^2 \boldsymbol{B}$. According to the functional formulation from Section 2, the corresponding stochastic problem assumes the form

$$\partial_t b_i + \partial_k (v_k b_i - v_i b_k) = \kappa_0 \partial^2 b_i + f_i^\theta, \tag{158}$$

where b_i is just the fluctuating component of the magnetic induction, κ_0 is the magnetic diffusion coefficient, and a stochastic source term f_i^θ has been added to the right side. This term is the random part of the curl of the current and stems from the intrinsic randomness of the magnetic field [31]. A detailed account of the MHD problem can be found in [73–75].

The random source f_i^θ in Equation (158) is assumed to be zero-mean Gaussian with the correlation function

$$\langle f_i^b(t, \boldsymbol{x}) f_j^b(t', \boldsymbol{x}') \rangle = \delta(t - t')\, C_{ij}(\boldsymbol{r}/L_\theta), \quad \boldsymbol{r} = \boldsymbol{x} - \boldsymbol{x}', \tag{159}$$

where $C_{ij}(\mathbf{r}/L_\theta)$ is a function, whose exact form is irrelevant. It must have a finite limit at $(\boldsymbol{r}/L_\theta) \to 0$ and vanish at $(\boldsymbol{r}/L_\theta) \to \infty$. The magnetic field b_i is solenoidal, therefore the terms $\partial_k(v_i b_k)$ and $(b_k \partial_k) v_i$ are equal.

In realistic scenarios, the stochastic NS equation (Equation (16)) has to be amended by a Lorentz term responsible for the backward influence of the magnetic field on the velocity field. This is brought about by a familiar force term in conducting fluid of form $\boldsymbol{v} \times \boldsymbol{B} \sim \boldsymbol{J} \sim (\boldsymbol{\nabla} \times \boldsymbol{B}) \times \boldsymbol{B}$. The most important consequence of this argument is an important physical effect known as the turbulent dynamo: generation of magnetic field at large scales by the turbulent motion. Let us make an important remark, which should clarify the fundamental difference between equilibrium statistical models and non-equilibrium ones. The turbulent dynamo might be explained by the mechanism of spontaneous symmetry breaking most spectacularly associated with the Higgs mechanism in particle physics. Here, "the ground state" of the turbulent gyrotropic fluid with vanishing mean $\boldsymbol{b} = 0$ of the magnetic field is unstable. It is stabilized by the spontaneous generation of a spatially uniform anomalous mean $\boldsymbol{b} \neq 0$. This happens in full analogy with the situation in a ferromagnetic material below T_c, in which magnetic order is stabilized by the appearance of spontaneous magnetization. The condition for total magnetization follows from an extremal condition imposed on free energy. On the other hand, for the dynamo problem, the resulting field is determined by the properties of the response function. In particular, it is required that all perturbations from a given stable state have to be damped out sufficiently quickly. Let us note that that the instability manifests itself not at the level of the action function but only at the following one-loop approximation.

Let us discuss the full MHD problem in more detail and introduce one specific generalization. From now on, we assume that the MHD problem is governed by two coupled equations [22]

$$\nabla_t \boldsymbol{v} = \nu_0 \boldsymbol{\nabla}^2 \boldsymbol{v} + (\boldsymbol{b} \cdot \boldsymbol{\nabla}) \boldsymbol{b} - \boldsymbol{\nabla} Q + \boldsymbol{f}^v, \tag{160}$$

$$\nabla_t \boldsymbol{b} = \nu_0 u_0 \boldsymbol{\nabla}^2 \boldsymbol{b} + A(\boldsymbol{b} \cdot \boldsymbol{\nabla}) \boldsymbol{v} - \boldsymbol{\nabla} P + \boldsymbol{f}^b, \tag{161}$$

where $\boldsymbol{b} \equiv \boldsymbol{b}(x)$ denotes a solenoidal vector field advected by a helical turbulent flow of an incompressible velocity field $\boldsymbol{v} \equiv \boldsymbol{v}(x)$. Both fields $\boldsymbol{v}$ and $\boldsymbol{b}$ are divergenceless. In other words, they satisfy the incompressibility condition

$$\boldsymbol{\nabla} \cdot \boldsymbol{b} = \boldsymbol{\nabla} \cdot \boldsymbol{v} = 0. \tag{162}$$

Further, u_0 is the bare inverse Prandtl number [23,26]. Functions $P \equiv P(x)$ and $Q \equiv Q(x)$ in Equations (160) and (161) representing pressure fields are not relevant in the following analysis. In fact, due to the solenoidal property in Equation (162), functions P and Q can be expressed in terms of a formal

Biot–Savart law [23,26]. A is a dimensionless parameter of the model. Three particular values of the parameter A have been analyzed in detail [35]. First, the value $A = 1$ yields the Kazantsev–Kraichnan kinematic dynamo model, where the pressure term in Equation (161) drops out. Second, the choice $A = 0$ leads to the model of passively advected vector admixture. Third, the value $A = -1$ corresponds to the linearized NS equation in a background field with given statistics. It is therefore convenient to retain A as a free parameter and analyze all three models together. The model in Equation (161) is often called the generic A model in the literature [35,82,83]. The complete model thus comprises two interconnected stochastic equations (Equations (160) and (161)).

Stochastic random sources $\boldsymbol{f}^v$ and $\boldsymbol{f}^b$ must be included in the coarse-grained setup of turbulence and MHD problems. As usual, it is assumed that the random variable $\boldsymbol{f}^b$ is Gaussian with zero mean and correlation function

$$D^b_{ij}(x;0) \equiv \langle f^b_i(x) f^b_j(0)\rangle = \delta(t) C_{ij}(|\boldsymbol{x}|/L). \tag{163}$$

Here, L is the integral scale of stirring of the magnetic field $\boldsymbol{b}$, C_{ij} is finite at $L \to \infty$ and at $|x| \gg L$ it should rapidly fall off. Apart from these restrictions, C_{ij} needs not be specified [35]. Further, $\boldsymbol{f}^v$ mimics the pumping of kinetic energy in the system at the largest spatial scales and is subject to the condition of real IR energy injection implemented through its specific power-like form suited to the RG approach [5]. However, the results obtained do not depend on specific forcing statistics because of the universality of fully developed turbulence. Hence, we only take advantage of the most suitable forcing in the RG analysis. Following this standard argument [5], we choose the pair correlation function of zero-mean Gaussian statistics

$$D^v_{ij}(x;0) \equiv \langle f^v_i(x) f^v_j(0)\rangle = \delta(t) \int \frac{\mathrm{d}^d k}{(2\pi)^d} D_0 k^{4-d-2\varepsilon} R_{ij}(\boldsymbol{k}) \mathrm{e}^{i\boldsymbol{k}\cdot\boldsymbol{x}}, \tag{164}$$

where $d = 3$ denotes the space dimension; $\boldsymbol{k}$ is the wavevector with the magnitude $k = |\boldsymbol{k}|$; and $D_0 \equiv g_0 \nu_0^3 > 0$ is the positive quantity, where g_0 is a coupling constant connected to the typical UV momentum scale Λ through $g_0 \simeq \Lambda^{2\varepsilon}$. The parameter ε specifies the power-like behavior of the energy pumping at large scales and assumes the value of 2 in the physically relevant IR energy pumping. In the RG analysis of the fully developed turbulence, ε is assumed small in calculations. Its physical value of 2 is inserted into perturbative expansions only as the last step [5]. The tensor quantity R_{ij} determines the spatial parity violation in the model at hand. In symmetric isotropic incompressible turbulent fluid, such as analyzed here, the tensor $R_{ij}(k)$ corresponds to the sum of the ordinary transverse projection operator $P_{ij}(\boldsymbol{k}) = \delta_{ij} - k_i k_j/k^2$ and a helical term $H_{ij}(\boldsymbol{k}) = i\rho\, \varepsilon_{ijl} k_l/k$, i.e.

$$R_{ij}(\boldsymbol{k}) = \delta_{ij}(\boldsymbol{k}) - \frac{k_i k_i}{\boldsymbol{k}^2} + i\rho\varepsilon_{ijl}\frac{k_l}{|\boldsymbol{k}|}, \tag{165}$$

where ε_{ijl} is the third-rank completely antisymmetric tensor. The real-valued helicity parameter ρ obeys the inequality $|\rho| \leq 1$ dictated by the condition that the correlation function is positive definite in Equation (164). Evidently, the value $\rho = 0$ corresponds to the isotropic (non-helical) case, while the value $\rho = 1$ corresponds to fully broken spatial parity.

Following the De Dominicis–Janssen approach (Section 2), the stochastic problem of Equations (161) and (160) is equivalent to a field-theoretic model with the set of fields $\Phi = \{\boldsymbol{v}, \boldsymbol{b}, \boldsymbol{v}', \boldsymbol{b}'\}$, where the fields with primes once more denote the response fields [5,45]. Thus, the field-theoretic model is given by the Dominicis–Janssen action

$$\mathcal{S}[\Phi] = \tfrac{1}{2}\left[v'_i D^v_{ij} v'_j + b'_i D^b_{ij} b'_j\right] + \boldsymbol{v}'[-\nabla_t \boldsymbol{v} + \nu_0 \boldsymbol{\nabla}^2 \boldsymbol{v} + (\boldsymbol{b}\cdot\boldsymbol{\nabla})\boldsymbol{b}] + \boldsymbol{b}'[-\nabla_t \boldsymbol{b} + \nu_0 u_0 \boldsymbol{\nabla}^2 \boldsymbol{b} + A(\boldsymbol{b}\cdot\boldsymbol{\nabla})\boldsymbol{v}], \tag{166}$$

where D^b_{ij} and D^v_{ij} are defined in Equations (163) and (164), respectively. As usual, summations over repeated indices $i, j \in 1, 2, 3$ are implied. As the original fields $\boldsymbol{v}$ and $\boldsymbol{b}$, the response fields are solenoidal, i.e., $\nabla \cdot \boldsymbol{v}' = \nabla \cdot \boldsymbol{b} = 0$. In the frequency–momentum representation, free propagators of the model in Equation (166) are

$$\langle b'_i b_j \rangle_0 = \langle b_i b'_j \rangle_0^* = \frac{P_{ij}(\mathbf{k})}{i\omega + \nu_0 u_0 k^2}, \qquad \langle b_i b_j \rangle_0 = \frac{C_{ij}(\mathbf{k})}{|-i\omega + \nu_0 u_0 k^2|^2}, \tag{167}$$

$$\langle v'_i v_j \rangle_0 = \langle v_i v'_j \rangle_0^* = \frac{P_{ij}(\mathbf{k})}{i\omega + \nu_0 k^2}, \qquad \langle v_i v_j \rangle_0 = \frac{g_0 \nu_0^3 k^{4-d-2\varepsilon} R_{ij}(\mathbf{k})}{|-i\omega + \nu_0 k^2|^2}. \tag{168}$$

The function $C_{ij}(k)$ is the Fourier transform of the function $C_{ij}(r/L)$ introduced in Equation (163). Moreover, the theory includes three interaction vertices

- $S_{b'bv}$: $b'_i(-v_j\partial_j b_i + Ab_j\partial_j v_i) = b'_i v_j V_{ijl} b_l$,
- $S_{v'vv}$:$-v'_i v_j \partial_j v_i = v'_i v_j W_{ijl} v_l/2$,
- $S_{v'bb}$:$v'_i b_j \partial_j b_i = v'_i v_j U_{ijl} v_l/2$,

In the momentum–frequency representation, they are associated with the vertex factors [5]

$$V_{ijl} = i(k_j\delta_{il} - Ak_l\delta_{ij}), \quad W_{ijl} = i(k_l\delta_{ij} + k_j\delta_{il}), \quad U_{ijl} = i(k_l\delta_{ij} + k_j\delta_{il}). \tag{169}$$

In all these vertices, momentum $\boldsymbol{k}$ is flowing into through the corresponding response field, i.e., in V_{ijl} it is the response field $\boldsymbol{b}'$ and in W_{ijl}, U_{ijl} the field $\boldsymbol{v}'$. A graphical representation of interaction vertices is displayed in Figure 4.

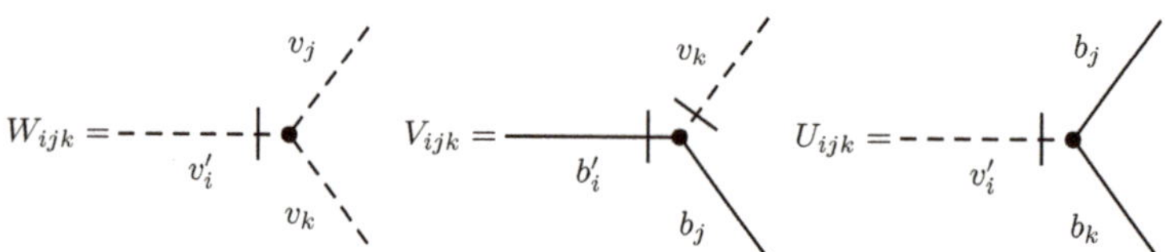

Figure 4. Graphical representation of all interaction vertices of the model related velocity non-linearities of the action (166).

The field theoretic model defined by the action in Equation (166) is intrinsically unstable due to the fact that 1PI graphs $\langle v'v \rangle_{1\text{-IR}}$ and $\langle b'b \rangle_{1\text{-IR}}$ are not UV finite. To ensure the stabilization of the advection–diffusion system, we assume—inspired by the argument in [84]—that the vector field $\boldsymbol{b}$ fluctuates around a spontaneously generated non-vanishing mean $\boldsymbol{B} = \langle \boldsymbol{b} \rangle \neq 0$ with the magnitude depending on the parameter A. The auxiliary field $\boldsymbol{b}'$ is still supposed to have vanishing mean. Technically, the spontaneous symmetry breaking is achieved by the substitution of the sum $\boldsymbol{B} + \boldsymbol{b}$ instead of the vector field $\boldsymbol{b}$ in the action in Equation (166). Such a change of variables gives rise to a new action functional with the free part now containing additional terms, whereas the interacting part remains intact. The action functional with broken symmetry is

$$\begin{aligned} \mathcal{S}[\Phi] = {} & \frac{1}{2}\left[v'_i D^v_{ij} v'_j + b'_i D^b_{ij} b'_j\right] + \boldsymbol{v}'[-\nabla_t \boldsymbol{v} + \nu_0 \boldsymbol{\nabla}^2 \boldsymbol{v} + (\boldsymbol{b}\cdot\boldsymbol{\nabla})\boldsymbol{b}] + \boldsymbol{b}'[-\nabla_t \boldsymbol{b} + \nu_0 u_0 \boldsymbol{\nabla}^2 \boldsymbol{b} \\ & + A(\boldsymbol{b}\cdot\boldsymbol{\nabla})\boldsymbol{v}] + \boldsymbol{v}'(\boldsymbol{B}\cdot\boldsymbol{\nabla})\boldsymbol{b} + A\boldsymbol{b}'(\boldsymbol{B}\cdot\boldsymbol{\nabla})\boldsymbol{v}. \end{aligned} \tag{170}$$

The quadratic part of this action determines propagators of the theory and now it has more involved structure than Equation (166). It is seen that the symmetry breaking brings about cross propagators $\langle \boldsymbol{v}\boldsymbol{b}' \rangle$, $\langle \boldsymbol{b}\boldsymbol{v}' \rangle$, $\langle \boldsymbol{b}\boldsymbol{v} \rangle$, and $\langle \boldsymbol{b}\boldsymbol{b} \rangle$. Furthermore, all propagators are more complicated and depend on the uniform magnetic field $\boldsymbol{B}$ explicitly. In the frequency–momentum representation, they are

$$
\begin{aligned}
\langle v_i v_j\rangle &= \frac{\beta(\boldsymbol{k})\beta^*(\boldsymbol{k})}{\xi(\boldsymbol{k})\xi^*(\boldsymbol{k})} D^v(\boldsymbol{k}) R_{ij}(\boldsymbol{k}), & \langle b_i b_j\rangle &= A^2 \frac{(\boldsymbol{B}\cdot\boldsymbol{k})^2}{\xi(\boldsymbol{k})\xi^*(\boldsymbol{k})} D^v(\boldsymbol{k}) R_{ij}(\boldsymbol{k}) \\
\langle v_i v_j'\rangle &= \frac{\beta^*(\boldsymbol{k})}{\xi^*(\boldsymbol{k})} P_{ij}(\boldsymbol{k}), & \langle b_i b_j'\rangle &= \frac{\alpha^*(\boldsymbol{k})}{\xi^*(\boldsymbol{k})} P_{ij}(\boldsymbol{k}), \qquad (171) \\
\langle b_i v_j'\rangle &= iA \frac{(\boldsymbol{B}\cdot\boldsymbol{k})}{\xi^*(\boldsymbol{k})} P_{ij}(\boldsymbol{k}), & \langle v_i b_j'\rangle &= i \frac{(\boldsymbol{B}\cdot\boldsymbol{k})}{\xi^*(\boldsymbol{k})} P_{ij}(\boldsymbol{k}), \qquad (172) \\
\langle b_i v_j\rangle &= iA \frac{\beta(\mathbf{k})(\boldsymbol{B}.\mathbf{k})}{\xi(\mathbf{k})\xi^*(\mathbf{k})} D_v(\mathbf{k}) R_{ij}(\mathbf{k}). & & \qquad (173)
\end{aligned}
$$

Here, following abbreviations have been introduced

$$
\alpha(\boldsymbol{k}) = i\omega + \nu k^2, \quad \beta(\boldsymbol{k}) = i\omega + u\nu k^2, \quad \xi(\boldsymbol{k}) = A(\boldsymbol{B}\cdot\boldsymbol{k})^2 + \alpha(\boldsymbol{k})\beta(\boldsymbol{k}) \tag{174}
$$

for brevity. Propagators are depicted in Figure 5.

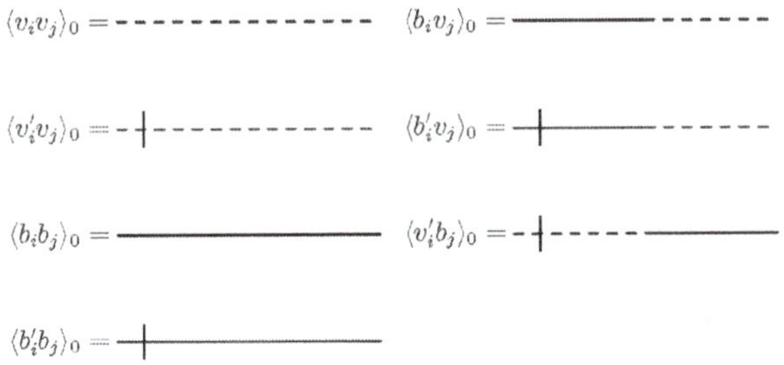

$\langle v_i v_j\rangle_0 =$ $\langle b_i v_j\rangle_0 =$

$\langle v_i' v_j\rangle_0 =$ $\langle b_i' v_j\rangle_0 =$

$\langle b_i b_j\rangle_0 =$ $\langle v_i' b_j\rangle_0 =$

$\langle b_i' b_j\rangle_0 =$

Figure 5. Graphical representation of all propagators of the model given by the quadratic part of the action (166).

Identification of all relevant UV divergences is carried out by the standard power counting. We omit the details, which are analogous to those in Ref. [82]. The model at hand is logarithmic at $\varepsilon = 0$. In the framework of the minimal subtraction (MS) scheme, this means that all possible UV divergences are poles in ε [5]. Following the analysis of Hnatič and Zalom [82], we arrive at the conclusion that all UV divergences are absorbed in counterterms $\boldsymbol{v}'\boldsymbol{\nabla}^2\boldsymbol{v}$ and $\boldsymbol{b}'\boldsymbol{\nabla}^2\boldsymbol{b}$. Thus, the model is multiplicatively renormalizable with renormalized parameters g_0, u_0, and ν_0

$$
\nu_0 = \nu Z_\nu, \quad g_0 = g\mu^{2\varepsilon} Z_g, \quad u_0 = u Z_u, \tag{175}
$$

where g, u, and ν are the corresponding renormalized parameters. The renormalization mass μ is introduced as a consequence of the analytic regularization used in calculations. The quantities $Z_i = Z_i(g, u; d, \rho; \varepsilon)$ contain poles in the exponent ε. The renormalized action functional can be written as follows

$$
\begin{aligned}
\mathcal{S}_R[\phi] = \frac{1}{2}\left[v_i' D_{ij}^v v_j' + b_i' D_{ij}^b b_j'\right] &+ \boldsymbol{v}'[-\nabla_t \boldsymbol{v} + \nu Z_1 \boldsymbol{\nabla}^2 \boldsymbol{v} + Z_3(\boldsymbol{b}\cdot\boldsymbol{\nabla})\boldsymbol{b}] + \boldsymbol{b}'[-\nabla_t \boldsymbol{b} + \nu u Z_2 \boldsymbol{\nabla}^2 \boldsymbol{b} \\
&+ A(\boldsymbol{b}\cdot\boldsymbol{\nabla})\boldsymbol{v}] + Z_3 \boldsymbol{v}'(\boldsymbol{B}\cdot\boldsymbol{\nabla})\boldsymbol{b} + A\boldsymbol{b}'(\boldsymbol{B}\cdot\boldsymbol{\nabla})\boldsymbol{v},
\end{aligned} \tag{176}
$$

where Z_1 and Z_2 are renormalization constants defined by relations

$$
Z_\nu = Z_1, \quad Z_g = Z_1^{-3}, \quad Z_u = Z_2 Z_1^{-1}. \tag{177}
$$

Both renormalization constants, Z_1 and Z_2, correspond to a different class of Feynman diagrams (details below). However, they reveal a common structure in the MS scheme: the nth order of perturbation theory is related to the nth power of g with the corresponding coefficient containing poles in ε of order n and less [5].

The relevant one-loop-order Feynman diagrams are shown in Figures 6 and 7.

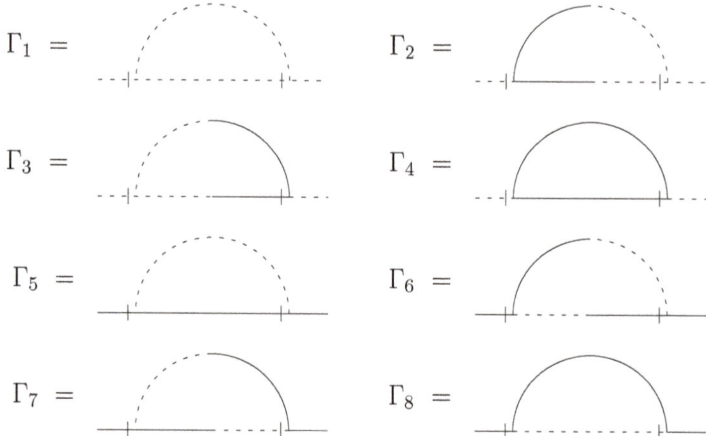

Figure 6. Graphical representation of all Feynman diagrams for two-point one-irreducible Green functions of the action (176). Graphs $\Gamma_1, \ldots, \Gamma_4$ represent perturbation expansion for $\Gamma_{v'v}$ function, and $\Gamma_5, \ldots, \Gamma_8$ for $\Gamma_{b'b}$ function.

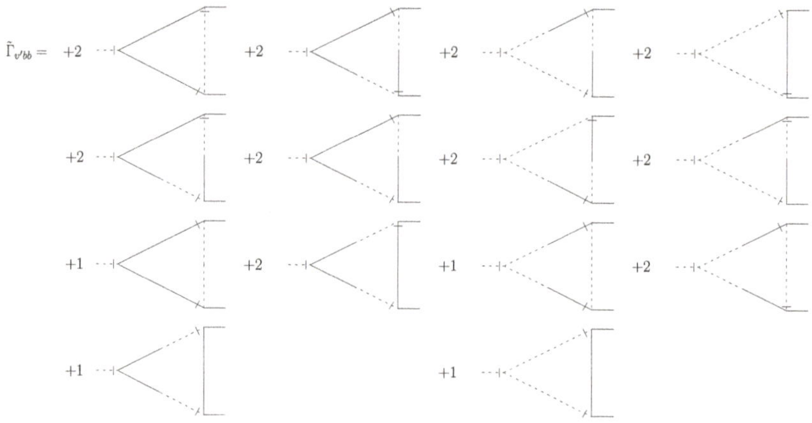

Figure 7. Graphical representation of all one-loop Feynman diagrams forces for one-irreducible Green function $\Gamma_{v'bb}$.

One-loop calculation [84] of self-energy graphs Σ_{ij} for the 1PI function $\Gamma_{b'_i b_j}$ yields the following expression

$$\Sigma_{ij} \sim \varepsilon_{isl} p_s T_{lj} \tag{178}$$

where

$$T_{lj} = a\Lambda\delta_{lj} - b|\boldsymbol{B}|(\delta_{lj} + e_l e_j), \quad \boldsymbol{e} \equiv \frac{\boldsymbol{B}}{|\boldsymbol{B}|}. \tag{179}$$

Here, e_i are components of the unit vector pointing in the direction of the spontaneous magnetic field. Terms proportional to δ_{ij} might in principle generate instabilities. In the literature, the term proportional to $e_l e_j$ is known as exotic. As detailed analysis [84] shows, δ_{ij} terms can be eliminated by imposing the following relation

$$|\boldsymbol{B}| = \frac{a\Lambda}{b} = \sqrt{\frac{1}{\pi|A|}\frac{\Gamma(d/2+3/2)}{\Gamma(d/2+1)}}u_* \nu\,\Lambda, \tag{180}$$

where the parameter A should not be equal to 0 or -1. At real space dimension $d = 3$, we get

$$|\boldsymbol{B}| = \frac{8}{3\pi\sqrt{|A|}}u_* \nu\,\Lambda. \tag{181}$$

To get insight into relevance of the last term on the right hand side of Equation (179), let us consider the approximate case of linearized MHD equations in polarized medium

$$\partial_t \boldsymbol{v} = \nu k^2 \boldsymbol{v} + i\gamma \boldsymbol{b}, \quad \partial_t \boldsymbol{b} = u\nu k^2 \boldsymbol{v} + i\gamma \boldsymbol{v} - i\mu[\boldsymbol{k}\times\boldsymbol{e}](\boldsymbol{e}\cdot\boldsymbol{b}), \tag{182}$$

where for brevity we have denoted $\gamma = \boldsymbol{B}\cdot\boldsymbol{k}$, introduced μ as notation for certain combination of model parameters and employed the time–momentum representation [84]. Solution is sought in the form of plane waves

$$\boldsymbol{v} \equiv \boldsymbol{v}(t,\boldsymbol{x}) = \boldsymbol{v}(t)\mathrm{e}^{i\boldsymbol{k}\cdot\boldsymbol{x}}, \qquad \boldsymbol{b} \equiv \boldsymbol{b}(t,\boldsymbol{x}) = \boldsymbol{b}(t)\mathrm{e}^{i\boldsymbol{k}\cdot\boldsymbol{x}}. \tag{183}$$

For an inviscid medium ($\nu = 0$) without the exotic term ($\mu = 0$), we get a solution known as Alfvén waves for which

$$\boldsymbol{v}(t) \sim \boldsymbol{b}(t) \sim \mathrm{e}^{-i\omega t}, \quad \omega \equiv \gamma. \tag{184}$$

The solution with the exotic term is more complicated. To find solutions with the exotic term ($\mu \neq 0$), let us first find a convenient orthonormal basis of three vectors $\boldsymbol{e}_1, \boldsymbol{e}_2, \boldsymbol{e}_3$. The following choice

$$\boldsymbol{e}_1 \equiv \frac{\boldsymbol{k}}{|\boldsymbol{k}|}, \quad \boldsymbol{e}_2 \equiv \frac{\boldsymbol{e} - \boldsymbol{e}_1\cos\theta}{\sin\theta}, \quad \boldsymbol{e}_3 \equiv [\boldsymbol{e}_1 \times \boldsymbol{e}_2] = \frac{\boldsymbol{e}_1\times\boldsymbol{e}}{\sin\theta} \tag{185}$$

turns out to be case and qualitatively is depicted in Figure 8.

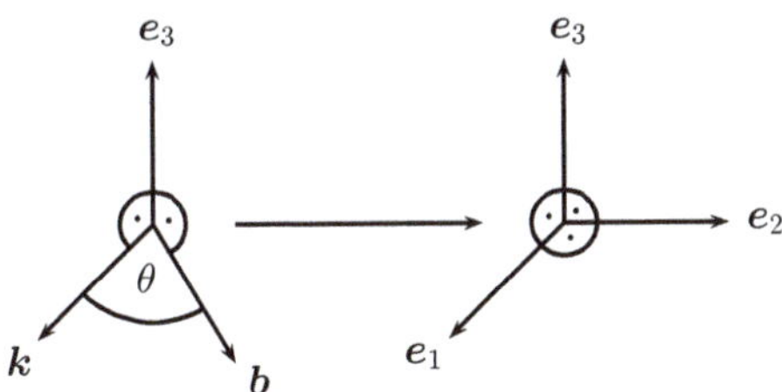

Figure 8. Construction of the proper orthonormal basis for a sought solution of linearized MHD equations.

The transverse velocity and magnetic vector fields are perpendicular to the wavevector $\boldsymbol{k}$ or, equivalently, to the basis vector $\boldsymbol{e}_1$. Let us decompose fields into perpendicular components $\boldsymbol{e}_2$ and $\boldsymbol{e}_3$ as follows

$$\boldsymbol{v}(t) = v_2(t)\boldsymbol{e}_2 + v_3(t)\boldsymbol{e}_3, \quad \boldsymbol{b}(t) = b_2(t)\boldsymbol{e}_2 + b_3(t)\boldsymbol{e}_3. \tag{186}$$

It is a straightforward exercise to check that modes $v_i, b_i; i = 2,3$ satisfy following equations

$$\partial_t v_2(t) = i\gamma b_2(t), \qquad \partial_t b_2(t) = i\gamma v_2(t), \tag{187}$$
$$\partial_t v_3(t) = i\gamma b_3(t), \qquad \partial_t b_3(t) = i\gamma v_3(t) + 2i\lambda b_2(t). \tag{188}$$

Solution of these equations can be represented in the form

$$\begin{aligned} cb_2(t) &= -v_2(t) = b_2 \mathrm{e}^{-i\gamma t}, \\ b_3(t) &= [b_3 + i\lambda b_2\, t]\mathrm{e}^{-i\gamma t}, \\ v_3(t) &= [-b_3 - \lambda/\gamma b_2 - i\lambda b_2\, t]\mathrm{e}^{-i\gamma t}. \end{aligned} \tag{189}$$

Thus, field-theoretic methods not only explain the emergence of turbulent dynamo, but also lead to an accompanying physical effect: appearance of disturbances in Alfvén waves perpendicular to the spontaneous field $\boldsymbol{B}$, which leads to their linear growth in time. This generates long-lived pulses $t\exp(-u\nu k^2 t)$—similar to Goldstone bosons.

An intriguing unsolved problem remains, however. Although gyrotropy is the cause of the dynamo effect, the value in Equation (180) calculated for the spontaneous magnetic field is independent of the parameter ρ. Thus, any however small gyrotropy causes emergence of a finite spontaneous field. It is thus not clear whether or not the dynamo will also occur in a normal fluid in which $\rho = 0$.

10. Effect of Strong Anisotropy

Another related problem that can be addressed is the effect of large-scale anisotropy on statistics of the velocity field, and the passively advected fields in the inertial range [34,36,85–99]. The classical Kolmogorov–Obukhov theory predicts that the anisotropy generated at large spatial scales by the forcing (boundary conditions, geometry of the mesh, etc.) fades away when the energy is transferred down to the smaller scales by means of the cascade mechanism [23,25]. This picture is corroborated in recent analyses for *even* correlation functions. Thus, they provide us with some backup to the hypothesis of the restored local isotropy of the turbulence for the velocity field and passively advected field in the inertial range [34,36,90–95,97–99]. More specifically, the exponents describing the scaling in the inertial range exhibit universality and hierarchy related to the state of anisotropy. In particular, the main contribution to an even function comes from the exponent from the isotropic shell [34,36,89,91–94,97–99]. However, the anisotropy survives in the inertial range being explicit in odd correlation functions. This is inconsistent with the behavior anticipated on grounds of the cascade picture. Further, the skewness factor decreases with the length scale much more slowly than expected [50,85–88,100–103], whereas the odd dimensionless ratios of structure functions of higher order (hyperskewness, etc.) increase. This hints at persistent anisotropy at small scales [34,36,90,92]. This appears a universal effect as it is observed for both the scalar [34,36] and vector [92] fields, advected by the Gaussian rapid-change velocity, as well as for the scalar advected by the two-dimensional velocity field generated by the stochastic Navier–Stokes equation [90].

Here, we demonstrate the anomalous scaling behavior of a passive scalar advected by the velocity field due to the Kraichnan model with strong anisotropy. In contrast with the studies in [34,36,50,85–88,103], where the velocity field was assumed isotropic and the anisotropy was generated at large scales by the imposed linear mean gradient, the uniaxial anisotropy in considered model is present at all scales. The anomalous exponents turn out to depend on the anisotropy parameters and are thus not universal.

The aim of this section is twofold. First, expressions for the structure functions and correlation functions of the scalar gradients are obtained and then the corresponding anomalous exponents are computed in the leading order of the ε expansion. Due to the anisotropy of the velocity flow, the composite operators of different ranks mix strongly with each other under renormalization. As a direct consequence, the corresponding anomalous exponents are given by the eigenvalues of matrices of generic structure (this is in contrast with the case of large-scale anisotropy, in which the matrices are diagonal or triangular). In terms of the zero-mode approach [93,94], this basically means that the $SO(d)$ decompositions of the correlation functions do not diagonalize differential operators in the corresponding exact equations.

As mentioned in Section 3, in the rapid-change model, the passive advection of a scalar field $\theta(x) \equiv \theta(t, \boldsymbol{x})$ is described by the stochastic equation (Equation (33)). The velocity $\boldsymbol{v}(x)$ correlator now instead of Equation (40) is assumed in the form

$$\langle v_i(x) v_j(x') \rangle = D_0 \frac{\delta(t-t')}{(2\pi)^d} \int \mathrm{d}^d k \, T_{ij}(\boldsymbol{k}) \, (k^2 + m^2)^{-d/2-\varepsilon/2} \exp[\mathrm{i}\boldsymbol{k} \cdot (\boldsymbol{x} - \boldsymbol{x}')], \tag{190}$$

where

$$\frac{D_0}{\nu_0} \equiv g_0 \equiv \Lambda^\varepsilon \tag{191}$$

and the relation $m = 1/L$ holds. In the isotropic case, the tensor quantity $T_{ij}(\boldsymbol{k})$ in Equation (190) is the usual transverse projector $T_{ij}(\boldsymbol{k}) = P_{ij}(\boldsymbol{k})$ (see Equation (18)). The velocity statistics is now assumed to be anisotropic in entire region of scales. The ordinary transverse projector is then replaced by the general transverse structure

$$T_{ij}(\boldsymbol{k}) = a(\psi) P_{ij}(\boldsymbol{k}) + b(\psi) \tilde{n}_i(\boldsymbol{k}) \tilde{n}_j(\boldsymbol{k}), \tag{192}$$

where the unit vector $\boldsymbol{n}$ denotes the singled out direction ($\boldsymbol{n}^2 = 1$),

$$\tilde{n}_i(\boldsymbol{k}) \equiv P_{ij}(\boldsymbol{k}) n_j, \tag{193}$$

and ψ is the angle between the vectors $\boldsymbol{k}$ and $\boldsymbol{n}$. In other words, $(\boldsymbol{n} \cdot \boldsymbol{k}) = k \cos\psi$ [note that $(\tilde{\boldsymbol{n}} \cdot \boldsymbol{k}) = 0$].

The positivity of the correlator in Equation (190) leads to the inequalities

$$a(\psi) > 0, \quad a(\psi) + b(\psi) \sin^2\psi > 0. \tag{194}$$

In practical calculations, one works with the special case

$$T_{ij}(\boldsymbol{k}) = \left[1 + \alpha_1 \frac{(\boldsymbol{n} \cdot \boldsymbol{k})^2}{k^2}\right] P_{ij}(\boldsymbol{k}) + \alpha_2 \tilde{n}_i(\boldsymbol{k}) \tilde{n}_j(\boldsymbol{k}). \tag{195}$$

Then, the inequalities in Equation (194) reduce to $\alpha_{1,2} > -1$.

Let us note that the quantities in Equations (192) and (195) are invariant with respect to transformation $\boldsymbol{n} \to -\boldsymbol{n}$. The anisotropy permits an introduction of the mixed correlator $\langle \boldsymbol{v} f \rangle \propto \boldsymbol{n}\delta(t-t') \, C'(r/L)$ with a function $C'(r/L)$ similar to $C(r/L)$ in Equation (34). This violates the evenness in $\boldsymbol{n}$ and gives rise to non-vanishing odd functions S_{2n+1}. However, this does not alter RG analysis [104].

The stochastic problem to be analyzed is tantamount to the field theoretic model of the set of fields $\phi \equiv \{\theta', \theta, \boldsymbol{v}\}$ with action

$$\mathcal{S}[\phi] = \frac{1}{2}\theta' D_\theta \theta' + \theta' \left[-\partial_t - (\boldsymbol{v} \cdot \boldsymbol{\nabla}) + \nu_0 \boldsymbol{\nabla}^2 + \frac{1}{2} D_{vij}(0) \partial_i \partial_j\right] \theta - \frac{1}{2} \boldsymbol{v} D_v^{-1} \boldsymbol{v}. \tag{196}$$

Here, D_θ and D_v are the correlators in Equations (34) and (190), respectively, and

$$D_{vij}(0) = D_0 \int \frac{\mathrm{d}^d q}{(2\pi)^d} \frac{T_{ij}(\boldsymbol{q})}{(q^2 + m^2)^{d/2+\varepsilon/2}} \tag{197}$$

is the diagonal term (in spatial variables) of the coefficient of the temporal δ function in the velocity pair correlation function in Equation (190)).

The model in Equatio (196) corresponds to the usual Feynman diagrammatic technique with the triple vertex in Equation (51), propagators in Equation (50) and

$$\langle \theta\theta \rangle_0 = C(k) \, (\omega^2 + \nu_0^2 k^4)^{-1}, \quad \langle \theta'\theta' \rangle_0 = 0, \tag{198}$$

where $C(k)$ is the Fourier transform of the function $C(r/L)$ in Equation 34). The bare propagator $\langle \boldsymbol{v}\boldsymbol{v}\rangle_0 \equiv \langle \boldsymbol{v}\boldsymbol{v}\rangle$ is defined by Equation (190) with the transverse projection operator from Equation (192) or Equation (195).

The pair correlation functions $\langle\phi\phi\rangle$ of the multicomponent field $\phi \equiv \{\theta', \theta, \boldsymbol{v}\}$ fulfill the Dyson equation. In terms of component fields, we arrive at the system of two equations (cf. [25])

$$G^{-1}(\omega, \boldsymbol{k}) = -\mathrm{i}\omega + \nu_0 k^2 - \Sigma_{\theta'\theta}(\omega, \boldsymbol{k}), \tag{199}$$

$$D(\omega, \boldsymbol{k}) = |G(\omega, \boldsymbol{k})|^2 \left[C(k) + \Sigma_{\theta'\theta'}(\omega, \boldsymbol{k})\right], \tag{200}$$

where $G(\omega, \boldsymbol{k}) \equiv \langle\theta\theta'\rangle$ and $D(\omega, \boldsymbol{k}) \equiv \langle\theta\theta\rangle$ are the exact response function and pair correlation function, respectively, and $\Sigma_{\theta'\theta}$ and $\Sigma_{\theta'\theta'}$ are self-energy operators represented by 1PI graphs. The rest of the self-energy functions $\Sigma_{\phi\phi}$ in the model in Equation (196) vanish identically.

It is a characteristic feature of models such as that in Equation (196) that all the skeleton multi-loop diagrams of self-energy functions $\Sigma_{\theta'\theta}$, $\Sigma_{\theta'\theta'}$ contain closed circuits of retarded propagators $\langle\theta\theta'\rangle$ (it is important that the propagator $\langle \boldsymbol{v}\boldsymbol{v}\rangle_0$ in Equation (190) is δ correlated in time) and hence vanish.

In the presence of anisotropy, a new counterterm of the form $\theta'(\boldsymbol{n}\cdot\boldsymbol{\nabla})^2\theta$ has to be introduced. There is no such term in the unrenormalized action functional in Equation (196). Thus, the model in Equation (196) in its original form is not multiplicatively renormalizable, and to employ RG techniques we have to extend the model by adding a contribution corresponding to the counterterm to the bare action

$$\mathcal{S}[\Phi] = \theta' D_\theta \theta'/2 + \theta' \left[-\partial_t - (\boldsymbol{v}\cdot\boldsymbol{\nabla}) + \nu_0 \boldsymbol{\nabla}^2 + \chi_0 \nu_0 (\boldsymbol{n}\cdot\boldsymbol{\nabla})^2\right]\theta - \boldsymbol{v} D_v^{-1} \boldsymbol{v}/2. \tag{201}$$

Here, χ_0 is a new dimensionless parameter. For stability of the system, positivity of the viscous contribution $\nu_0 k^2 + \chi_0 \nu_0 (\boldsymbol{n}\cdot\boldsymbol{k})^2$ is required, which leads to the inequality $\chi_0 > -1$. The real physical value of the new parameter χ_0 is zero. However, this does not hinder the application of the RG techniques. It is first assumed to be arbitrary, and the equality $\chi_0 = 0$ is imposed as the initial condition in the solution of equations for invariant variables. The vanishing χ_0 corresponds to some nonzero value of its renormalized counterpart, which can be explicitly calculated.

For the action in Equation (201), the nontrivial bare propagators in Equation (198) are replaced by

$$\langle\theta\theta'\rangle_0 = \langle\theta'\theta\rangle_0^* = \frac{1}{-\mathrm{i}\omega + \nu_0 k^2 + \chi_0 \nu_0 (\boldsymbol{n}\cdot\boldsymbol{k})^2}, \tag{202}$$

$$\langle\theta\theta\rangle_0 = \frac{C(k)}{|-\mathrm{i}\omega + \nu_0 k^2 + \chi_0 \nu_0 (\boldsymbol{n}\cdot\boldsymbol{k})^2|^2}. \tag{203}$$

Once properly extended, the model becomes multiplicatively renormalizable: due to generation of counterterms, two independent renormalization constants $Z_{1,2}$ are introduced as coefficients of the counterterms. This yields the renormalized action in the form

$$\mathcal{S}_R[\Phi] = \theta' D_\theta \theta'/2 + \theta' \left[-\partial_t - (\boldsymbol{v}\cdot\boldsymbol{\nabla}) + \nu Z_1 \boldsymbol{\nabla}^2 + \chi\nu Z_2 (\boldsymbol{n}\cdot\boldsymbol{\nabla})^2\right]\theta - \boldsymbol{v} D_v^{-1} \boldsymbol{v}/2, \tag{204}$$

or to the multiplicative renormalization of all the parameters ν_0, g_0 and χ_0 of the action in Equation (201):

$$\nu_0 = \nu Z_\nu, \quad g_0 = g\mu^\varepsilon Z_g, \quad \chi_0 = \chi Z_\chi. \tag{205}$$

The correlator (Equation (190)) in Equation (204) is expressed in terms of renormalized variables using Equations (205). From direct comparison of Equations (201), (204), and (205), we obtain the relations

$$Z_1 = Z_\nu, \quad Z_2 = Z_\chi Z_\nu, \quad Z_g = Z_\nu^{-1}. \tag{206}$$

The beta functions are given by

$$\beta_g(g,\alpha) \equiv \widetilde{\mathcal{D}}_\mu g = g\,(-\varepsilon - \gamma_g) = g\,(-\varepsilon + \gamma_\nu) = g\,(-\varepsilon + \gamma_1), \tag{207}$$

$$\beta_\chi(g,\chi) \equiv \widetilde{\mathcal{D}}_\mu \chi = -\chi\gamma_\chi = \chi(\gamma_1 - \gamma_2). \tag{208}$$

The relation between β_g and γ_ν in Equation (207) is a consequence of the definitions and the last equality in Equation (206).

One-loop calculation [104] leads to the following expressions for the anomalous dimension $\gamma_1(g)$

$$\gamma_1(g) = \frac{g\bar{S}_d}{2d(d+2)}\Big[(d-1)(d+2) + \alpha_1(d+1) + \alpha_2\Big], \tag{209}$$

and for $\gamma_2(g,\alpha)$

$$\gamma_2(g,\alpha) = \frac{g\bar{S}_d}{2d(d+2)\chi}[-2\alpha_1 + \alpha_2(d^2-2)], \tag{210}$$

respectively. Let us note that Equations (209)–(210) are exact [104].

From explicit Equations (209) and (210), we see that the RG equations have just one IR stable fixed point

$$g^*\bar{S}_d = \frac{2d(d+2)\varepsilon}{(d-1)(d+2)+\alpha_1(d+1)+\alpha_2}, \quad \chi^* = \frac{-2\alpha_1 + \alpha_2(d^2-2)}{(d-1)(d+2)+\alpha_1(d+1)+\alpha_2}. \tag{211}$$

At this point, both eigenvalues of the matrix Ω are equal to ε; the values $\gamma_1^* = \gamma_2^* = \gamma_\nu^* = \varepsilon$ are exact as a consequence of Equations (207)–(208) [here and below, $\gamma_F^* \equiv \gamma_F(g^*,\chi^*)$]. The fixed point in Equation (211) is degenerate. By this we mean that its coordinates depend explicitly on the anisotropy parameters $\alpha_{1,2}$.

Let us consider the solution of the RG equation in the example of the even two-time structure functions

$$S_{2n}(\boldsymbol{r},\tau) \equiv \langle[\theta(t,\boldsymbol{x}) - \theta(t',\boldsymbol{x}')]^{2n}\rangle, \quad \boldsymbol{r} \equiv \boldsymbol{x} - \boldsymbol{x}', \quad \tau \equiv t - t'. \tag{212}$$

They obey the RG equation $\mathcal{D}_{RG}S_{2n} = 0$ with the differential operator $\mathcal{D}_{RG} = \mathcal{D}_\mu + \beta_g\partial_g + \beta_\chi\partial_\chi - \gamma_{nu}\mathcal{D}_\nu$.

In renormalized variables, a dimensional argument leads to

$$S_{2n}(\boldsymbol{r},\tau) = \nu^{-n}r^{2n}\tilde{R}_{2n}(\mu r, \tau\nu/r^2, r/L, g, \chi). \tag{213}$$

Here, $\tilde{R}_{2n}$ is a scaling function of dimensionless variables (dependence on d, ε, $\alpha_{1,2}$ and the angle between the vectors $\boldsymbol{r}$ and $\boldsymbol{n}$ is implied). From the RG equation, the representation

$$S_{2n}(\boldsymbol{r},\tau) = (\bar{\nu})^{-n}r^{2n}\tilde{R}_{2n}(1, \tau\bar{\nu}/r^2, r/L, \bar{g}, \bar{\chi}), \tag{214}$$

is easily derived in terms of the invariant variables $\bar{e} = \bar{e}(\mu r, e)$. The identity $\bar{L} \equiv L$ is a consequence of the fact that L is not renormalized. The relation between the bare and invariant variables takes the form

$$\nu_0 = \bar{\nu}Z_\nu(\bar{g}), \quad g_0 = \bar{g}r^{-\varepsilon}Z_g(\bar{g}), \quad \chi_0 = \bar{\chi}Z_\chi(\bar{g},\bar{\chi}). \tag{215}$$

Equation (215) defines the invariant variables as functions of the bare parameters in an implicit form. It is valid because expressions on both sides satisfy the same RG equation, and at $\mu r = 1$ Equation (215) coincides with Equation (205).

The asymptotic behavior at large μr of the invariant variables is governed by the IR stable fixed point: $\bar{g} \to g^*$, $\bar{\chi} \to \chi^*$ for $\mu r \to \infty$. However, in multi-charge problems, the possibility must be considered that, even when the stable IR point exists, not every phase trajectory approaches it in the limit $\mu r \to \infty$. The phase trajectory may first pass outside the natural region of stability (physical region is given by the inequalities $g > 0$, $\chi > -1$) or go to infinity within this region. It can be easily verified that the RG flow reaches the fixed point in Equation (211) for arbitrary initial conditions $g_0 > 0$, $\chi_0 > -1$, including the physical case $\chi_0 = 0$. Moreover, the large μr behavior of the invariant viscosity $\bar{\nu}$ can be extracted explicitly from Equation (215) together with the last relation in Equation (206): $\bar{\nu} = D_0 r^{\varepsilon}/\bar{g} \to D_0 r^{\varepsilon}/g^*$ (we remind the reader that $D_0 = g_0 \nu_0$). Then, for $\mu r \to \infty$ and any fixed mr, we get

$$S_{2n}(\boldsymbol{r},\tau) = D_0^{-n} r^{n(2-\varepsilon)} g^{*n}\, R_{2n}(\tau D_0 r^{\Delta_t}, r/L), \tag{216}$$

where the scaling function

$$R_{2n}(D_0 \tau r^{\Delta_t}, r/L) \equiv \tilde{R}_{2n}(1, D_0 \tau r^{\Delta_t}, r/L, g_*, \alpha_*), \tag{217}$$

has appeared and $\Delta_t \equiv -2 + \gamma_\nu^* = -2 + \varepsilon$ is the critical dimension of time. For the equal-time structure function, we have

$$S_{2n}(\boldsymbol{r}) = D_0^{-n} r^{n(2-\varepsilon)} g^{*n}\, R_{2n}(r/L). \tag{218}$$

Here, the definition of R_{2n} is evident from Equation (217). It should be noted that Equations (216)–(218) comprise a proof of the independence of the structure functions in the IR range (large μr and any r/L) of the viscosity coefficient or, in other words, of the UV scale: the parameters g_0 and ν_0 appear in Equation (216) only in the combination $D_0 = g_0 \nu_0$. A similar property has been found for the stochastic Navier–Stokes equation [105].

In contrast to the previously mentioned models, the scaling function $\tilde{R}$ in Equation (214) contains two different scales—corresponding to spatial and time differences, respectively. The information about its behavior can again be obtained using the OPE method from Section 5. Therefore, let $F(r,\tau)$ stands for some multiplicatively renormalized quantity. Dimensionality considerations imply

$$F_R(r,\tau) = \nu^{d_F^\omega} r^{-d_F} \tilde{R}_F(\mu r, \tau\nu/r^2, r/L, g, \alpha), \tag{219}$$

where d_F^ω and d_F are the canonical dimension in frequency and the total canonical dimension of F, respectively (see Section 4 or [5]), and R_F is a function of completely dimensionless variables. The analog of Equation (214) has the form

$$F(r,\tau) = Z_F(g,\alpha) F_R = Z_F(\bar{g},\bar{\alpha})\,(\bar{\nu})^{d_F^\omega} r^{-d_F} \tilde{R}_F(1, \tau\bar{\nu}/r^2, r/L, \bar{g}, \bar{\alpha}). \tag{220}$$

In the large μr limit, one has $Z_F(\bar{g},\bar{\alpha}) \simeq \text{const}\,(\Lambda r)^{-\gamma_F^*}$ [106]. The UV scale comes into this relation from Equation (191). In the IR range ($\Lambda r \sim \mu r$ large, r/L arbitrary), Equation (220) takes on the form

$$F(r,\tau) \simeq \text{const}\,\Lambda^{-\gamma_F^*} D_0^{d_F^\omega}\, r^{-\Delta[F]} R_F(D_0 \tau r^{\Delta_t}, r/L). \tag{221}$$

Here,

$$\Delta[F] \equiv \Delta_F = d_F^k - \Delta_t d_F^\omega + \gamma_F^*, \quad \Delta_t = -2 + \varepsilon \tag{222}$$

is the critical dimension of the function F. The scaling function R_F is related to $\tilde{R}_F$ as in Equation (217). For nonvanishing γ_F^*, the function F in the IR range exhibits dependence on Λ or, equivalently, on ν_0.

As a detailed analysis reveals [104], the operator θ^N requires no counterterms at all, i.e., it is in fact UV finite, $\theta^N = Z\,[\theta^N]^R$ with $Z = 1$. As a straightforward consequence, the critical dimension of $\theta^N(x)$ is given by Equation (222) with no correction from γ_F^* and reduces to the sum of the critical dimensions of the factors:

$$\Delta[\theta^N] = N\Delta[\theta] = N(-1 + \varepsilon/2). \tag{223}$$

The structure functions in Equation (212) are linear combinations of pair correlators involving the operators θ^N, therefore Equation (223) shows that they indeed satisfy the RG equation of the form in Equation (65), as discussed in Section 4.

Tensor composite operators $\partial_{i_1}\theta \cdots \partial_{i_p}\theta\,(\partial_i\theta\partial_i\theta)^n$ consisting solely of scalar gradients play an an important role. It is technically convenient to work with the scalar operators obtained by contracting the tensors with the vectors $\boldsymbol{n}$ in the fashion

$$F[N,p] \equiv [(\boldsymbol{n}\cdot\boldsymbol{\nabla})\theta]^p(\partial_i\theta\partial_i\theta)^n, \quad N \equiv 2n+p. \tag{224}$$

Canonical dimensions of these operators depend on the total number of the fields θ and have the following form $d_F = 0$, $d_F^\omega = -N$. These operators (Equation (224)) mix with each other only in the renormalization procedure. As a consequence, the corresponding infinite renormalization matrix

$$F[N,p] = \sum_{N',p'} Z_{[N,p]\,[N',p']}\,F^R[N',p'] \tag{225}$$

is block-triangular, i.e., $Z_{[N,p]\,[N',p']} = 0$ for $N' > N$. It is then clear that the critical dimensions associated with the operators $F[N,p]$ are fully determined by the eigenvalues of the finite subblocks with $N' = N$.

In the isotropic case, as well as in the presence of large-scale anisotropy, the elements $Z_{[N,p]\,[N,p']}$ vanish for $p < p'$, and the block $Z_{[N,p]\,[N,p']}$ is triangular together with the related blocks of the matrices U_F and Δ_F in Equations (77) and (222). Moreover, it can be diagonalized in terms of to irreducible operators (scalars, vectors, and traceless tensors), but even for nonvanishing imposed gradient its eigenvalues coincide with those of the isotropic case. Hence, the introduction of large-scale anisotropy has no effect on critical dimensions of operators in Equation (224) (see [34,36]). In the case of small-scale anisotropy, the operators with different values of p mix in renormalization resulting in the matrix $Z_{[N,p]\,[N,p']}$ is of generic form (neither diagonal nor triangula).

The calculation of the renormalization constants $Z_{[N,p]\,[N,p']}$ can be illustrated within the one-loop approximation. Let $\Gamma(x;\theta)$ be the generating functional of the 1PI Green functions with the insertion of just one composite operator $F[N,p]$ of Equation (224) containing any number of fields θ. Here, x is the argument of the composite operator and θ is the functional argument, the "classical counterpart" of the random field θ. The general interest is in the Nth term of the expansion of $\Gamma(x;\theta)$ in θ, which is denoted as $\Gamma_N(x;\theta)$; it has the form

$$\Gamma_N(x;\theta) = \frac{1}{N!}\int dx_1 \cdots \int dx_N\,\theta(x_1)\cdots\theta(x_N)\,\langle F[N,p](x)\theta(x_1)\cdots\theta(x_N)\rangle_{1-\text{ir}}. \tag{226}$$

The matrix of critical dimensions in Equation (222) is given in the one-loop approximation by the expression

$$\Delta_{[N,p][N,p']} = N\varepsilon/2 + \gamma^*_{[N,p][N,p']}, \tag{227}$$

where the asterisk implies the substitution in Equation (211). Details of calculation of $\gamma_{[N,p][N,p']}$ can be found in [104].

As already mentioned, the critical dimensions are determined by the eigenvalues of the matrix in Equation (227). It is readily checked that, in the isotropic case ($\alpha_{1,2} = 0$), elements of the matrix with $p' > p$ vanish, it becomes triangular, and its eigenvalues are the diagonal elements $\Delta[N,p] \equiv \Delta_{[N,p][N,p]}$:

$$\Delta[N,p] = N\varepsilon/2 + \frac{2p(p-1) - (d-1)(N-p)(d+N+p)}{2(d-1)(d+2)}\,\varepsilon + O(\varepsilon^2). \tag{228}$$

From this equation, we observe that, for fixed N and arbitrary $d \geq 2$, the dimension $\Delta[N,p]$ monotonically decreases with p. It reaches a minimum at the minimal allowed value of $p = p_N$, i.e., $p_N = 0$ if N is even and $p_N = 1$ if N is odd:

$$\Delta[N,p] > \Delta[N,p'] \quad \text{if} \quad p > p'. \tag{229}$$

This minimal value $\Delta[N, p_N]$ decreases monotonically as N increases separately for even and odd values of N, viz.

$$0 \geq \Delta[2n, 0] > \Delta[2n+2, 0], \quad \Delta[2n+1, 1] > \Delta[2n+3, 1]. \tag{230}$$

A similar hierarchy is demonstrated by the critical dimensions of certain tensor operators in the stirred Navier–Stokes turbulence (see Ref. [107] and Section 2.3 of [31]). However, no clear hierarchy is demonstrated by neighboring even and odd dimensions: relations

$$\Delta[2n+1, 1] - \Delta[2n, 0] = \frac{\varepsilon(d+2-4n)}{2(d+2)}, \quad \Delta[2n+2, 0] - \Delta[2n+1, 1] = \frac{\varepsilon(2-d)}{2(d+2)} \tag{231}$$

lead to inequality $\Delta[2n+1, 1] > \Delta[2n+2, 0]$ for any $d > 2$. On the other hand, the relation $\Delta[2n, 0] > \Delta[2n+1, 1]$ is satisfied only if n is large enough, $n > (d+2)/4$.

Let us denote $\Delta[N, p]$ the eigenvalue of the matrix in Equation (227), which coincides with Equation (228) for $\alpha_{1,2} = 0$. Because the eigenvalues depend continuously on parameters $\alpha_{1,2}$, at least for small enough values of $\alpha_{1,2}$, this notation is unambiguous.

The dimension $\Delta[2, 0]$ vanishes identically for any $\alpha_{1,2}$ in all orders in ε. As in the isotropic model, this is proved employing the Schwinger equation

$$\int \mathcal{D}\Phi \frac{\delta}{\delta\theta'(x)} \left[\theta(x) e^{\mathcal{S}_R[\Phi] + A\Phi}\right] = 0. \tag{232}$$

It can be easily checked [104] that $\Delta[2, 0] \equiv 0$, while the to the leading order in ε one obtains

$$\begin{aligned}\Delta[2,2]/\varepsilon = 2 + \Big\{&-(d-2)d(d+2)(d+4)F_0^* - (d+2)(d+4)(2+(d-2)\alpha_1 \\ &+ d\alpha_2)F_1^* + +3(d+4)(d-2\alpha_1+2d\alpha_2)F_2^* + 15d(\alpha_1-\alpha_2)F_3^*\Big\} \Big/ \Big\{(d-1) \\ &\times (d+4)[(d-1)(d+2) + (d+1)\alpha_1 + \alpha_2]\Big\},\end{aligned} \tag{233}$$

where $F_n^* \equiv F(1, 1/2+n; d/2+n; -\alpha_*)$ with α_* from Equation (211).

For $N > 2$, the eigenvalues may be found in closed form only within an expansion in $\alpha_{1,2}$. The explicit expressions can be found in [104]. They illustrate two features which appear to hold for all N:

- The most relevant anisotropy correction is of order $(\alpha_{1,2})$ for $p \neq 0$ and $(\alpha_{1,2}^2)$ for $p = 0$. This means that $\gamma^*[N, 0]$ are anisotropy independent in the *linear* approximation.
- This leading contribution depends on $\alpha_{1,2}$ only through the combination $\alpha_3 \equiv 2\alpha_1 + d\alpha_2$.

This conjecture is supported by the following expressions for $N = 6, 8$ and $p = 0$:

$$\gamma^*[6,0]/\varepsilon = \frac{-2(d+6)}{(d+2)} - \frac{12(d-2)^2(d+1)(d^2+14d+48)\alpha_3^2}{(d-1)^2 d(d+2)^4(d+4)^2}, \tag{234}$$

$$\gamma^*[8,0]/\varepsilon = \frac{-4(d+8)}{(d+2)} - \frac{24(d-2)^2(d+1)(d^2+18d+80)\alpha_3^2}{(d-1)^2 d(d+2)^4(d+4)^2}. \tag{235}$$

Using the operator product expansion (Section 5), we infer for the scaling function $R(r/L)$ in the representation in Equation (218) for the correlator $\langle F_1(x)F_2(x')\rangle$: the expression

$$R(r/L) = \sum_F A_F \left(\frac{r}{L}\right)^{\Delta_F}, \quad r \ll L, \tag{236}$$

where the coefficients A_F are regular in $(r/L)^2$.

Now, let us discuss the equal-time structure functions S_N in Equation (52). From this, it is assumed that the mixed correlator $\langle \boldsymbol{v} f \rangle$ differs from zero; this does not influence the critical dimensions. However, it gives rise to non-vanishing odd structure functions. As a rule, the operators entering into the OPE are those generated by Taylor expansions together with all admissible operators which admix to them in the renormalization procedure [5,6]. The leading term of the Taylor expansion of S_N is the operator $F[N, N]$ from Equation (224). Renormalization brings about all the operators $F[N', p]$ with $N' \leq N$ and all allowed values of p. The operators with $N' > N$ (whose contributions would be more relevant) are absent in Equation (236), because they do not appear in the Taylor expansion of S_N and do not admix to the terms therein. Hence, the RG representation in Equation (216) together with the OPE representation in Equation (236) leads in the inertial range to the asymptotic expression for the structure function:

$$S_N(\boldsymbol{r}) = D_0^{-N/2} r^{N(1-\varepsilon/2)} \sum_{N' \leq N} \sum_{p} \Big\{ C_{N',p} \, (r/L)^{\Delta[N',p]} + \cdots \Big\}. \tag{237}$$

The second sum is taken over all values of p, allowed for a given N', numerical coefficients $C_{N',p}$ depend on ε, d, $\alpha_{1,2}$ and the angle ϑ between $\boldsymbol{r}$ and $\boldsymbol{n}$. The ellipsis denotes the contributions of the operators other than $F[N, p]$, e.g., $\partial^2\theta\partial^2\theta$. They generate terms of order $(r/L)^{2+(\varepsilon)}$ and higher and are discarded in the following.

A few general remarks are in order:

- If the mixed correlator $\langle \boldsymbol{v} f \rangle$ is missing, the odd structure functions vanish. At the same time, the contributions to even functions are given only by the operators with even values of N'. Only contributions with $p = 0$ remain in the isotropic case ($\alpha_{1,2} = 0$) [104,108]. When anisotropy is present, $\alpha_{1,2} \neq 0$, the operators with $p \neq 0$ assume nonvanishing means, and their dimensions $\Delta[N', p]$ enter the right hand side of Equation (237).
- The most relevant term of the small r/L behavior is obviously related to the contribution with the smallest value of the exponent $\Delta[N', p]$. Now, we employ the hierarchy relations in Equations (229) and (230), which hold for $\alpha_{1,2} = 0$ and consequently remain valid at least for $\alpha_{1,2} \ll 1$. We conclude that, for sufficiently weak anisotropy, the leading term in (237) is given by the dimension $\Delta[N, 0]$ for any S_N. In all particular cases considered, this hierarchy remains for finite values of the anisotropy parameters as well. The contribution with $\Delta[N, 0]$ stays leading for such N and $\alpha_{1,2}$.
- It is possible that the inequalities in Equations (229) and (230) may be violated for some values of n, d and $\alpha_{1,2}$, so that the leading contribution to Equation (237) is given by a dimension with $N' \neq N$ and/or $p > 0$.
- The introduction of the mixed correlator $\langle \boldsymbol{v} f \rangle \propto \boldsymbol{n}\delta(t - t')\, C'(r/L)$ explicitly violates the evenness in $\boldsymbol{n}$ and generates non-vanishing odd functions S_{2n+1} and leads to to the contributions with odd N' to the expansion in Equation (237) for even functions. If the relations in Equations (229) and (230) are satisfied, in even functions, the leading term is still given by the contribution with $\Delta[N, 0]$. If the relations in Equation (231) are fulfilled, in the odd function S_{2n+1}, the leading term is given by the dimension $\Delta[2n, 0]$ for $n < (d+2)/4$ and by $\Delta[2n+1, 1]$ for $n > (d+2)/4$. Let us note that, for the model with an imposed gradient, for all n, the leading terms of S_{2n+1} are given by the dimensions $\Delta[2n+1, 1]$ [34,36].

Furthermore, it is permissible that the matrix in Equation (227) for some $\alpha_{1,2}$ has a pair of complex conjugate eigenvalues, Δ and Δ^*. In this case, the small r/L behavior of the scaling function $\xi(r/L)$ in Equation (237) would contain oscillating terms of the form

$$(r/L)^{\mathrm{Re}\,\Delta} \Big\{ C_1 \cos\big[\mathrm{Im}\,\Delta\,(r/L)\big] + C_2 \sin\big[\mathrm{Im}\,\Delta\,(r/L)\big] \Big\},$$

with constant factors C_i.

Representations similar to Equations (218) and (237) can be readily written for arbitrary equal-time pair correlation functions, provided their canonical and critical dimensions are known. In particular, for the operators $F[N, p]$ in the IR region ($\Lambda r \to \infty$, r/L fixed), we arrive at

$$\langle F[N_1, p_1] F[N_2, p_2] \rangle = \nu_0^{-(N_1+N_2)/2} \sum_{N,p} \sum_{N',p'} (\Lambda r)^{-\Delta_{[N,p]} - \Delta_{[N',p']}} R_{N,p;N',p'}(r/L). \tag{238}$$

Here, the indices N and N' satisfy the inequalities $N \leq N_1$ and $N' \leq N_2$. The indices p and p' assume all allowed values for given N and N'. The small r/L behavior of the functions $R_{N,p;N',p'}(r/L)$ is of the form

$$\xi_{N,p;N',p'}(mr) = \sum_{N'',p''} C_{N'',p''} (r/L)^{\Delta[N'',p'']}, \tag{239}$$

with the constraint $N'' \leq N + N'$ and corresponding values of p''; $C_{N'',p''}$ are numerical coefficients.

Thus far, we have considered the particular case of the velocity correlator given by Equations (190) and (195). Explicit calculations [104] show that contributions to the renormalization constants (as a direct consequence, to the coordinates of the fixed point and the anomalous dimensions as well) come solely from the even polynomials in the expansion in Equation (192). This is why odd polynomials were absent in Equation (192) from the very beginning. Further, from the explicit form [104] of self-energy graph $\Sigma_{\theta'\theta}$, it can be shown that only the coefficients a_l with $l = 0, 1$ and b_l with $l = 0, 1, 2$ contribute to the constants $Z_{1,2}$ in Equation (204) and, consequently, to the basic RG functions in Equations (207) as well as Equation (208) to the coordinates of the fixed point in Equation (211). Hence, the fixed point in the generic model in Equation (192) is fully parameterized by these five coefficients. The higher coefficients appear only through the positivity conditions in Equation (194).

Further, for $\chi = 0$, only coefficients a_l with $l \leq 2$ and b_l with $l \leq 3$ might contribute to the integrals H_n and, therefore, to the one-loop critical dimensions in Equation (227). As a consequence, calculation of the latter is significantly simplified in the special case $a_0 = 1$, $a_1 = 0$ and $b_l = 0$ for $l \leq 2$ in Equation (192). The coordinates of the fixed point in Equation (211) coincide with those in the isotropic model. In particular, $\alpha_* = 0$, and the anomalous exponents will depend on the two parameters a_2 and b_3 solely. In all analyzed cases, the general picture is akin to that outlined above for the case in Equation (195). For instance, the hierarchy of the critical dimensions, stated in the inequalities in Equations (229)–(231), persists also to this case. We conclude that the special case in Equation (195) represents adequately the main features of the generic model in Equation (192).

The exponents are related to the critical dimensions of composite operators in Equation (224) constructed from the scalar gradients. In contrast to the isotropic flow, these operators in the model under consideration mix in renormalization in such a way that the matrices of their critical dimensions are neither of diagonal nor triangular form. These matrices are calculated explicitly to the order (ε) However, their eigenvalues (anomalous exponents) can be calculated only as perturbative series in $\alpha_{1,2}$ (Equations (234) and (235)) or in a numerical fashion.

In the limit of zero anisotropy, the exponents can be related to definite tensor composite operators constructed from the scalar gradients, and exhibit certain hierarchy connected to the degree of anisotropy. It can be summarized as follows: the lower is the rank, the lower is the dimension and, therefore, the contribution to the inertial-range behavior is more important.

Leading terms of the even (odd) structure functions are caused by the scalar (vector) operators. For the case of finite anisotropy, the exponents cannot be related to individual operators (which are mixed in renormalization procedure), but the aforementioned hierarchy is present for all the studied cases.

A short comment about the second-order structure function $S_2(\boldsymbol{r})$ is appropriate here. It can be studied employing the RG and zero-mode techniques [104]; as in the isotropic case [39,109–111]. Its leading term is $S_2 \propto r^{2-\varepsilon}$, but the amplitude now depends on $\alpha_{1,2}$ and the angle between the vectors $\boldsymbol{r}$ and $\boldsymbol{n}$ from Equation (195). The first correction due to anisotropy is $(r/L)^{\Delta[2,2]}$ with the exponent $\Delta[2,2] = O(\varepsilon)$ from Equation (233).

In isotropic velocity field, the anisotropy inserted at large scales by external forcing or imposed mean gradient, persists in the inertial range and shows in *odd* correlation functions: the skewness factor $S_3/S_2^{3/2}$ decreases at $r/L \to 0$ but slowly (see [50,85–88,100–103]), while the higher-order ratios $S_{2n+1}/S_2^{n+1/2}$ increase (see, e.g., [34,36,90,92]).

In the studied model, the inertial-range behavior of the skewness factor is described by the expression $S_3/S_2^{3/2} \propto (r/L)^{\Delta[3,1]}$. For $\alpha_{1,2} \to 0$, the exponent $\Delta[3,1]$ is given by Equation (228) with $n = 3$ and $p = 1$. It is positive and coincides with the previous result [85,86]. By numerical means, it can be shown that, if the anisotropy becomes strong enough, the exponent $\Delta[3,1]$ becomes negative and the skewness factor increases towards the depth of the inertial interval. Let us note that the higher-order odd ratios increase already when the anisotropy is weak.

11. Conclusions

The crucial problem in many phenomena encountered in physics is the proper identification of underlying symmetry and the mechanism that leads to its violation. The solution to such a task is by no means obvious and requires deep understanding of the physical problem. In non-equilibrium statistical physics, which deals with systems containing many interacting degrees of freedom, symmetries are present at different levels of theoretical description.

In this article, our main aim is to summarize and elucidate approaches suitable for theoretical analysis of symmetries and their violations in problems related to various aspects of fluid dynamics. In particular, we concentrate on an overall analysis of symmetries emerging in fully developed turbulence and related problems. All studied problems are of classical nature and share the common property that they cannot be described within the standard equilibrium statistical physics and thus constitute systems far from the thermal equilibrium. Nevertheless, it is possible to study them with the use of powerful and versatile methods borrowed from high energy physics. We demonstrate how to construct a functional-integral representation of classical stochastic problems based on the introduction of an action functional analogous to that of particle physics. Use of the action functional is especially illuminating in revealing symmetries of models and their consequences. The functional representation is analyzed using Feynman diagrammatic technique and renormalization group approach. The latter is especially convenient for an analysis of large-scale behavior. To this end, it is necessary to find R stable fixed point(s) of the RG β functions of the model at hand. The fundamental difference between the RG in high-energy physics and statistical systems is that only the opposite spatial scales are of interest. In high-energy physics, small spatial scales are studied and the asymptotic behavior of models at large momenta is vital. In classical statistical systems, there is a maximal spatial scale L. The goal of the theory is then to determine how much L affects relevant physical quantities. We apply field-theoretic methods to classical non-equilibrium systems and as a paradigmatic example present the stochastic approach to fully developed turbulence. We review the basic setup, and briefly describe the RG method and the related topic of operator product expansion. Examples are given of the use of the latter as an important tool in the analysis of experimentally measurable quantities. The important Galilei symmetry is described in the functional representation and its consequences derived. Consequences of breaking of Galilei symmetry in the stochastic Navier–Stokes model with colored noise are discussed with the use of the RG. Important effects of parity violation and strong anisotropy in turbulent flows are analyzed in detail.

Author Contributions: All the authors contributed equally to all the parts of this work. Funding acquisition M.H.

Funding: The work was supported by VEGA grant No. 1/0345/17 of the Ministry of Education, Science, Research and Sport of the Slovak Republic, and the grant of the Slovak Research and Development Agency under the contract No. APVV-16-0186.

Acknowledgments: The authors thank Loran Ts. Adzhemyan, Nikolay V. Antonov, Michail M. Kompaniets, Mikhail Yu. Nalimov and Nikolay M. Gulitskiy for many illuminating and fruitful discussions.

Conflicts of Interest: The authors declare no conflict of interest.

Abbreviations

The following abbreviations are used in this manuscript:

NS	Navier–Stokes
RG	Renormalization Group
SDE	Stochastic Differential Equation
UV	Ultraviolet
IR	Infrared
OPE	Operator Product Expansion
1PI	one-particle irreducible
MHD	Magnetohydrodynamics

References

1. Bogoliubov, N.N.; Shirkov, D.V. *Introduction to the Theory of Quantized Fields*; Interscience: New York, NY, USA, 1980.
2. Peskin, M.; Schroeder, D.V. *An Introduction to Quantum Field Theory*; Addison-Wesley: Boston, MA, USA, 1995.
3. Landau, L.D.; Lifshitz, E.M. *Statistical Physics, Part 1*; Pergamon Press: Oxford, UK, 1980.
4. Nazmitdinov, R.G. From Chaos to Order in Mesoscopic Systems. *Phys. Part. Nucl. Lett.* **2019**, *16*, 159.
5. Vasil'ev, A.N. *The Field Theoretic Renormalization Group in Critical Behavior Theory and Stochastic Dynamics*; Chapman Hall/CRC: Boca Raton, FL, USA, 2004.
6. Zinn-Justin, J. *Quantum Field Theory and Critical Phenomena*, 4th ed.; Oxford University Press: Oxford, UK, 2002.
7. Stückelberg, E.; Petermann, A. La normalisation des constantes dans la théorie des quanta. *Helv. Phys. Acta* **1953**, *26*, 499.
8. Shirkov, D.V. Die Renormierungsgruppe für zwei Ladungen in der pseudoskalaren Mesontheorie. *DAN ZSSR* **1955**, *105*, 972.
9. Gell-Mann, M.; Low, F.E. Quantum Electrodynamics at Small Distances. *Phys. Rev.* **1954**, *95*, 1300.
10. Bogoliubov, N.N.; Shirkov, D.V. On renormalization groups in quantum electrodynamics. *DAN ZSSR* **1955**, *103*, 203.
11. Bogoliubov, N.N.; Shirkov, D.V. Application of the renormalization group to improve the formulae of perturbation theory. *DAN ZSSR* **1955**, *103*, 391.
12. Wilson, K.G.; Kogut, J. The renormalization group and the ε expansion. *Phys. Rep.* **1974**, *12*, 75.
13. Wegner, F.J. The Critical State, General Aspects. In *Phase Transitions and Critical Phenomena*; Domb, C.; Green, M.S., Eds.; Academic Press Inc.: New York, NY, USA, 1976; pp. 7–124.
14. Kadanoff, L.P. *Statistical Physics: Statics, Dynamics and Renormalization*; World Scientific Publishing Co.: Singapore, 2000.
15. Täuber, U.C. *Critical Dynamics: A Field Theory Approach to Equilibrium and Non-Equilibrium Scaling Behavior*; Cambridge University Press: New York, NY, USA, 2014.
16. Hohenberg, P.C.; Halperin, B.I. Theory of dynamic critical phenomena. *Rev. Mod. Phys.* **1977**, *49*, 435.
17. Van Kampen, N.G. *Stochastic Processes in Physics and Chemistry*; North-Holland: Amsterdam, The Netherlands, 2007.
18. Blaizot, J.P.; Iancu, E.; Mehtar-Tani, Y. Medium-induced qcd cascade: Democratic branching and wave turbulence. *Phys. Rev. Lett.* **2013**, *111*, 052001.
19. Sabbah, H.N.; Stein, P.D. Turbulent blood flow in humans: Its primary role in the production of ejection murmurs. *Circ. Res.* **1976**, *38*, 513–525.
20. Goldstein, M.L.; Wicks, R.T.; Perri, S.; Sahraoui, F. Kinetic scale turbulence and dissipation in the solar wind: Key observational results and future outlook. *Phil. Trans. R. Soc. A* **2015**, *373*, 20140147.
21. Zhuravleva, I.; Churazov, E.; Schekochihin, A.A.; Allen, S.W.; Arévalo, P.; Fabian, A.C.; Forman, W.R.; Sanders, J.S.; Simionescu, A.; Sunyaev, R.; et al. Turbulent heating in galaxy clusters brightest in X-rays. *Nature* **2014**, *515*, 85–87.
22. Landau, L.D.; Lifshitz, E.M. *Fluid Mechanics*; Pergamon Press: Oxford, UK, 1959.
23. Frisch, U. *Turbulence: The Legacy of A. N. Kolmogorov*; Cambridge University Press: Cambridge, UK, 1995.
24. McComb, W.D. *The Physics of Fluid Turbulence*; Clarendon: Oxford, UK, 1990.
25. Monin, A.S.; Yaglom, A.M. *Statistical Fluid Mechanics: Vol 2*; MIT Press: Cambridge, MA, USA, 1975.

26. Davidson, P.A. *Turbulence: An Introduction for Scientists and Engineers*; Oxford University Press: Oxford, UK, 2015.
27. Forster, D.; Nelson, D.R.; Stephen, M.J. Large-distance and long-time properties of a randomly stirred fluid. *Phys. Rev. A* **1977**, *16*, 732.
28. Dominicis, C.D.; Martin, P.C. energy spectra of certain randomly-stirred fluids. *Phys. Rev. A* **1979**, *19*, 419.
29. Yakhot, V.; Orszag, S.A. Renormalization Group Analysis of Turbulence. I. Basic Theory. *J. Sci. Comput.* **1986**, *1*, 3.
30. Smith, L.M.; Woodruff, S.L. Renormalization-group analysis of turbulence. *Annu. Rev. Fluid Mech.* **1998**, *30*, 275.
31. Adzhemyan, L.T.; Antonov, N.V.; Vasil'ev, A.N. *The Field Theoretic Renormalization Group in Fully Developed Turbulence*; Gordon & Breach: London, UK, 1999.
32. Adzhemyan, L.T.; Vasil'ev, A.N.; Pis'mak, Y.M. Renormalization-group approach in the theory of turbulence: The dimensions of composite operators. *Theor. Math. Phys.* **1983**, *57*, 1131.
33. Falkovich, G.; Gawędzki, K.; Vergassola, M. Particles and fields in fluid turbulence. *Rev. Mod. Phys.* **2001**, *73*, 913.
34. Antonov, N.V. Anomalous scaling regimes of a passive scalar advected by the synthetic velocity field. *Phys. Rev. E* **1999**, *60*, 6691.
35. Antonov, N.V. Renormalization group, operator product expansion and anomalous scaling in models of turbulent advection. *J. Phys. A* **2006**, *39*, 7825.
36. Antonov, N.V. Anomalous scaling of a passive scalar advected by the synthetic compressible flow. *J. Phys. D* **2000**, *144*, 370.
37. Gawędzki, K.; Vergassola, M. Phase Transition in the Passive Scalar Advection. *J. Phys. D* **2000**, *138*, 63.
38. Adzhemyan, L.T.; Antonov, N.V. Renormalization group and anomalous scaling in a simple model of passive scalar advection in compressible flow. *Phys. Rev. E* **1998**, *58*, 7381.
39. Kraichnan, R.H. Small-Scale Structure of a Scalar Field Convected by Turbulence. *Phys. Fluids* **1968**, *11*, 945.
40. Adzhemyan, L.T.; Antonov, N.V.; Honkonen, J. Anomalous scaling of a passive scalar advected by the turbulent velocity field with finite correlation time: Two-loop approximation. *Phys. Rev. E* **2002**, *66*, 036313.
41. Antonov, N.V.; Hnatich, M.; Honkonen, J.; Jurčišin, M. Turbulence with Pressure: Anomalous Scaling of a Passive Vector Field. *Phys. Rev. E* **2003**, *68*, 046306.
42. Folk, R.; Moser, G. Critical dynamics: A field-theoretical approach. *J. Phys. A Math. Gen.* **2006**, *39*, R207.
43. Janssen, H.K. On a Lagrangean for Classical Field Dynamics and Renormalization Group Calculations of Dynamical Critical Properties. *Z. Phys. B* **1976**, *23*, 377.
44. Dominicis, C.D. Techniques de renormalisation de la théroe des champs et dynamique des phénoménes critiques. *J. Phys. Colloq. Fr.* **1976**, *37*, C1–247.
45. Martin, P.C.; Siggia, E.D.; Rose, H.A. Statistical Dynamics of Classical Systems. *Phys. Rev. A* **1973**, *8*, 423.
46. Amit, D.J.; Martín-Mayor, V. *Field Theory, the Renormalization Group and Critical Phenomena*; World Scientific: Singapore, 2005.
47. Vasil'ev, A.N. *Functional Methods in Quantum Field Theory and Statistical Physics*; Gordon and Breach: Amsterdam, The Netherlands, 1998.
48. Bec, J.; Frisch, U. Les Houches 2000: New Trends in Turbulence. In *Burgulence*; Lesieur, M., Yaglom, A., David, F., Eds.; Springer: Dodrecht, The Netherlands, 2001.
49. Bohr, T.; Jensen, M.; Paladin, G.; Vulpiani, A. *Dynamical Systems Approach to Turbulence*; Cambridge University: Cambridge, UK, 1998.
50. Holzer, M.; Siggia, E.D. Turbulent mixing of a passive scalar. *Phys. Fluids* **1994**, *6*, 1820.
51. Avellaneda, M.; Majda, A. Mathematical models with exact renormalization for turbulent transport. *Commun. Math. Phys.* **1990**, *131*, 381.
52. Avellaneda, M.; Majda, A. Mathematical models with exact renormalization for turbulent transport, II: Fractal interfaces, non-Gaussian statistics and the sweeping effect. *Commun. Math. Phys.* **1992**, *146*, 139.
53. Majda, A. Explicit inertial range renormalization theory in a model for turbulent diffusion. *J. Stat. Phys.* **1993**, *73*, 515.
54. Horntrop, D.; Majda, A. Subtle statistical behavior in simple models for random advection-diffusion. *J. Math. Sci. Univ. Tokyo* **1994**, *1*, 23.
55. Zhang, Q.; Glimm, J. Inertial range scaling of laminar shear flow as a model of turbulent transport. *Commun. Math. Phys.* **1992**, *146*, 217.

56. Chertkov, M.; Falkovich, G.; Lebedev, V. Nonuniversality of the Scaling Exponents of a Passive Scalar Convected by a Random Flow. *Phys. Rev. Lett.* **1996**, *76*, 3707.
57. Eyink, G. Intermittency and anomalous scaling of passive scalars in any space dimension. *Phys. Rev. E* **1996**, *54*, 1497.
58. Kraichnan, R.H. Relation between Lagrangian and Eulerian correlation times of a turbulent velocity field. *Phys. Fluids* **1964**, *7*, 142.
59. Chen, S.; Kraichnan, R.H. Sweeping decorrelation in isotropic turbulence. *Phys. Fluids A* **1989**, *1*, 2019.
60. L'vov, V.S. Scale invariant theory of fully developed hydrodynamic turbulence-Hamiltonian approach. *Phys. Rep.* **1991**, *207*, 1.
61. Honkonen, J.; Karjalainen, E. Diffusion in a random medium with long-range correlations. *J. Phys. A Math. Gen.* **1988**, *21*, 4217.
62. Bouchaud, J.P.; Georges, A. Anomalous diffusion in disordered media: Statistical mechanisms, models and physical applications. *Phys. Rep.* **1990**, *195*, 127.
63. 't Hooft, G. Dimensional regularization and the renormalization group. *Nucl. Phys. B* **1973**, *61*, 455.
64. Antonia, R.A.; Satyaprakash, B.R.; Hussian, A.K. Statistics of fine-scale velocity in turbulent plane and circular jets. *J. Fluid. Mech.* **1982**, *119*, 55.
65. Anselmet, F.; Gagne, Y.; Hopfinger, E.; Antonia, R.A. High-order velocity structure functions in turbulent shear flows. *J. Fluid. Mech.* **1984**, *140*, 63.
66. Collins, J. *Renormalization*; Cambridge University Press: Cambridge, UK, 1985.
67. Adzhemyan, L.T.; Antonov, N.V.; Vasil'ev, A.N. Quantum field renormalization group in the theory of fully developed turbulence. *Phys. Usp.* **1996**, *39*, 1193.
68. Adzhemyan, L.T.; Antonov, N.V.; Vasil'ev, A.N. Infrared divergences and the renormalization group in the theory of fully developed turbulence. *Zh. Eksp. Teor. Fiz.* **1989**, *95*, 1272.
69. Adzhemyan, L.T.; Vasil'ev, A.N.; Hnatich, M. Renormalization-group approach in the theory of turbulence: renormalization and critical dimensions of the composite operators of the energy-momentum tensor. *Theor. Math. Phys.* **1988**, *74*, 115.
70. Landau, L.D.; Lifshitz, E.M. *Mechanics*; Pergamon Press: Oxford, UK, 1960.
71. Antonov, N.V.; Gulitskiy, N.M.; Kostenko, M.M.; Malyshev, A.V. Statistical symmetry restoration in fully developed turbulence: Renormalization group analysis of two models. *Phys. Rev. E* **2018**, *97*, 033101.
72. Gardiner, C.W. *Handbook of Stochastic Methods: For Physics, Chemistry, and the Natural Sciences*; Springer: Berlin/Heidelberg, Germany, 2009.
73. Moffatt, H.K. *Magnetic Field Generation in Electrically Conducting Fluids*; Cambridge University Press: Cambridge, UK, 1978.
74. Biskamp, D. *Magnetohydrodynamic Turbulence*; Cambridge University Press: Cambridge, UK, 2003.
75. Shore, S.N. *Astrophysical Hydrodynamics: An Introduction*; Wiley Vch: Weinheim, Germany, 2007.
76. Priest, E. *Magnetohydrodynamics of the sun*; Cambridge University Press: Cambridge, UK, 2014.
77. Tu, C.Y.; Marsch, E. MHD structures, waves and turbulence in the solar wind: Observations and theories. *Space Sci. Rev.* **1995**, *73*, 1.
78. Balbus, A.S.; Hawley, J.F. Instability, turbulence, and enhanced transport in accretion disks. *Rev. Mod. Phys.* **1998**, *70*, 1.
79. Chabrier, G. Galactic Stellar and Substellar Initial Mass Function. *Publ. Astron. Soc. Pac.* **2003**, *115*, 763.
80. Elmegreen, B.G.; Scalo, J. Interstellar Turbulence I: Observations and Processes. *Annu. Rev. Astron. Astrophys.* **2004**, *42*, 211.
81. Federrath, C. The origin of physical variations in the star formation law. *Mon. Not. R. Astron. Soc.* **2013**, *436*, 1245.
82. Hnatič, M.; Zalom, P. Helical turbulent Prandtl number in the A model of passive vector advection. *Phys. Rev. E* **2016**, *94*, 053113.
83. Hnatič, M.; Zalom, P. Field-theoretical Model of Helical Turbulence. *Nonlin. Phenom. Complex Syst.* **2017**, *20*, 238.
84. Adzhemyan, L.T.; Vasil'ev, A.N.; Hnatic, M. Turbulent dynamo as spontaneous symmetry breaking. *Theor. Math. Phys.* **1987**, *72*, 940.
85. Pumir, A. Anomalous scaling behaviour of a passive scalar in the presence of a mean gradient. *Europhys. Lett.* **1996**, *34*, 25.

86. Pumir, A. Structure of the three-point correlation function of a passive scalar in the presence of a mean gradient. *Phys. Rev. E.* **1998**, *57*, 2914.
87. Schraiman, B.I.; Siggia, E.D. Symmetry and Scaling of Turbulent Mixing. *Phys. Rev. Lett.* **1996**, *77*, 2463.
88. Pumir, A.; Schraiman, B.I.; Siggia, E.D. Perturbation theory for the δ-correlated model of passive scalar advection near the Batchelor limit. *Phys. Rev. E* **1996**, *55*, R1263.
89. Wiese, K.J. The passive polymer problem. *J. Stat. Phys.* **2000**, *101*, 843.
90. Celani, A.; Lanotte, A.; Mazzino, A.; Vergassola, M. Universality and Saturation of Intermittency in Passive Scalar Turbulence. *Phys. Rev. Lett.* **2000**, *84*, 2385.
91. Lanotte, A.; Mazzino, A. Anisotropic nonperturbative zero modes for passively advected magnetic fields. *Phys. Rev. E* **1999**, *60*, R3483.
92. Antonov, N.V.; Lanotte, A.; Mazzino, A. Persistence of small-scale anisotropies and anomalous scaling in a model of magnetohydrodynamics turbulence. *Phys. Rev. E* **2000**, *61*, 6586.
93. Arad, I.; Biferale, L.; Procaccia, I. Nonperturbative spectrum of anomalous scaling exponents in the anisotropic sectors of passively advected magnetic fields. *Phys. Rev. E* **2000**, *61*, 2654.
94. Arad, I.; L'vov, V.S.; Podivilov, E.; Procaccia, I. Anomalous scaling in the anisotropic sectors of the Kraichnan model of passive scalar advection. *Phys. Rev. E* **2000**, *62*, 4904.
95. Borue, V.; Orszag, S.A. Numerical study of three-dimensional Kolmogorov flow at high Reynolds numbers. *J. Fluid. Mech.* **1996**, *306*, 293.
96. Saddoughi, S.G.; Veeravalli, S.V. Local isotropy in turbulent boundary layers at high Reynolds number. *J. Fluid. Mech.* **1994**, *268*, 333.
97. Arad, I.; Dhruva, B.; Kurien, S.; L'vov, V.S.; Procaccia, I.; Sreenivasan, K.R. Extraction of Anisotropic Contributions in Turbulent Flows. *Phys. Rev. Lett* **1998**, *81*, 5330.
98. Arad, I.; Biferale, L.; Mazzitelli, I.; Procaccia, I. Disentangling Scaling Properties in Anisotropic and Inhomogeneous Turbulence. *Phys. Rev. Lett,* **1999**, *82*, 5040.
99. Kurien, S.; L'vov, V.S.; Procaccia, I.; Sreenivasan, K.R. Scaling structure of the velocity statistics in atmospheric boundary layers. *Phys. Rev. E* **2000**, *61*, 407.
100. Antonia, R.A.; Hopfinger, E.J.; Gagne, Y.; Anselmet, F. Temperature structure functions in turbulent shear flows. *Phys. Rev. A* **1984**, *30*, 2704.
101. Sreenivasan, K.R. On local isotropy of passive scalars in turbulent shear flows. *Proc. R. Soc. Lond. Ser. A* **1991**, *434*, 165.
102. Sreenivasan, K.R.; Antonia, R.A. The phenomenology of small-scale turbulence. *Annu. Rev. Fluid. Mech.* **1997**, *29*, 435.
103. Tong, C.; Warhaft, Z. On passive scalar derivative statistics in grid turbulence. *Phys. Fluids.* **1994**, *6*, 2165.
104. Adzhemyan, L.T.; Antonov, N.V.; Hnatich, M.; Novikov, S.V. Anomalous scaling of a passive scalar in the presence of strong anisotropy. *Phys. Rev. E* **2000**, *63*, 016309.
105. Fournier, J.D.; Frisch, U. Remarks on the renormalization group in statistical fluid dynamics. *Phys. Rev. A* **1983**, *28*, 1000.
106. Antonov, N.V.; Vasil'ev, A.N. Renormalization group in the theory of developed turbulence. The problem of justifying the Kolmogorov hypotheses for composite operators. *Theor. Math. Phys.* **1997**, *110*, 97.
107. Adzhemyan, L.T.; Borisenok, S.V.; Girina, V.I. Renormalization group approach and short-distance expansion in theory of developed turbulence: Asymptotics of the triplex equal-time correlation function. *Theor. Math. Phys.* **1995**, *105*, 1556.
108. Adzhemyan, L.T.; Antonov, N.V.; Vasil'ev, A.N. Renormalization Group, Operator Product Expansion, and Anomalous Scaling in a Model of Advected Passive Scalar. *Phys. Rev. E* **1998**, *58*, 1823.
109. Kraichnan, R.H. Anomalous scaling of a randomly advected passive scalar. *Phys. Rev. Lett.* **1994**, *72*, 1016.
110. Kraichnan, R.H. Convection of a passive scalar by a quasi-uniform random straining field. *J. Fluid. Mech.* **1974**, *64*, 737.
111. Kraichnan, R.H. Passive Scalar: Scaling Exponents and Realizability. *Phys. Rev. Lett.* **1997**, *78*, 4922.

MDPI
St. Alban-Anlage 66
4052 Basel
Switzerland
Tel. +41 61 683 77 34
Fax +41 61 302 89 18
www.mdpi.com

Symmetry Editorial Office
E-mail: symmetry@mdpi.com
www.mdpi.com/journal/symmetry

www.ingramcontent.com/pod-product-compliance
Lightning Source LLC
LaVergne TN
LVHW070107170726
843515LV00007B/1631

* 9 7 8 3 0 3 6 5 2 7 5 9 8 *